Some Results That Are Neutral

- SAS, ASA, AAS, SSS, H-L congruence conditions for triangles
- A point is on the perpendicular bisector of a line segment if and only if it is equidistant from the endpoints of the segment
- A point is on the bisector of an angle if and only if it is equidi⸱ from the sides of the angle
- The Isosceles Triangle Theorem and its converse
- The Vertical Angles Theorem
- The Exterior Angle Theorem for Triangl⸱
- The Saccheri-Legendre Theorem
- The Triangle Inequality
- The Hinge Theorem
- The Alternate Interior Angle Theorem
- The Midpoint Connection Theorem for triangles (concerning parallelism)

Some Results That Are Strictly Euclidean

- The Converse of the Alternate Interior Angle Theorem
- The Exterior Angle Theorem for Triangles (strong version)
- Angle Sum Theorem for Triangles (strong version)
- Transitivity of Parallelism
- Rectangles exist
- Pairs of similar, but non-congruent, triangles exist
- The opposite sides of a parallelogram are congruent
- The Pythagorean Theorem
- The Midpoint Connection Theorem for Triangles (concerning length)
- The Theorems of Menelaus and Ceva
- The concurrence of the perpendicular bisectors of the sides of a triangle
- The Nine-Point Circle Theorem

Some Results That Are Strictly Hyperbolic

- The Angle of Parallelism is acute
- Rectangles do not exist
- The summit angles of a Saccheri Quadrilateral are acute
- The fourth angle of a Lambert Quadrilateral is acute
- The sum of the interior angles of a triangle is less than $180°$
- Parallel lines are *not* everywhere equidistant
- Triangles exist that cannot be circumscribed
- The AAA congruence condition for triangles
- There is a maximal area for triangles

ROADS
TO GEOMETRY

ROADS
TO GEOMETRY

Edward C. Wallace
Stephen F. West

The State University of New York
College of Arts and Science at Geneseo

Prentice Hall, Englewood Cliffs, New Jersey 07632

Library of Congress Cataloging-in-Publication Data

Wallace, Edward C.,
  Roads to geometry / Edward C. Wallace and Stephen F. West.
    p.   cm.
  Includes index.
  ISBN 0-13-781725-8
  1. Geometry. I. West, Stephen F.,   . II. Title.
QA445.W35 1992   91-11464
516—dc20   CIP

Acquisitions Editor: Priscilla McGeehon
Editorial/Production Supervision
  and Interior Design: Valerie Zaborski
Copy Editor: Carol Dean
Cover Designer: Joe DiDomenico
Prepress Buyer: Paula Massenaro
Manufacturing Buyer: Lori Bulwin
Editorial Assistant: Marisol L. Torres

© 1992 by Prentice-Hall, Inc.
A Simon and Schuster Company
Englewood Cliffs, New Jersey 07632

Printed in the United States of America

10 9 8 7 6 5 4 3

ISBN 0-13-781725-8

Prentice-Hall International (UK) Limited, *London*
Prentice-Hall of Australia Pty. Limited, *Sydney*
Prentice-Hall Canada Inc., *Toronto*
Prentice-Hall Hispanoamericana, S.A., *Mexico*
Prentice-Hall of India Private Limited, *New Delhi*
Prentice-Hall of Japan, Inc., *Tokyo*
Simon & Schuster Asia Pte. Ltd., *Singapore*
Editora Prentice-Hall do Brasil, Ltda., *Rio de Janeiro*

CONTENTS

3   Traveling Together (Neutral Geometry)   65

4   One Way to Go (Euclidean Geometry of
     the Plane)   98

5 Side Trips (Analytic and Transformational Geometry)   185

6 Other Ways to Go (Non-Euclidean Geometries)   251

PREFACE

The goal of this book is to provide a geometric experience which clarifies, extends, and unifies concepts generally discussed in traditional high school geometry courses and to present additional topics that assist in gaining a better understanding of elementary geometry. As its title *Roads to Geometry* indicates, this book is designed to provide the reader with a "map" for a voyage through plane geometry and its various branches. As prerequisites, this book assumes only a prior course in high school geometry and the mathematical maturity usually provided by a semester of calculus or discrete mathematics.

Preparations for this voyage begin in Chapter 1 with a discussion of the "Rules of the Road" in which the reader is familiarized with the properties of axiomatic systems and application of the axiomatic method to investigations of these systems. A discussion of several examples of finite and incidence geometries provides a framework within which we may investigate plane geometry.

With these preparations complete, the voyage commences in Chapter 2 where we are confronted with "Many Ways to Go." Here, within a historical perspective, we travel a variety of "roads" through geometry by investigating different axiomatic approaches to the study of Euclidean plane geometry. Axiomatic developments of geometry as proposed by Euclid, David Hilbert, G. D. Birkhoff, and the School Mathematics Study Group (SMSG) are compared and contrasted.

In Chapter 3, "Traveling Together," we investigate the content of neutral geometry. The SMSG postulates provide our pedagogical choice for a "main arterial" as we prepare ourselves for the choice between the Euclidean and non-Euclidean "exits" that appear on the horizon.

Chapter 4 provides "One Way to Go" as we explore the Euclidean plane. In this chapter we extend ideas developed in neutral geometry and provide a traditional look at the geometric topics of congruence, area, similarity, circles, and constructions from a Euclidean perspective.

While still within the Euclidean plane, Chapter 5 provides two "Side Trips" through analytical and transformational approaches to geometry. The real numbers, algebra, isometries, similarities, analytical transformations, and inversion and their applications to geometric theorem proving are discussed.

In Chapter 6 we consider "Other Ways to Go." We return briefly to neutral geometry in preparation for our venture into the non-Euclidean plane. In addition to a discussion of hyperbolic geometry, this chapter contains a detailed description of the Poincaré disk model and a brief excursion into elliptic geometry.

Finally, in Chapter 7, "All Roads Lead To . . ." projective geometry. Here we delve into a more general geometry than we have studied in previous chapters as we investigate the real projective plane and the ideas of duality, perspectivity, and projective transformations.

This text is appropriate for several kinds of students. Preservice teachers of geometry are provided with a rigorous yet accessible treatment of plane geometry in a historical context. Mathematics majors will find its axiomatic development sufficiently rigorous to provide a foundation for further study in the areas of Euclidean and non-Euclidean geometry. Through the choice of the SMSG postulate set as a basis for the development of plane geometry, this book avoids the pitfalls of many "foundations of geometry" texts which encumber the reader with such a detailed development of preliminary results that many other substantive and elegant results are inaccessible in a one-semester course.

The chapters of this book separate nicely into independent units. The material in Chapters 1 and 2 provides preliminary groundwork for the study of geometry. Instructors who feel that their classes are exceptionally well prepared can omit these chapters in the interest of freeing time for material presented later in the book. Instructors teaching more typical classes will find the discussion of axiomatics in Chapter 1 and the comparisons of the various axiom sets in Chapter 2 very helpful in conveying the notion of mathematical rigor. Instructors can teach a semester of Euclidean geometry using Chapters 1 through 5, while those instructors more interested in non-Euclidean Geometries can opt to cover Chapters 1, 2, 3, and 6.

At the end of each section is an ample collection of exercises of varying difficulty which provide problems that both extend and clarify results of the section as well as problems that apply those results. At the end of each of Chapters 3 through 7 is a summary listing all the new definitions and theorems of the chapter. In addition, a "tear-out" page listing the SMSG axioms is included in the back cover so that the student does not have to turn to an appendix each time an axiom is invoked.

The authors hope that *Roads to Geometry* will in some way encourage the reader to more fully appreciate the marvelous worlds of Euclidean and non-Euclidean plane geometry and to that end we wish you bon voyage.

E. C. Wallace
S. F. West

1

RULES
OF THE ROAD

Axiomatic Systems

1.1 INTRODUCTION

The word "geometry" comes from the Greek words meaning "earth measure," which when taken literally imply that geometry involves measuring earthly things. Ancient geometry, in part, had its beginnings in the practical mensuration necessary for the agriculture of the Babylonians and Egyptians. These civilizations were "known for their engineering prowess in marsh drainage, irrigation, flood control, and the erection of great edifices and structures."[1] Much of Egyptian and Babylonian geometry was restricted to computation of the lengths of line segments, areas, and volumes. These results, which were usually presented as a sequence of arithmetical instructions, were derived empirically and in some instances were incorrect. For example, the Egyptians used the formula $A = \frac{1}{4}(a + c)(b + d)$ to calculate the area of an arbitrary quadrilateral with successive sides of lengths a, b, c, and d. This formula proves to be correct for rectangles but not for quadrilaterals in general. It wasn't until the first half of the sixth century B.C. that mathematicians began to question whether or not these empirical results were always true.

Recent developments in the formal study of geometry bear little resemblance to its historical beginnings. In fact, as you will see in later chapters, the contemporary study of geometry does not necessarily require that we measure anything *or* even restrict ourselves to the earth.

[1] Howard Eves, "The History of Geometry," in *Historical Topics for the Mathematics Classroom* (Washington, D.C.: National Council of Teachers of Mathematics, 1969), p. 168.

1

The remainder of this section briefly introduces several of the great philosopher/mathematicians of antiquity and their roles in the birth of "demonstrative geometry."

Thales of Miletus

The transformation of the study of geometry from a purely practical science (namely, surveying) to a branch of pure mathematics was undertaken by Greek scholars and took place over a number of centuries. The individual most often credited with initiating the formal study of demonstrative geometry as a discipline is Thales of Miletus (c. 640–546 B.C.). In his early days Thales was a merchant, and in this capacity he traveled to Egypt and the Middle East. He returned to Greece with a knowledge of the measurement techniques used by the Egyptians at that time. The greatest contribution made by Thales to the study of geometry was his ability to abstract the ideas of the Egyptians from a physical context to a mental one. The propositions for which he has been given credit are among the simplest in plane geometry. For example, Proclus, in his "Eudemian Summary,"[2] stated that Thales was "the first to demonstrate that the circle is bisected by the diameter."[3] As one can see, this rather simple assertion is not noteworthy by virtue of its profound content. The significance of Thales' contribution lies not in the content of the propositions themselves but in his use of logical reasoning to argue in favor of them. Proclus describes an indirect "proof" (presumably due to Thales) to support the circle bisection theorem. While Thales' proof is not acceptable by today's standards (and was even avoided by Euclid) it showed, for the first time, an attempt to justify geometric statements using reason instead of intuition and experimentation.

D. E. Smith speaks about the importance of Thales' work in geometry: "Without Thales there would not have been a Pythagoras—or such a Pythagoras; and without Pythagoras there would not have been a Plato—or such a Plato."[4]

Pythagoras

Pythagoras (c. 572 B.C.) was born on the Greek Island of Samos before Thales' death and was probably a student of Thales. Pythagoras traveled widely throughout the Mediterranean region, and it is very possible that his

---

[2] The "Eudemian Summary" is a small part of Proclus' *Commentary on the First Book of Euclid* in which he describes the history of Thales as given by Eudemus of Rhomes in a work available to Proclus but which has since been lost to us.

[3] Glenn R. Morrow, (trans.), *Proclus—A Commentary on the First Book of Euclid's Elements* (Princeton, N.J.: Princeton University Press, 1970), p. 124.

[4] D. E. Smith, *History of Mathematics* (New York: Dover Publications, Inc., 1958), I, p. 68.

journeys took him to India, since his philosophical orientation was more closely aligned with the Indian civilization than with the Greek. On returning to Europe, Pythagoras migrated to Croton, a Greek colony in southern Italy, and established a quasi-religious brotherhood called the Pythagoreans. It is likely that much of the mathematics attributed to Pythagoras was actually developed by members of this brotherhood during the 200 or so years of its existence. The Pythagoreans took mathematical thought a step beyond the point to which Thales had brought it. Whereas Thales had formalized a portion of the geometry that he encountered, the Pythagorean philosophy was to develop mathematical results exclusively as the result of deduction. It was during this time that "chains of propositions were developed in which each successive proposition was derived from earlier ones."[5] The Pythagorean school set the tone for all the Greek mathematics that was to follow, and since the ideas of Plato were largely committed to mathematics, one could say that Pythagoras had a major effect on all of Greek philosophy.

Plato

Plato's role in the development of geometry (and of mathematics in general) is often overshadowed by his preeminent status in Greek philosophy in general. The Academy of Plato, which was established about 387 B.C., attracted the most famous scholars of the time. At the Academy the study of mathematics was confined to pure mathematics, with the emphasis placed on soundness of reason. One of Plato's most famous students was Aristotle (c. 584 B.C.), who in his work *Analytica Posteriora* did much to systematize the classical logic that formed a basis for all Greek mathematics. By about 400 B.C. Greek civilization had developed to the point where intellectual pursuits were valued for their intrinsic virtue. Plato was of the mind that "mathematics purifies and elevates the soul."[6] Since there was no need for mathematicians at the Academy to concern themselves with applications of their work, the emphasis could be placed on the processes involved in the development of mathematical thought rather than on worldly products of that thought. Thus mathematics had by 350 B.C. taken on the nature of a "pure science."

Euclid

The name most often associated with ancient Greek geometry is that of Euclid. Not a great deal is known about Euclid's background. He may have been born in Greece, or he may have been an Egyptian who went to Alexan-

---

[5] Eves, in *Historical Topics*, p. 172.
[6] Morrow, *Proclus—A Commentary*, p. 25.

dria to study and teach. He is believed to have been the first mathematics professor at the great University of Alexandria. His lifetime overlapped that of Plato, and he may have been a student at Plato's academy. Proclus, in his *Commentary*, tells us that Euclid was influenced by Plato's philosophy, but there is no direct evidence that the two ever met.

By Euclid's time (c. 325 B.C.) the development of rational thought had progressed sufficiently to allow for, and even demand, a systematic study of geometry. Euclid's monumental work, *Elements of Geometry*, a single chain of 465 propositions which in part encompasses plane and solid geometry, has for over 2000 years, remained as the most widely known example of a formal axiomatic system. As we will see in subsequent chapters, Euclid's work was far from flawless. Still its strengths far outnumber its weaknesses, as attested by the fact that it overshadowed and replaced all previous writings in this area.

In order to place Euclid's historic effort in context, we shall in the next section discuss what is meant by an axiomatic system and investigate the properties that axiomatic systems possess.

EXERCISE SET 1.1

1. As indicated earlier, Egyptian geometers used the formula $A = \frac{1}{4}(a + c)(b + d)$ to calculate the area of any quadrilateral whose successive sides have lengths a, b, c, and d.
   (a) Does this formula work for squares? For rectangles that are not squares?
   (b) If you choose specific lengths for the sides of an isosceles trapezoid, how does the result compare to the actual area? Repeat for two other isosceles trapezoids. Do the same for three specific parallelograms.
   (c) Generalize your results for part (b).

2. If a and b are the lengths of the legs of a right triangle and c is the length of the hypotenuse, Babylonian geometers approximated the length of the hypotenuse by the formula $c = b + (a^2/2b)$.
   (a) How does this approximation compare to the actual result when $a = 3$ and $b = 4$? When $a = 5$ and $b = 12$? When $a = 12$ and $b = 5$?
   (b) Give an algebraic argument demonstrating that this formula results in an approximation that is too large.

3. The following was translated from a Babylonian tablet created about 2600 B.C. Explain what it means.

   60 is the circumference, 2 is the perpendicular, find the chord. Double 2 and get 4, do you see? Take 4 from 20 and get 16. Square 20, and you get 400. Square 16, and you get 256. Take 256 from 400 and you get 144. Find the square root of 144. 12, the square root, is the chord. This is the procedure.[7]

   ---

   [7] Howard Eves, *A Survey of Geometry* (Boston: Allyn and Bacon, 1965), p. 7, Problem 1.2-2.

4. The Moscow Papyrus (c. 1850 B.C.) contains the following problem:

> If you are told: A truncated pyramid of 6 for the vertical height by 4 on the base by 2 on the top. You are to square this 4, result 16. You are to double 4, result 8. You are to square 2, result 4. You are to add the 16, the 8, and the 4, result 28. You are to take one third of 6, result 2. You are to take 28 twice, result 56. See, it is 56. You will find it right.

Show that this is a special case of the general formula, $V = \frac{1}{3}h(a^2 + ab + b^2)$, for the volume of the frustum of a pyramid whose bases are squares, whose sides are a and b, respectively, and whose height is h.

− **5.** An Egyptian document, the Rhind Papyrus (c. 1650 B.C.), states that the area of a circle can be determined by finding the area of a square whose side is $\frac{8}{9}$ of the diameter of the circle. Is this correct? What value of π is implied by this technique?

6. It is said that Thales indirectly measured the distance from a point on shore to a ship at sea using the equivalent of angle-side-angle (ASA) triangle congruence theorem. Make a diagram that could be used to accomplish this feat.

− **7.** Eratosthenes (c. 275 B.C.), a scholar and librarian at the University at Alexandria, is credited with calculating the circumference of the earth using the following method: Eratosthenes observed that on the summer solstice the sun was directly overhead at noon in Syene (the present site of Aswan), while at the same time in Alexandria, which was due north, the rays of the sun were inclined 7°12′, thus indicating that Alexandria was 7°12′ north of Syene along the earth's surface. Using the known distance between the two cities of 5000 stades (approximately 530 miles), he was able to approximate the circumference of the earth. Make a diagram that depicts this method and calculate the circumference in stades and in miles. How does this result compare to present-day estimates?

1.2 AXIOMATIC SYSTEMS AND THEIR PROPERTIES

The Axiomatic Method

As we begin our study of geometry, it is important that we have a basic understanding of the *axiomatic method* used in the development of all of modern mathematics. The axiomatic method is a procedure by which we demonstrate or prove that results (theorems, and so on) discovered by experimentation, observation, trial and error, or even by "intuitive insight," are indeed correct. Little is known about the origins of the axiomatic method. Most historians, relying on accounts given by Proclus in his "Eudemian Summary," indicate that the method seems to have begun its evolution during the time of the Pythagoreans as a further development and refinement of various early deductive procedures.

In an axiomatic system the proof of a specific result is simply a sequence of statements, each of which follows logically from the ones before and leads from a statement that is known to be true to the statement which is to be proven. First, for a proof to be convincing, it is necessary to establish ground rules for determining when one statement follows logically from another. For the purposes of this development, our rules of logic will consist of the standard two-valued logic studied in most introductory logic courses. Second, it is important that all readers of the proof have a clear understanding of the terms and statements used in the discussion. To ensure this clarity, we might try to define each of the terms in our discussion. If, however, one of our definitions contains an unfamiliar term, then the reader has the right to expect a definition of this term. Thus a chain of definitions is created. This chain must be circular or linear (think of a concrete model). Since circularity is unacceptable in any logical development, we may assume that the chain of definitions is linear. Now this linear chain may be an infinite sequence of definitions, or it must stop at some point. An unending sequence of definitions is at best unsatisfying, so the collection of definitions must end at some point and one or more of the terms will remain undefined. These terms are known as the *undefined* or *primitive terms* of our axiomatic system.

The primitive terms and definitions can now be combined into the statements or *theorems* of our axiomatic system. For these theorems to be of mathematical value, we must supply logically deduced proofs of their validity. We now need additional statements to prove these theorems which in turn require proof. As before, we form a chain of statements that leads us to the conclusion that, to avoid circularity, one or more of these statements must remain unproven. These statements, called *axioms* or *postulates*,[8] must be assumed, and they form the fundamental truths[9] of our axiomatic system.

To summarize, any logical development of an axiomatic system must therefore conform to the pattern represented in Table 1.2.1.

To illustrate an axiomatic system and the relationships among its undefined terms, axioms, and theorems we will consider the following example.

Example 1.2.1

A simple abstract axiomatic system. Undefined terms: Fe's, Fo's, and the relation, "belongs to."[10]

---

[8] Today, the words "axiom" and "postulate" are used interchangeably. Historically, the word "postulate" has been used to represent an assumed truth confined to a particular subject area, while "axiom" represents a more universal truth applicable to all areas of mathematics.

[9] The truth of these axioms is not at issue—just the reader's willingness to accept them as true.

[10] We will occasionally use the terminology "a Fe is on a Fo" or "a Fo contains a Fe," and by this we mean that the Fe "belongs to" the Fo.

TABLE 1.2.1   The Axiomatic Method

1. Any axiomatic system must contain a set of technical terms that are deliberately chosen as undefined terms and are subject to the interpretation of the reader.

2. All other technical terms of the system are ultimately defined by means of the undefined terms. These terms are the definitions of the system.

3. The axiomatic system contains a set of statements, dealing with undefined terms and definitions, that are chosen to remain unproven. These are the axioms of the system.

4. All other statements of the system must be logical consequences[11] of the axioms. These derived statements are called the theorems of the axiomatic system.

AXIOM 1.   There exist exactly three distinct Fe's in this system.

AXIOM 2.   Two distinct Fe's belong to exactly one Fo.

AXIOM 3.   Not all Fe's belong to the same Fo.

AXIOM 4.   Any two distinct Fo's contain at least one Fe which belongs to both.

FE-FO THEOREM 1.   Two distinct Fo's contain exactly one Fe.

Proof. Since Axiom 4 states that two distinct Fo's contain at least one Fe, we need only show that these two Fo's contain no more than one Fe. For this purpose we will use an indirect proof and assume that two Fo's share more than one Fe. The simplest case of "more than one" is two. Now each of these two Fe's belong to two distinct Fo's, but that in turn contradicts Axiom 2, and we are done.

FE-FO THEOREM 2.   There are exactly three Fo's.

Proof. Axiom 2 tells us that each pair of Fe's is on exactly one Fo. Axiom 1 provides us with exactly three Fe's, and by counting distinct pairs of Fe's, we find that we have at least three Fo's. Now suppose that there exists a distinct fourth Fo. Theorem 1 tells us that the fourth Fo must share a Fe with each of the other Fo's. Therefore it must contain at least one of the two of the existing three Fe's, but Axiom 2 prohibits this. Therefore a fourth Fo cannot exist, and there are exactly three Fo's.

[11] Recall that it is presumed that underlying the axiomatic system is some type of logical structure on which valid arguments are based.

FE-FO THEOREM 3.   Each Fo has exactly two Fe's which belong to it.

Proof. By Theorem 2, we have exactly three Fo's. Now Axiom 4 provides that each Fo has at least one Fe and prevents it from containing exactly one, while Axiom 1 and Axiom 3 prevent a Fo from containing more than two Fe's.

In the previous example the undefined terms are truly undefined, and the axioms and theorems are nonsensical in nature. The real value in its study lies in the fact that, as in any axiomatic system, the truth of its theorems comes not from what they say but from the fact that they have been logically deduced from the axioms. Additional practice proving theorems in this system and in another simple system can be found in the exercises.

In the following section we will investigate the consequences of giving some type of meaning or interpretation to the undefined terms of our axiomatic system.

Models

In the previous section we discussed the axiomatic method and an example of an abstract axiomatic system. Each axiomatic system contains a number of undefined terms. Since these terms are truly undefined, they have no inherent meaning and each reader may choose to interpret them in his or her own way. By giving each undefined term in a system a particular meaning, we create an *interpretation* of that system. If for a given interpretation of a system, all of the axioms are "correct" statements, we call the interpretation a *model*.

Example 1.2.2

In Example 1.2.1 we can designate the Fe's as people and the Fo's as committees, and the axioms become

AXIOM 1.   There are exactly three people.

AXIOM 2.   Two distinct people belong to exactly one committee.

AXIOM 3.   Not all people belong to the same committee.

AXIOM 4.   Any two distinct committees contain one person who belongs to both.

Let the people be Bob, Ted, and Carol, and the committees Entertainment (Bob and Ted), Finance (Ted and Carol), and Refreshments (Bob and Carol) (Figure 1.2.1.) We can see that as a collection these axioms are "correct" statements, and therefore this interpretation is an example of a model.

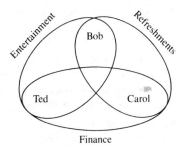

Figure 1.2.1

Example 1.2.3

In Example 1.2.1 we can designate the Fe's as books, the Fo's as horizontal shelves, and the relation "belongs to" as "is on" (Figure 1.2.2), and the axioms become

AXIOM 1.   There are exactly three books.

AXIOM 2.   Two books are on exactly one shelf.

AXIOM 3.   Not all books are on the same shelf.

AXIOM 4.   Any two distinct shelves contain one book which is on both.

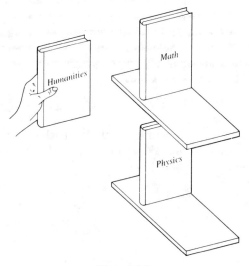

Figure 1.2.2

Since Axiom 4 is not a "correct" statement, this interpretation is not a model for the system.

Since the theorems of any axiomatic system are logical consequences of the axioms of the system, their validity is independent of any interpretation of the undefined terms. One of the main properties of any model of an axiomatic system is that all theorems of the system are correct statements in the model. Therefore the theorems proven in previous sections (and those proven by you in the exercises) become correct statements under the interpretation in Example 1.2.2. For example, Fe-Fo Theorem 1 now reads, Each committee contains exactly two people. However, in the interpretation found in Example 1.2.3, the theorems need not be correct statements since the axioms are not satisfied.

Models play an important role in the study of axiomatic systems. Suppose that in our system we have a statement for which we are unsure of the existence of a proof (i.e., we don't know whether the statement is a theorem or not). By investigating models, we can gain insight into the correctness of the statement. In particular, if one model exists in which the statement is incorrect, we can be assured that no proof of the statement exists. The reader is cautioned, however, that the reverse is not always true. Simply because a statement is correct in one model does not ensure its correctness in all models, and thus it may not be a theorem in the axiomatic system. It is always a good practice to create several different models when investigating axiomatic systems.

When creating models, it is possible to produce two seemingly different models which on closer inspection are essentially the same. By "essentially the same" we mean that there exists a one-to-one correspondence between the interpretation of each set of undefined terms such that any relationship between the undefined terms in one model is preserved, under that one-to-one correspondence, in the second model. Two models that are essentially the same are said to be *isomorphic*, and the one-to-one correspondence is called an *isomorphism*.

Example 1.2.4

Consider the axiomatic system in Example 1.2.1 where Fe's are interpreted as trees and Fo's are interpreted as rows (Figure 1.2.3). This model is isomorphic to the model in Example 1.2.2.

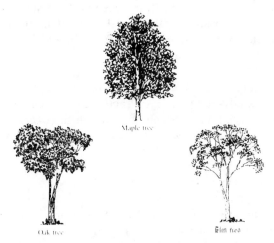

Figure 1.2.3

Example 1.2.5

Let the Fe's of our axiomatic system be interpreted as the letters in the set $S = \{P, Q, R\}$, and let the Fo's be interpreted as all of the two element subsets of S. This model is isomorphic to the model in Example 1.2.2 under the one-to-one correspondence

$$P \leftrightarrow \text{Bob} \qquad \{P, Q\} \leftrightarrow \text{Entertainment Committee}$$

$$Q \leftrightarrow \text{Ted} \qquad \{P, R\} \leftrightarrow \text{Refreshments Committee}$$

$$R \leftrightarrow \text{Carol} \qquad \{Q, R\} \leftrightarrow \text{Finance Committee}$$

This model is also isomorphic to the model in Example 1.2.4. Why?

To illustrate the fact that every one-to-one correspondence between two different interpretations of the undefined terms of some axiomatic system is not necessarily an isomorphism, consider the following example in which the relationship between the undefined terms is not preserved.

Example 1.2.6

Let the Fo's be the elements of the set $T = \{x, y, z\}$, and let each Fe be represented by the pair of Fo's that share it. ($\{x, y\}$, $\{x, z\}$, and $\{y, z\}$ are the Fe's of our model). Now consider the following one-to-one correspondence between this model and the model in Example 1.2.5:

$$\{x, y\} \leftrightarrow P \qquad \{P, Q\} \leftrightarrow z$$

$$\{y, z\} \leftrightarrow Q \qquad \{Q, R\} \leftrightarrow y$$

$$\{x, z\} \leftrightarrow R \qquad \{P, R\} \leftrightarrow x$$

Distinct Fo's $\{Q, R\}$ and $\{P, R\}$ in Example 1.2.5 share the Fe R, while the corresponding Fo's x and y in the model above share the Fe $\{x, y\}$ and not $\{x, z\}$ as one-to-one correspondence indicates. Therefore the illustrated one-to-one correspondence is not an isomorphism.

We will soon see that the creation of models will play a very important role in our future study of geometry. Practice in creating models and determining isomorphic relationships can be found in the exercises.

In the following section we shall briefly investigate several important properties that can be exhibited by axiomatic systems.

Properties of Axiomatic Systems

The most important and most fundamental property of an axiomatic system is *consistency*. A set of axioms is said to be consistent if it is impossible to deduce from these axioms a theorem that contradicts any axiom or previously proven theorem. Without this property, an axiomatic system has no mathematical value, and further study of its properties is useless. To establish the consistency of an axiomatic system, we will make use of models similar to those discussed in the last section.

Models may be separated into two categories: (1) *concrete* models, where interpretations of the undefined terms are objects or relations adapted from the real world, and (2) *abstract* models, where interpretations of the undefined terms are taken from some other axiomatic system. The model in Example 1.2.4 is an example of a concrete model, whereas the one in Example 1.2.5 is an abstract model.

When a concrete model has been produced, we claim to have established the *absolute* consistency of our axiomatic system. Otherwise, contradictory theorems deduced from the axioms would have contradictory counterparts in the real world, which we accept as impossible.

We shall see, however, that the establishment of absolute consistency is not always possible, since it may not be possible to establish a concrete model. For example, should an axiomatic system similar to the one discussed in Example 1.2.1 require an infinite number of Fe's, the creation of a concrete model would be impossible since, in the real world, there does not exist an infinite collection of "things" that would serve as an interpretation. In such cases we establish a model using concepts from some other axiomatic system whose consistency we are willing to assume (namely, the real numbers). In other words, the new axiomatic system is consistent if the one in which we have chosen our model is consistent. When this situation arises, we claim to have established the *relative* consistency of our axiomatic system, and the two axiomatic systems are said to be *relatively consistent*.

The next two properties of axiomatic systems that we will discuss, independence and completeness, have a nature that is distinctly different from the property of consistency. This difference lies in the fact that, unlike the consistency property, we do not require that axiomatic systems possess these properties to be useful (worthy of study). In fact, later on it may prove advantageous for us to employ systems that do not possess these properties.

An individual axiom is said to be *independent* if it cannot be logically deduced from the other axioms in the system. The entire set of axioms is said to be independent if each of its axioms is independent. It should be clear to the reader that an axiomatic system should not be invalidated simply because its axioms are not independent. The worst that can be said is that any nonindependent system has some redundancies, since one or more of its axioms could also appear as theorems. In general, a mathematician would prefer an independent axiom set, since any axiomatic system is best built on a minimum number of assumptions. It is often advantageous, however, to use a set of axioms that is not independent. For example, the axiomatic development studied in most high school geometry classes is based on a set of axioms that is not independent. The reasoning behind this, which is primarily pedagogical, is that often a very important and useful theorem occurs early in the development, and the difficulty of its proof, using an independent axiom set, may preclude its use for all but an experienced mathematician. In such cases the theorem may be included as an axiom, thus making it available to students with less mathematical maturity.

To demonstrate the independence of an axiom, we will again make use of models. By producing a model in which one axiom is incorrect and the remaining axioms are correct, we ensure the independence of the first axiom since only correct statements may be logically deduced from correct statements.

Example 1.2.7

Let the Fe's in Example 1.2.1 be the elements in the set $\{A, B, C, D\}$, and the Fo's be the subsets $\{A, B\}$, $\{A, C\}$, $\{A, D\}$, and $\{B, C, D\}$. Clearly, Axiom 1 is incorrect in this model since it contains four Fe's, but careful investigation shows that Axioms 2 through 4 are all correct. Therefore Axiom 1 is independent of the remaining three. The reader is encouraged to find models that demonstrate the independence of Axioms 2 through 4.

The property of completeness is also concerned with the size of the axiom set. Whereas independence guaranteed that our set of axioms was not too large, completeness guarantees that our chosen axioms are sufficient in number to prove or disprove any statement that arises concerning our collection of undefined terms. We say that an axiom set is of sufficient size, or *complete*, if it is impossible to add an additional consistent and independent axiom without adding additional undefined terms.

Testing for completeness is another question altogether. As with consistency, the failure to find a new consistent, independent axiom does not eliminate the possibility of its existence and therefore is an insufficient procedure by which to prove completeness. We can, however, use the isomorphism of models to demonstrate a form of completeness. If all models of a given axiomatic system are isomorphic, then the set of axioms is said to be *categorical*. The property of categoricalness can be shown to imply completeness; the proof, however, is beyond the scope of this discussion.[12]

In the next section, we will illustrate the properties of axiomatic systems in a geometric context.

EXERCISE SET 1.2

The Axiomatic Method

To answer Problems 1 through 4, use the axiomatic system outlined in Example 1.2.1.

1. Prove: A Fo cannot contain three distinct Fe's.
2. Prove: There exists a set of two Fo's that contains all the Fe's of the system.
3. Prove: For every set of two distinct Fe's, every Fo in the system must contain at least one of them.
4. Prove: All three Fo's cannot contain the same Fe.
5. Consider the following axiom set in which x's, y's, and "on" are the undefined terms:

    AXIOM 1.   There exist exactly five x's.

    AXIOM 2.   Any two distinct x's have exactly one y on both of them.

    AXIOM 3.   Each y is on exactly two x's.

How many y's are there in the system? Prove your result.

To answer Problems 6 through 9, use the axiom set in Problem 5.

6. Prove that any two y's have at most one x on both.
7. Prove that not all x's are on the same y.
8. Prove that there exist exactly four y's on each x.
9. Prove that for any y_1 and any x_1 not on that y_1 there exist exactly two other distinct y's on x_1 that do not contain any of the x's on y_1.

Models

10. Verify that the axioms in Example 1.2.2 are "correct" statements.
11. Verify that Axioms 1 and 3 in Example 1.2.3 are "correct" statements and explain why Axioms 2 and 4 are not correct.
12. Verify that the model in Example 1.2.5 is isomorphic to the model in Example 1.2.2.

---

[12] For a detailed discussion of completeness and categoricalness, see Howard Eves, *Foundations and Fundamental Concepts of Mathematics* (Boston: PWS-KENT Publishing Co., 1990), pp. 160–162.

13. Devise a one-to-one correspondence between the undefined terms in the models in Examples 1.2.5 and 1.2.6 that is an isomorphism and verify your result.

14. Devise another model that is isomorphic to the one in Example 1.2.2. Find a model that is not isomorphic, if possible.

15. Devise a model for the axiom system described in Problem 5.

16. Consider an infinite set of undefined elements S and the undefined relation R which satisfies the following axioms:

  AXIOM 1.   If $a, b \in S$ and a R b, then $a \neq b$.

  AXIOM 2.   If $a, b, c \in S$, a R b, and b R c, then a R c.

  (a) Show that an interpretation with S as the set of integers and a R b as "a is less than b" is a model for the system.
  (b) Would $S =$ the set of integers and a R b interpreted as "a is greater than b" also be a model?
  (c) Are the models in parts (a) and (b) isomorphic?
  (d) Would $S =$ the set of real numbers and a R b interpreted as "a is less than b" be another model?
  (e) Is the model in part (d) isomorphic to the model in part (a)?

Properties of Axiomatic Systems

17. Devise two additional concrete models for the axiom system in Example 1.2.1. Are these models isomorphic? Justify your result.

18. Devise two additional abstract models for the axiom system in Example 1.2.1.

19. Are the models in Problem 16 concrete or abstract? Explain your answer.

20. Explain why it is not possible to devise a concrete model of the real number system.

21. Demonstrate the independence of Axioms 2 through 4 in Example 1.2.1.

22. Devise two concrete models for the axiom set in Problem 5. Are these models isomorphic? Are they isomorphic to the model found in Problem 15? Justify your results.

23. Devise an abstract model for the axiom set in Problem 5. Is this model isomorphic to those in Problem 22? Justify your results.

24. Demonstrate the independence of the three axioms in Problem 5.

25. Consider a set of undefined elements, S, and the undefined relation R which satisfies the following axioms:

  AXIOM 1.   If $a \in S$, then a R a.

  AXIOM 2.   If $a, b \in S$ and a R b, then b R a.

  AXIOM 3.   If $a, b, c \in S$, a R b, and b R c, then a R c.

  (a) Devise a concrete model for the axiom system.
  (b) Devise an abstract model for this system.
  (c) Are the axioms independent?

1.3 FINITE GEOMETRIES

In this section we will investigate several geometries, each having a small number of axioms and theorems and only a finite number of points. These *finite geometries*, as we call them, afford us the opportunity to study a geometry of relatively simple structure using the axiomatic method. In these investigations we will begin to recognize the importance of relying on our axioms and on the underlying logical structure in determining the "truth" of our theorems.[13]

Four-Point Geometry

The *Four-point geometry*, which, as you will see, derives its name from its first axiom, has as its undefined terms point, line, and "on".[14] The following set of three axioms will be assumed:

AXIOM 1.   There exist exactly four points.

AXIOM 2.   Any two distinct points have exactly one line on both of them.

AXIOM 3.   Each line is on exactly two points.

Now as mentioned in a previous section, models often offer insight into an axiomatic system. If points are interpreted as dots on the paper and lines as pencil lines, a model of the four-point geometry can be represented by many drawings, three of which are shown in Figure 1.3.1.

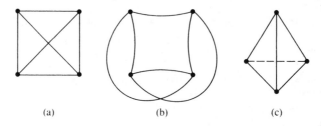

          (a)   (b)   (c)

Figure 1.3.1

The reader should verify that all three axioms apply to all three figures.

DEFINITION 1.3.1.   Two lines on the same point are said to *intersect* and are called *intersecting lines*.

---

[13] The reader is reminded that by our willingness to assume the axioms, we establish the axiom set as the only basis for determining "truth" in our system, and where intuition runs contrary, the axioms are the ultimate truth.

[14] Various expressions can also be used to represent "on." For example, each of the following has the same meaning: A point is "on" a line; a line "contains" a point; or a line "goes through" a point.

DEFINITION 1.3.2.   Two lines that do not intersect are called *parallel lines*.

FOUR-POINT THEOREM 1.   In the Four-point geometry, if two distinct lines intersect, then they have exactly one point in common.

Proof. By Definition 1.3.1, two distinct intersecting lines have at least one point in common, and Axiom 2 prohibits them from having more than one in common.

FOUR-POINT THEOREM 2.   The Four-point geometry has exactly six lines.

Proof. From Axiom 2, each pair of points has exactly one line on both of them, and Axiom 1 provides four points. Thus, by simple combinatorics, there must exist six pairs of points, hence six lines. Axiom 3 guarantees no more and no less.

FOUR-POINT THEOREM 3.   Each point of the Four-point geometry has exactly three lines on it.

Proof. By Axiom 2, each point has a line in common with each of the other three points. Therefore we have at least three lines on each point. Suppose a fourth line was on one of the given points, then, by Axiom 3, it must be on one of the other points, but this would violate Axiom 2. Therefore there are exactly three lines on each point.

FOUR-POINT THEOREM 4.   In the Four-point geometry, each distinct line has exactly one line parallel to it.

Proof. Axioms 1 and 3 provide us with a line *l* and a point *P* not on *l*. Four-point Theorem 3 tells us that there are exactly three lines on *P*, and Axiom 2 tells us that two of them must intersect *l*. Therefore we have at least one line parallel to *l*. Suppose that there was a second line parallel to *l*. This line could not contain *P* without violating Four-point Theorem 3, and since it is parallel to *l*, it cannot contain either of the points on *l*. Now either the second parallel contains only one point, which violates Axiom 3, or there exists a fifth point, which violates Axiom 1. Therefore the second parallel cannot exist, and there exists exactly one.

Alternate Proof. Since this geometry is finite, it is possible to examine every possible case of points and lines. By using Figure 1.3.2, where the points are represented by the letters *A*, *B*, *C*, and *D* and the lines by columns of letters, we may check directly to see that two distinct lines intersect in exactly one point, that there must be exactly six lines, that each point has

$$l_1 \quad l_2 \quad l_3 \quad l_4 \quad l_5 \quad l_6$$

$$A \quad A \quad A \quad B \quad B \quad C$$

$$B \quad C \quad D \quad C \quad D \quad D \quad \textbf{Figure 1.3.2}$$

exactly three lines on it, and that each line has exactly one line parallel to it.[15]

The Geometries of Fano and Young

The next two finite geometries we will investigate take on a slightly different character than those of the last section, in that their axioms do not explicitly state the number of points or the number of lines in the geometry. We must make use of the interrelationships among the axioms to discover and prove the exact number of points and lines as theorems.

The first geometry is of historical significance, since it was Gino Fano who first initiated the study of finite geometries. In 1892, Fano considered a finite three-dimensional geometry consisting of 15 points, 35 lines, and 15 planes. One of those planes yields the finite geometry presented here. The following five axioms completely characterize what we will call *Fano's geometry*, with point, line, and "on" serving as the undefined terms:

AXIOM 1.   There exists at least one line.

AXIOM 2.   There are exactly three points on every line.

AXIOM 3.   Not all points are on the same line.

AXIOM 4.   There is exactly one line on any two distinct points.

AXIOM 5.   There is at least one point on any two distinct lines.

At this time, the reader is encouraged to use the axioms to devise a model for Fano's geometry before progressing further. Two representations of a model for Fano's geometry can be found in Figures 1.3.3a and 1.3.3b. As we begin to make conjectures about Fano's geometry, we may find the representation contained in Figure 1.3.3b to be of greater assistance. Since not every set of three points forms a line in the geometry, Figure

---

[15] This method is allowable, since the geometry is finite and therefore the number of cases must be finite. Should a proof involve a geometry that has an infinite number of points or an infinite number of lines, the number of cases necessary to check would in turn be infinite and thus impossible to accomplish.

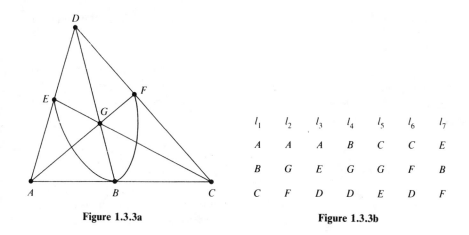

l_1	l_2	l_3	l_4	l_5	l_6	l_7
A	A	A	B	C	C	E
B	G	E	G	G	F	B
C	F	D	D	E	D	F

Figure 1.3.3a   **Figure 1.3.3b**

1.3.3a is subject to misinterpretation, whereas in Figure 1.3.3b the lines are explicitly stated.

FANO'S THEOREM 1.   In Fano's geometry, two distinct lines have exactly one point in common.

Proof. Figure 1.3.3b is easily checked for this result. However, the reader is encouraged to provide a proof without directly referring to the model. (See Exercise Set 1.3, Problem 13.)

FANO'S THEOREM 2.   Fano's geometry contains exactly seven points and seven lines.

Proof. Axioms 1 through 3 provide us with at least four points, three of which are on a line *l*, and one point *P* which is not on *l*. Now by Axiom 4, *P* and each point on line *l* must determine a distinct line, and by Axiom 2, each of these lines must contain three points. These three points cannot be any of the original four points without violating Axiom 4; therefore we have at least seven points. We must now show that there cannot be more than seven points. Assume that there exists a distinct eighth point *Q*. Points *P* and *Q* must determine a line *m*, which by Axiom 5 must intersect *l*. The point of intersection cannot be any of the three points on *l* without violating Axiom 4; therefore *l* must contain a fourth point, which contradicts Axiom 2. The proof that Fano's geometry has exactly seven lines is left for the reader. (See Exercise Set 1.3, Problem 14.)

It should be noted that if we define parallel lines as before, Fano's geometry is an example of a geometry that has no parallel lines. Additional theorems in Fano's geometry can be found in the exercises at the end of the section.

The second geometry we will investigate provides an example of how changing only one axiom can create a distinctly different system. The new geometry, called *Young's geometry*, assumes the first four axioms of Fano's geometry along with the following replacement for Axiom 5.

AXIOM 5.   For each line *l* and each point *P* not on *l*, there exists exactly one line on *P* that does not contain any points on *l*.

Figure 1.3.4 depicts a model of Young's geometry in which the points are the letters *A* through *I* and the lines are columns of letters:

l_1	l_2	l_3	l_4	l_5	l_6	l_7	l_8	l_9	l_{10}	l_{11}	l_{12}
A	A	A	B	B	B	C	C	D	D	G	H
B	D	E	E	D	F	F	E	E	H	H	F
C	G	I	H	I	G	I	G	F	C	I	A

Figure 1.3.4

Closer inspection of Axiom 5 in Young's geometry indicates that this geometry does indeed have parallel lines.

YOUNG'S THEOREM 1.   Every point in Young's geometry is on at least four lines.

Proof. Let *P* be any point, and let *l* be any line that does not contain *P*. By Axiom 2, *l* contains exactly three points, and by Axiom 4, *P* and each point on line *l* must determine a distinct line; therefore we have at least three lines. Now by Axiom 5, there must be a line that contains *P* but contains no points on *l*; thus we have at least four lines.

YOUNG'S THEOREM 2.   Young's geometry contains exactly nine points. The proof is left as an exercise. See exercise set 1.3, problem 23.

YOUNG'S THEOREM 3.   Young's geometry contains exactly 12 lines. The proof is left as an exercise. See exercise set 1.3, problem 24.

In the next section we will investigate another characteristic type of axiomatic geometry that is not necessarily finite in nature.

EXERCISE SET 1.3

Finite Geometries

1. Verify that Four-point Theorems 1 through 4 are "correct" in each of the drawings in Figure 1.3.1.
2. Devise a concrete model for the Four-point geometry.
3. Devise an abstract model for the Four-point geometry.
4. Prove that there exists a set of two lines in the Four-point geometry that contains all the points of the geometry.
5. Rewrite each of the axioms of the Four-point geometry interchanging the words "point" and "line" (and making appropriate adjustments in the grammar and definitions[16]). Each of these new axioms is called a *plane dual*; the resulting geometry is called the *Four-line geometry*.
6. Rewrite each of the theorems of the Four-point geometry interchanging the words "point" and "line." These new statements are theorems in the Four-line geometry.[17]
7. If points are interpreted as dots on the paper and lines as dashes, make a drawing representing a model of the Four-line geometry.
8. Without using the principle of duality, prove that the Four-line geometry has exactly six points.
9. Without using the principle of duality, prove that each line of the Four-line geometry has exactly three points on it.
10. Without using the principle of duality, prove that in the Four-line geometry a set of two lines cannot contain all the points of the geometry.
11. Suppose that Axiom 1 of the Four-point geometry is changed to read as follows: There exist exactly five points. Also suppose that Axioms 2 and 3 remain the same.
    (a) Make a drawing representing the new five-point geometry.
    (b) State and prove at least two theorems in the five-point geometry. (*Hint:* See Exercise Set 1.2, Problems 5 through 9.)

The Geometries of Fano and Young

12. Devise a model that demonstrates the absolute consistency of Fano's geometry.
13. Prove Fano's Theorem 1 without using the models.
14. Complete the proof of Fano's Theorem 2 by proving that Fano's geometry has exactly seven lines.
15. In Fano's geometry prove that each point is on exactly three lines.
16. In Fano's geometry prove that the set of all lines on any point contains all the points of the geometry.
17. In Fano's geometry prove that for any pair of points there exists exactly two lines containing neither point.
18. In Fano's geometry prove that for any set of three nonconcurrent lines there exists exactly one point not on any of the three lines.
19. Show that each of the axioms in Fano's geometry is independent.

---

[16] For example, parallel points are points that do not lie on the same line.

[17] The concept of planar duality assures us that the plane dual of any valid theorem in one geometry is also a valid theorem in the new geometry.

20. Write the plane dual for each of the axioms of Fano's geometry and draw a representation of a model satisfying these axioms.
21. Devise a dot-and-pencil-line model for Young's geometry.
22. Prove the stronger version of Young's Theorem 1 in which each point must lie on *exactly* four lines.
23. Prove Young's Theorem 2.
24. Prove Young's Theorem 3.
25. In Young's geometry, prove that every line has exactly two lines parallel to it.
26. In Young's geometry, prove that two lines parallel to a third line are parallel to each other.
27. In Young's geometry, suppose that Axiom 2 is changed to read as follows: There are exactly two points on every line. How many points and lines would the geometry have? What if every line had exactly four points? Generalize your result for the case where each line contains exactly *n* points (*n* being some positive integer).
28. Describe the similarities and/or differences between Young's geometry and Fano's geometry.

1.4 AXIOMS FOR INCIDENCE GEOMETRY

In the previous two sections we applied the axiomatic method in a geometric setting to prove results in several examples of finite geometries. Our past experience should indicate that geometries exist that do not have a finite number of points and lines. In this section we will investigate a set of axioms that does not explicitly state that the number of points or lines is finite. In particular, we will find that the axioms apply to both finite and infinite geometries.

As before, our undefined terms will consist of point, line, and "on", and we will assume the following four axioms as a basis for our geometry.

INCIDENCE AXIOM 1. For each two distinct points there exists a unique line on both of them.

INCIDENCE AXIOM 2. For every line there exist at least two distinct points on it.

INCIDENCE AXIOM 3. There exist at least three distinct points.

INCIDENCE AXIOM 4. Not all points lie on the same line.

Any geometry that satisfies all four incidence axioms will be called an *incidence geometry*.

Example 1.4.1

Consider the Four-point geometry in Section 1.3. Axiom 2 is essentially the same as Incidence Axiom 1, Axiom 3 implies Incidence Axiom 2, and Axioms 1 and 2 together imply Incidence Axioms 3 and 4. Therefore the Four-point geometry is an incidence geometry.

Example 1.4.2

Fano's geometry and Young's geometry are examples of incidence geometries.

Example 1.4.3

Recall your high school geometry and consider a fixed circle in the plane. Suppose that we interpret "point" as any point in the interior of the circle, and interpret "line" as an open chord of the circle. This model represents an incidence geometry containing an infinite number of points.

Example 1.4.4

The Four-line geometry discussed in Problem 5 in Exercise Set 1.3 is not an incidence geometry since Incidence Axiom 1 is not satisfied.

The following theorems represent only a sample of the types of theorems that can be proven using only the axioms of incidence geometry. Additional theorems in a variety of different incidence geometries can be found in the exercises at the end of this section.

INCIDENCE THEOREM 1.   If two distinct lines intersect, then the intersection is exactly one point.

Proof. If lines l and m intersect, then by definition the intersection is at least one point P. Now if we assume that l and m share a second distinct point Q, we have different lines each containing distinct points P and Q. But this violates Incidence Axiom 1.

INCIDENCE THEOREM 2.   For each point there exist at least two lines containing it.

Proof. First we observe, as a consequence of Incidence Axioms 3 and 4, that for every point P there is at least one line l not containing P. Now Incidence Axiom 2 provides that l must contain at least two points, and by Incidence Axiom 1, P and each of these points determines a unique line. Therefore there are at least two lines on P.

INCIDENCE THEOREM 3.   There exist three lines that do not share a common point.

Proof. By Axioms 3 and 4, our incidence geometry must have three noncollinear points. These three points must pairwise determine distinct lines. Therefore there are at least three lines, and these lines cannot all share the same point.

At this stage we may wish to consider the question of the existence of parallel lines (nonintersecting lines) in incidence geometry. Our incidence axioms do not explicitly state that parallel lines exist, and therefore, we may ask, Can we prove that they exist? It is here that investigating models of incidence geometry can prove to be helpful. Since Fano's geometry is a model of incidence geometry that has no parallel lines, it should be clear that the existence of parallel lines cannot be deduced from the axioms. Therefore if we are to prove that a model of incidence geometry has parallel lines, they must be the result of an axiom or its consequences. For example, Young's geometry is an incidence geometry that has parallel lines as a consequence of its Axiom 5.

If we consider any line l and any point P, where P is not on l, then three possibilities exist for a parallel axiom.

1. There exist no lines on P that are parallel to l.
2. There exists exactly one line on P that is parallel to l, or
3. There exists more than one line on P parallel to l.

As we will discover in subsequent chapters, an incidence geometry that assumes alternative 2 above or its axioms imply some equivalent statement is said to be *Euclidean* or to have the *Euclidean parallel property*. If the incidence geometry assumes alternative 1 or 3, or its axioms imply some equivalent statement, then it is said to be *non-Euclidean*.

Example 1.4.5

The Four-point geometry has the Euclidean parallel property since its axioms imply alternative 2.

Example 1.4.6

Young's geometry has the Euclidean parallel property since its Axiom 5 is equivalent to alternative 2.

Example 1.4.7

Fano's geometry has the non-Euclidean property since its Axiom 5 is equivalent to alternative 1.

Example 1.4.8

The five-point geometry discussed in Problem 11 in Exercise Set 1.3.1 has the non-Euclidean property, since its axioms imply that for any line l and any point P not on line l there exist two lines through P and parallel to l, which satisfies alternative 3.

Example 1.4.9

The geometry in Example 1.4.3 has the non-Euclidean property since it satisfies alternative 3.

In the first section of this chapter we made a brief reference to the systematic study of geometry made by Euclid in the *Elements*. The question that naturally arises seems to be, How does Euclid's development fare when viewed as an axiomatic system in the context of this chapter? In the next chapter we shall attempt to answer this question and in doing so gain further insight into our own investigation of geometry.

EXERCISE SET 1.4

1. Verify that Fano's geometry and Young's geometry are both incidence geometries.
2. Show that the axioms of incidence geometry are independent.
3. Determine which of the following interpretations of the undefined terms are models of incidence geometry and indicate which parallel alternative is exhibited.
   (a) Points are points on a Euclidean plane, and lines are nondegenerate circles in the Euclidean plane.
   (b) Points are points on a Euclidean plane, and lines are all those lines on the plane that pass through a given fixed point P.
   (c) Points are points on a Euclidean plane, and lines are concentric circles all having the same fixed center.
   (d) Points are Euclidean points in the interior of a fixed circle, and lines are the parts of Euclidean lines that intersect the interior of the circle.
   (e) Points are points on the surface of a Euclidean sphere, and lines are great circles on the surface of that sphere.
   (f) The same as part (e) except that any two points that lie on opposite ends of a diameter are identified as the same point.
   (g) Points are points on a Euclidean hemisphere (not including those points on the great circle that define the hemisphere) and lines are great semicircles (i.e., those points on a great circle that intersect with the hemisphere).
   (h) In Euclidean 3-space, points are interpreted as lines and lines are interpreted as Euclidean planes.
   (i) Points are lines in Euclidean 3-space, and lines are Euclidean planes all of which contain the same fixed line.
4. An incidence geometry that exhibits the Euclidean parallel property is called an *affine geometry*. Which of the geometries discussed in Sections 1.3 and 1.4 are affine geometries?
5. Prove that in a finite affine geometry all lines must contain the same number of points.
6. In an affine geometry prove that if a line is parallel to one of two intersecting lines, then it must intersect the other.

7. In a finite affine geometry prove that if every line contains exactly n points, then every point has exactly $(n + 1)$ lines on it.

8. In a finite affine geometry prove that if every line contains exactly n points, then there are exactly n^2 points and $n(n + 1)$ lines in the geometry. (See Exercise Set 1.3, Problem 27.)

9. An incidence geometry having no parallel lines (parallel alternative 1) and in which each line has at least three points is called a *projective geometry*. Which of the geometries discussed in Sections 1.3 and 1.4 are projective geometries?

10. Consider Fano's geometry with Axiom 2 replaced by the following: There are exactly four points on every line. Devise a model for this geometry. Is this new geometry affine or projective? Prove your result.

MANY WAYS TO GO

Axiom Sets for Geometry

2.1 INTRODUCTION

Historically, the development of geometry can be characterized as a transition from inductive discoveries to increasingly sophisticated deductive processes. Early geometers were interested in results only insofar as they were useful to everyday projects. This was true for other areas of mathematics as well. The ancient Babylonians had, for their time, an advanced arithmetic system that evolved because of their extensive involvement in trade and commerce. For much the same reason, the Egyptians made progress in geometry because of practical considerations (e.g., the regular flooding of the Nile River which washed away property lines, requiring frequent surveys for reconstruction of those lines). The Egyptians had little or no concern for the theory behind their geometry. It was of no interest to them *why* their geometry worked. They cared only that the methods were successful, and they had an abundance of empirical evidence to support the validity of their techniques. Much of Egyptian geometry was based on trial and error, and many of their results were approximations. Since the Egyptians were interested only in the application of geometric ideas, none of this was viewed as a flaw.

As Greek mathematicians and philosophers became more influential in the Mediterranean region, the study of geometry underwent significant changes. Geometry was viewed as an abstraction of the real world, an idealized model, that served as a setting in which pure reason could be exercised. Practical considerations were of no importance, and the significance of the subject lay in the soundness of the logical processes employed. To a large extent all of this had occurred by about 250 B.C., when Euclid produced his monumental work, the *Elements*. The *Elements* endured as the most influential work in geometry for more than 2000 years until rather recently when modern mathematicians began to raise and answer questions concerning issues that had been overlooked, avoided, or ignored during that time span.

In this chapter we will informally assess Euclid's *Elements* as a formal axiomatic system and compare it with alternative systems that have been offered during the last century or so in preparation for a more formal development of Euclidean and non-Euclidean geometries in Chapters 3 through 6. Since the material is mostly preparatory in nature, the few proofs offered will be given rather loosely. Similarly, the exercises given in this chapter need not be approached with extreme rigor. There is time enough in Chapters 3 through 7 to provide a more rigorous development of the geometries we shall encounter along the roads to geometry. Since, chronologically, the first important deductive presentation of geometry was given in the *Elements*, we begin with a look at geometry as it must have been perceived by Euclid.

2.2 EUCLID'S GEOMETRY AND EUCLID'S *ELEMENTS*

The Greek mathematicians of Euclid's time thought of geometry as an abstract model of the world in which they lived. The notions of point, line, plane (or surface) and so on, were meant to be consistent with human perceptions. The postulates and axioms (or common notions) were for the most part commonsense ideas that provoked little argument (with one major and one minor exception which we will discuss in detail later in this and other chapters). Euclid's task, as he saw it, was to define all the necessary terms, state the necessary postulates, and apply sound logic to derive the theorems that constitute the geometry. More recent geometers (such as David Hilbert and G. D. Birkhoff) have offered approaches to Euclid's geometry (i.e., the set of theorems derived by Euclid) that use different axiom sets and different approaches to the definition of terms.

In contrast, we shall see that other mathematicians have constructed non-Euclidean geometries (i.e., geometric systems that include theorems that contradict theorems proved by Euclid). This is possible only because in these geometries postulates are assumed that contradict some of those posited by Euclid. We begin the survey with a discussion of Euclid's geometry as he presented it 2200 years ago in the *Elements*.

One does not have to read past Book I of the *Elements* in order to obtain an accurate feeling for the entire work. Although it may be a bit unfair to do so, we will, in this section, evaluate Book I of the *Elements* as an axiomatic system using the criteria discussed in Chapter 1.

Euclid begins his exposition by listing 23 definitions, 2 of which are shown here (a complete list of Euclid's definitions for Book I may be found in Appendix A):

DEFINITIONS

1. A *point* is that which has no part.
2. A *line* is breadthless length.

In Euclid's time the need for undefined terms was not yet recognized (this is evident from Definition 1). If a point is to be defined as "that which has no part," the term "part" is left undefined. If in trying to overcome this problem we give a definition for "part," we will either introduce a new undefined term (a part is that which occupies *area*) or give a circular definition that involves, in some way, the term "point" (a part is a collection of points). This dilemma points up the need to leave some geometric terms undefined. Most would agree that we have a better intuitive feeling for the geometric concept of a point than of a part. Because of this, most modern geometers have chosen to leave "point" as an undefined term. A similar problem arises in Definition 2 above, since it refers to breadth and length, neither of which has been defined.

From this we see that Euclid failed to recognize the need for certain terms to be left undefined. This is not surprising. The early Greek approach to axiomatics did not call for a listing of undefined terms (or primitives). Euclid's perception of a point was a physical point "shrunk to nothing" which could then be thought of in real terms. Consequently, his definition attempted to portray the idea of a "sizeless physical entity." This and other definitions can be more accurately called descriptions than definitions, although the Greeks of Euclid's time did not recognize the need to distinguish between these two ideas. It wasn't until about 2000 years later that mathematicians agreed on the requirements of the axiomatic method that we discussed in Chapter 1. The notion of an undefined term is actually a relatively recent idea.

We can easily pardon Euclid for failing to recognize the need for undefined terms (he was, after all, a product of his environment). And we should give him credit for recognizing the need for a mutual understanding of the terms used in developing the system. Let us then direct our attention toward another essential component of a formal axiomatic system: the axioms (or postulates).

Euclid actually made a distinction between two types of assumptions.

He used the term "postulate" to refer to assumptions that were specific to geometry. Common notions (often called axioms), on the other hand, were assumptions used throughout mathematics and not specifically linked to geometry. Both sets of assumptions may be found in Appendix A, but for convenience the postulates are listed here.

POSTULATES

Let the following be postulated:

1. To draw a straight line from any point to any point.
2. To produce a finite straight line continuously in a straight line.
3. To describe a circle with any center and distance.
4. That all right angles are equal to one another.
5. That, if a straight line falling on two straight lines makes the interior angles on the same side less than two right angles, the two straight lines, if produced indefinitely, meet on that side on which the angles are less than two right angles.

We will now consider the following question: How does this axiom set fare in view of the contemporary requirements that an axiom set be consistent, independent, and complete? We begin with a discussion of completeness.

As discussed in Chapter 1, an axiom system is considered complete if it is impossible to add a new independent axiom that is consistent with the other axioms and contains no new undefined terms. With this in mind consider the following "sixth postulate" for Euclidean geometry.

6. There exist at least three points that are not on the same line.

Is this assumption independent of the others? Since existence of points is mentioned nowhere in Postulates 1 through 5, those postulates could not be used to establish the existence of points, collinear or not. An implication of this independence is that Euclid's set of axioms is not complete.

Is Postulate 6 consistent with the others? If so, we should be able to construct a model displaying all six axioms simultaneously. To show consistency we will use (as a model) what is normally thought of as the Euclidean plane: a piece of infinitely long and wide paper. If we note that triangles exist in the plane and that their vertices are not on a single line, we clearly have a model that exhibits all six postulates at once. From this we may conclude that the sixth postulate is consistent with the others.

Does this axiom involve any new undefined terms? Clearly, no.[1] As a result of this discussion, we may conclude that the axiom set offered by Euclid is *not* complete, in the sense that we may add a sixth independent postulate that is consistent with the others without using new undefined terms. This is not to say that the addition of Postulate 6 makes the set complete. But the preceding discussion does point out an oversight made by Euclid that has been noted by many recent geometers: Unless the existence of points is explicitly postulated, the existence of points may not be assumed and, as a result, geometries lacking an existence postulate could be vacuous. When proposing a geometric system, modern geometers always postulate the existence of points in some way in order to guarantee that the theorems apply to a nontrivial model.

Euclid's failure to begin with a complete axiom set was a major shortcoming of his work. In the absence of various postulates (such as the existence of points), Euclid often tacitly assumed things he felt were obvious. The problem, of course, is not that these assumptions were invalid, but rather that they were not stated explicitly.

For example, Proposition (or Theorem) 1 of Book I states that it is possible to construct an equilateral triangle on any given line segment. Euclid's proof for this proposition can be summarized as follows (refer to Figure 2.2.1).

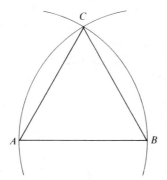

Figure 2.2.1

In order to construct an equilateral triangle with $\overline{AB}$ as a side, Euclid first constructs a circle using point A as the center and $\overline{AB}$ as a radius[2] (we will call this circle A). Likewise, he constructs circle B with its center at B

---

[1] Postulate 6, in one sense, does contain a new term—"on" (or incidence of point to line). Euclid failed to provide a definition for the incidence relation, although an understanding of this relation is assumed throughout.

[2] The term "radius" is used in two ways. $\overline{AB}$ *can* serve as *a* radius of the circle, while the length of $\overline{AB}$ is *the* radius of the circle. The intent is generally clear from the context.

and radius $\overline{AB}$. C is used to denote the point of intersection of circles A and B. Since C is on circle A, $AC = AB$.[3] Since C is also on circle B, $BC = AB$. Consequently, $AB = AC = BC$, so that $\triangle ABC$ is equilateral. Is the reasoning sound? Yes, providing the geometric model conforms to our intuition. In other words, this argument is reasonable if we are working in the traditional Euclidean plane. Unfortunately, the postulates listed by Euclid are not sufficient to guarantee that the geometry is Euclidean in nature. Suppose, for example, we take as the model a coordinatized plane (e.g., a piece of graph paper) consisting only of points with rational coordinates. Now suppose that A is the origin $(0,0)$ and B is $(1,0)$. With some work we can show that the coordinates of point C are $(\frac{1}{2}, \sqrt{3}/2)$. Unfortunately, $\sqrt{3}/2$ is irrational, so C is not a point in this plane. This means that circles A and B *do not* intersect, so that the proof fails by assuming the existence of a point that does not exist in the model.

Of course, you may object by saying that the plane in which we are working *does* contain points with irrational coordinates. This is true if we are working in what we usually think of as Euclidean geometry. However, there is nothing in the postulate set that guarantees that we are. If we are to reason solely from the postulates, and not at all from intuition, postulates are needed that give substance to our intuitions. Identifying the needed postulates isn't as easy as you might at first think. We'll talk more about this when we discuss the postulate sets of Hilbert and Birkhoff later in this chapter.

Euclid made other tacit assumptions in some of his proofs. Proposition 2 in Book I proposes the following way: "To place at a given point (as an extremity) a straight line equal to a given straight line." Euclid's proof for this theorem is summarized as follows (refer to Figure 2.2.2).

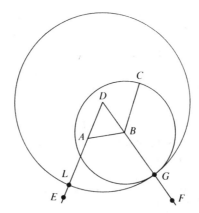

Figure 2.2.2

---

[3] The notation AB will be used to denote the distance between A and B, or the length of line segment $\overline{AB}$.

$\overline{BC}$ is the given straight line that we will replicate at point A (the given point). To do this Euclid draws $\overline{AB}$ and then applies Proposition 1 to construct equilateral triangle $\triangle ABD$. He then extends rays[4] $\overrightarrow{DA}$ and $\overrightarrow{DB}$ arbitrarily to points E and F. Next he constructs a circle with its center at B and with a radius of length BC (which postulate assures us that this can be done?) and uses G to represent the intersection of this circle and $\overrightarrow{DB}$. Euclid then constructs a second circle with its center at D and radius $\overline{DG}$. The intersection of this circle and $\overrightarrow{DA}$ is called L. Since $BC = BG$, $DL = DG$, and $DA = BD$, we may conclude that $AL = BG = BC$. Therefore $\overline{AL}$ is a segment with an extremity at A and with length equal to BC, completing the proof.

The proof is sound for the configuration as shown here, but what if the points are located differently? Proclus, in his *Commentary*, may have been the first to raise an objection to this: "Our geometer has taken a point lying off the line and at one side; but for the sake of patience we must consider all cases. . . ."[5] For instance, the proof as given by Euclid does not directly apply if point A is on $\overline{BC}$. Other possibilities include the case where A, B, and C are collinear, with B between A and C, or C between B and A. In the *Elements*, Euclid offered only one proof, using the configuration shown in Figure 2.2.2. If the proposition is to be proved for all cases, a more general demonstration is needed.

It turns out that with some modification the argument will apply to the choice of any three points A, B, and C and that the theorem can be proved regardless of the placement of these points. This is not always the case, however, when we allow pictures to guide us through a proof. Consider the following proof that all triangles are isosceles.

In $\triangle ABC$, allow the bisector of $\angle A$ to meet the perpendicular bisector of side $\overline{BC}$ at a point we shall call D. Construct $\overline{DF}$ and $\overline{DE}$, the perpendiculars from D to $\overline{AB}$ and $\overline{AC}$, and draw $\overline{DA}$, $\overline{DB}$, and $\overline{DC}$ (Figure 2.2.3). Using

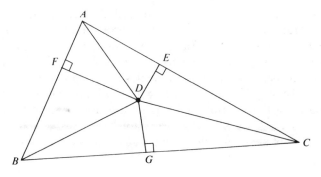

Figure 2.2.3

---

[4] The notation $\overrightarrow{DA}$ is used to denote the ray with endpoint D that contains the point A, which is distinct from D.

[5] Morrow, *Proclus—A Commentary*, p. 225.

the usual Euclidean congruence requirements, we may show that $\triangle ADF \cong \triangle ADE$ and $\triangle BGD \cong \triangle CGD$, hence $\triangle BDF \cong \triangle CDE$. If we then add the lengths of corresponding parts of congruent triangles, we find that $AB = AC$, and the triangle with which we started, and about which we assumed nothing (?), is isosceles. The flaw in this proof lies not so much in the logic but rather in assumptions made by referring to the figure. The hidden assumption is subtle and identifying it is left as an exercise.

In summary then we see that Euclid's presentation of plane geometry was flawed on at least three counts:

1. Failure to recognize the need for undefined terms or primitives.
2. Use of subtle but unstated postulates in the proof of theorems.
3. Reliance on diagrams to guide the logic in the construction of proofs.

We should not be too hard on Euclid, however. His work was a very early attempt to systematize geometry and is a remarkably good effort given its chronological location. The rules concerning the structure of acceptable axiomatic systems have changed greatly since 250 B.C., and it is a bit unfair to judge a work from that time period using the currently accepted standards. It wasn't until the late 1800s that modern geometers were able to construct consistent[6] and independent axiomatic sets for Euclid's geometry. We'll discuss a couple of these in the next two sections.

Another aspect of Euclid's *Elements* that needs further discussion is the nature of Postulate 5. A brief look at the five postulates from Book I should be sufficient for one to notice that Postulate 5 is far more complex than any other postulate. While Postulates 1 through 4 are self-evident to most, Postulate 5 usually requires several readings before the meaning becomes clear.

The statement of Postulate 5 says, directly, that two lines will meet to the right of a transversal that crosses them providing the sum of the measures of the interior angles to the right of the transversal is less than two right angles (180°). Similarly, the lines will meet on the left of the transversal if the sum of the measures of the interior angles on the left of the transversal is less than 180°. By implication, no intersection will take place when the sum of the measures of the interior angles is *exactly* 180°. The ramifications of this postulate are, as we will see, enormously important. Several of the direct (although not always immediate) consequences of this postulate are as follows:

1. Through a given point only one parallel can be drawn to a given line.

---

[6] The question of consistency of a set of Euclidean axioms is difficult. Relative consistency (see Chapter 1) is the best we can hope for. This issue is discussed in depth in Chapter 6.

2. There exists at least one triangle in which the sum of the measures of the interior angles is 180°.
3. There exists a pair of similar, but not congruent, triangles.
4. There exists a pair of straight lines that are everywhere equidistant from one another.
5. Every triangle can be circumscribed.
6. The sum of the measures of the interior angles of a triangle is the same for all triangles.

Because of the complexity of Postulate 5, many mathematicians thought that it could be shown to be a consequence of the other Euclidean postulates which were thought to be much more obvious. For nearly 2000 years after Euclid, geometers tried to show that Postulate 5 was a consequence of Postulates 1 through 4 by attempting to prove it using those postulates. In fact, if any of the direct consequences 1 through 6 shown above could be proved without Postulate 5, we could, with some effort, prove Postulate 5 by working backward and thereby reduce the number of Euclidean postulates to four. With this in mind, consider the following proof that the sum of the measures of the interior angles of a triangle is 180°.

THEOREM.   The sum of the measures of the interior angles of a triangle is 180°.

Proof. Let x represent the sum of the measures of the interior angles of a triangle (Figure 2.2.4).
Then,

$$m\angle 3 + m\angle 4 + m\angle 5 = x$$

and

$$m\angle 1 + m\angle 2 + m\angle 6 = x$$

Now

$$m\angle 1 + m\angle 2 + m\angle 3 + m\angle 4 + m\angle 5 + m\angle 6 = 2x$$

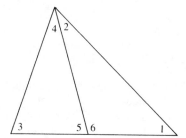

Figure 2.2.4

But

$$m\angle 1 + m\angle 2 + m\angle 4 + m\angle 3 = x$$

and

$$m\angle 5 + m\angle 6 = 180°$$

Therefore,

$$x + 180° = 2x \text{ and } x = 180°.$$

Note that nowhere in this proof is any mention made of parallel lines. In fact, it appears that this proof is in no way based on the parallel postulate. But it is. In fact, it subtly uses one of the six direct consequences of Postulate 5 listed earlier. What is that hidden assumption? (See Exercise Set 2.2, Problem 12.)

EXERCISE SET 2.2

1. Find a model of a geometry in which Euclid's 5th postulate is not valid. (*Hint:* Review some of the finite geometries in Chapter 1.)
2. What is the subtle flaw in the proof that all triangles are isosceles?
3. Proposition 3 in Book I of the *Elements* states the following: "Given two unequal straight lines, to cut off from the greater a straight line equal to the less."
   (a) Euclid refers to "unequal straight lines." In current terminology, two lines are equal if and only if they are the same set of points. What do you think Euclid is referring to when he talks of "unequal straight lines?"
   (b) Restate Proposition 3 using current terminology.
   (c) Provide a proof for Proposition 3.
4. Suppose the terms "point," "line," "between" (for points), and "congruent" are taken as undefined terms. Give a definition for each of the following terms. Are there other terms that need to be defined first? What are they and how might you define them?
   (a) Parallel lines
   (b) Perpendicular lines
   (c) Line segment
   (d) Ray
   (e) Circle
   (f) Square
5. Euclid's third postulate asserts that a circle can be drawn using any point as its center and any distance as its radius. Unfortunately, the term "distance" is not defined in the set of definitions.
   (a) Try to provide a satisfactory definition for distance.
   (b) Try to provide a satisfactory definition for a circle that does not require a definition for distance.
   (c) Try to restate Postulate 3 without using the term "distance."
6. Restate Euclid's fifth postulate in more understandable terms.
7. Euclid's fourth proposition in Book I states the SAS congruence condition. The proof proceeds by placing the first triangle on top of the second in such a way

that the congruent sides and angles coincide. Then he argues that the remaining side and angles must also coincide, hence be equal in measure. Is this a valid argument? Explain why or why not.

8. Earlier we saw that a sixth, independent, consistent, postulate could be appended to Euclid's postulate set without introducing new undefined terms, showing that Euclid's set of postulates is not complete. Which of the following postulates, if any, could be used as a sixth postulate to show that Euclid's postulate set is not complete?

   (a) There exist at least two points.
   (b) Not all points are collinear.
   (c) Between each pair of distinct points exists a point distinct from the two.
   (d) Every line contains at least two points.
   (e) Given a line and a point not on that line, there exists a unique line through the point that is parallel to the line.

9. Consider the following "postulate": Given any two distinct points A and B, there exists a third distinct point C which is between A and B.

   (a) Does this postulate contain any new undefined terms?
   (b) Do you think that this postulate is independent of Euclid's five postulates? Explain.

10. Does Postulate 5 imply the existence of parallel lines? Explain why or why not.

11. Explain why you think the following statement is (or is not) equivalent to Postulate 5: Given a line l and a point P not on l, there exists a line through P that is parallel to l.

12. What is the hidden assumption in the proof that the sum of the measures of the interior angles of a triangle is $180°$?

2.3 AN INTRODUCTION TO MODERN EUCLIDEAN GEOMETRIES

There is evidence that even Euclid himself recognized at least one of the shortcomings of the *Elements*. As one looks at the five postulates stated in Book I, it is clear that Euclid's fifth, or parallel, postulate is by far the most difficult to comprehend. Euclid implies a skepticism of Postulate 5 by avoiding its use until Proposition 29 when it is employed for the first time.

Proclus, in his *Commentary*, takes the position that the parallel postulate should be dropped from the list of postulates and added as a theorem. The same endeavor was undertaken by many other mathematicians over the course of the 2000 years following the introduction of the *Elements*. Nearly everyone who studies Postulate 5 notices that it lacks the obvious nature of Euclid's other four postulates. One reason for this is that Postulate 5 requires the reader to imagine the behavior of lines as they approach infinity. Another reason is the roundabout way that it addresses the concept of parallelism. It has long been a mystery why Euclid chose this form for the fifth postulate, although many conjectures concerning his rationale have been made. It will be shown in Chapter 3 that Euclid's fifth postulate is logically equivalent to the following, more straightforward, statement attributed to John Playfair (1795), which is commonly referred to as *Playfair's postulate*.

PLAYFAIR'S POSTULATE.   For every line *l* and every point *P* not on *l* there exists a unique line *m* that contains *P* and is parallel to *l*.

Most people find this statement easier to understand than Euclid's Postulate 5. Nonetheless, it can be shown that each implies the other, so that either can be used to construct a model in which all the theorems in the *Elements* are valid. Even Playfair's postulate, however, is less obvious than Euclid's other four, since it too involves imagining the behavior of lines as they extend without bound. Consequently, while Playfair helped clarify Euclid's parallel postulate, his contribution did not resolve the question of whether or not Postulate 5 is independent of the others. Mathematicians of Playfair's era continued to search for a proof of the parallel postulate using Postulates 1 through 4, in each instance without success.

One technique used in several of these attempts was an indirect approach in which one assumes that Postulate 5 is false and then attempts to show that this assumption leads to a contradiction. This result implies that Postulate 5 (or its equivalent) is the only consistent statement concerning parallels for a geometry that assumes Postulates 1 through 4.

We will discuss these attempts in greater detail in Chapter 6, but for now it will suffice to say that it was eventually shown that Postulate 5 *is* independent of the others and that consistent geometries can be derived using either Postulate 5 (i.e., Playfair's postulate) or a suitable negation of it. Of course, the use of contradictory parallel postulates results in geometries that contradict one another. Nevertheless, the geometries that result can and, in Chapter 6, will be shown to be consistent axiomatic systems.

Since most of us have intuitive notions that coincide with the Euclidean model of plane geometry, some valid theorems in a geometric model that negates Postulate 5 appear contradictory to us. However, within the axiomatic system there are no contradictions, so that the system is mathematically valid.

During the remainder of this chapter we will briefly discuss five axiom sets for geometry: three that produce Euclidean models and two that are non-Euclidean in nature since they negate Euclid's fifth postulate. We begin with a rather recent effort (less than 100 years old) that successfully preserves the flavor of the *Elements* but which corrects the flaws mentioned earlier.

EXERCISE SET 2.3

1. Write the negation of Playfair's postulate. Explain how this statement allows for two different non-Euclidean geometries.
2. Do you think that any of the propositions given by Euclid are false? If so, which ones? If not, explain why modern geometers felt the need to redo his historic work.

2.4 HILBERT'S MODEL FOR EUCLIDEAN GEOMETRY

During the nineteenth century, it became evident to mathematicians interested in geometry that there was no single axiom set that resulted in a single (i.e., universal) model of geometry. Rather, there evolved a consensus that the validity of a geometric axiomatic system is dependent on the consistency, independence, and completeness of the axiom set upon which it is built. To be sure, it was known that different axiom sets would result in different models, but the study of geometry was by then no longer restricted by the notion that every resulting model had to be consistent with the Euclidean model. Still the Euclidean model for geometry was by far the most intuitive, and since it was the model with the broadest historical basis, mathematicians took on the task of constructing an axiom set that would result in the theorems of Euclid but without the shortcomings mentioned earlier in this chapter. The most well known of these works, *Grundlagen der Geometrie (Foundations of Geometry)*, which is true to the spirit within which Euclid worked, was published in 1899 by a German named David Hilbert, arguably the most prominent mathematician of the era.

Unlike Euclid, Hilbert was well versed in the requirements of modern axiomatics. For undefined terms, Hilbert chose "point," "line," "plane," "on" (incidence of point to line), "between" (as a relation concerning three distinct points), and "congruence." It is a tribute to Hilbert's genius (and to his perseverance) that he was able to recognize that all the necessary terms of Euclidean plane geometry could be defined starting with this simple set of primitives. For example, Hilbert was able to distinguish between lines and line segments (something Euclid neglected to do) by using the following definition:

DEFINITION.   *Line Segment $\overline{AB}$*. The set of all points that are between point A and point B. Points A and B are called the *endpoints* of the segment.[7]

Hilbert partitioned his axioms into five groups: axioms of connection (or incidence), axioms of order, axioms of congruence, the axiom of parallels (Playfair's postulate), and axioms of continuity. The entire axiom set may be found in Appendix B, but for now let's consider just the axioms of connection.

Group I   Axioms of Connection

I-1. Through any two distinct points A, B, there is always a line m.

I-2. Through any two distinct points A, B, there is not more than one line m.

---

[7] Most recent geometers have defined the term "line segment" in such a way as to include the endpoints. Hilbert chose to exclude the endpoints.

1-3. On every line there exist at least two distinct points. There exist at least three points which are not on the same line.

1-4. Through any three points not on the same line, there is one and only one plane.

It would be difficult to fail to notice the contrast between Hilbert's Postulates I-1 and I-2 and Euclid's first postulate: "To draw a line from any point to any point."

Euclid certainly intended to imply the same idea but failed to do so as completely. Hilbert not only postulates the existence of a line between any two distinct points but also postulates the uniqueness of that line. (Euclid assumes this throughout but fails to postulate the idea.) In addition, since Postulate I-3 describes the existence of points, we are assured of having lines with which to work, a formality Euclid overlooked.

Since Hilbert's axiom set is intended to provide the basis for traditional Euclidean geometry, we should expect that all our intuitive ideas about plane geometry are valid in his axiomatic system. For example, suppose that *l* and *m* are lines. Is it possible that *l* and *m* intersect in more than one point? Since our intuition says no, but no axiom assures that intersecting lines intersect just once, we should hope that this property can be proved as a theorem. With this in mind, consider the following.

THEOREM. Two distinct lines cannot intersect in more than one point.

Proof. We will proceed indirectly and assume otherwise: Suppose that *l* and *m* intersect in two distinct points called *A* and *B*. Under this hypothesis *A* and *B* are distinct points contained by two distinct lines, a contradiction of Postulate I-2. Consequently, the assumption that *l* and *m* intersect in more than one point cannot be valid, so that we may conclude that pairs of distinct lines *do not* intersect in more than one point. Two points need to be made concerning this proof.

First, the theorem is so obviously true in Euclidean geometry that it might seem to qualify as an axiom. However, the very fact that the statement can be proved indicates that it is *not* independent of the others. Had Hilbert included it as an axiom, the set of axioms would not be independent. It is unfortunately true that things are complicated by having to prove obvious statements. Still the elegance of constructing an independent axiom set outweighs, for mathematicians of Hilbert's status, the added bother.

Second, notice the indirect nature of this proof: Assume otherwise—derive contradiction—conclude theorem. This technique was applied in many of the proofs from the finite geometries discussed in Chapter 1. If you are still uncomfortable reasoning in this fashion, you should work hard to

overcome this problem, since proofs of this kind are at the core of many of
the arguments in both Euclidean and non-Euclidean geometries. The exercise set at the end of this section requires a number of indirect proofs such as
this.

While Hilbert's axioms of connection are straightforward, some of the
other axioms and the implications that may be deduced from them are less
obvious. In particular, read the following four axioms of order (those concerning the undefined term "between"):

Group II   Axioms of Order

II-1. If point B is between points A and C (we will denote this by A-B-C),
then A, B, and C are distinct points on the same line and C-B-A.

II-2. For any two distinct points A and C there is at least one point B on the
line $\overleftrightarrow{AC}$ such that A-C-B.

II-3. If A, B, and C are three points on the same line, then exactly one is
between the other two.

II-4. Let A, B, and C be three points that are not on the same line, and let m
be a line in the plane containing A, B, and C that does not contain any
of the points A, B, or C. Then, if m contains a point of the segment $\overline{AB}$
it will also contain a point of segment $\overline{AC}$ or a point of segment $\overline{BC}$.

The purpose of the axioms of order is to give meaning to the undefined
term "between." Since "between" is a primitive, we have nothing to guarantee that the term fits our notion of betweenness except the axioms of the
system. While it may not be obvious at first, the betweenness relation implicitly defined by these axioms is consistent with our intuition.

The axioms of order imply many results we expect in a Euclidean
geometry. For example, Axiom II-2 allows us to conclude that if we can
locate two points A and C on a line (which we can by Axioms I-3 and I-1), we
can find a third point B on the line that is farther[8] from A than C is. Of
course, having assured the existence of B, we may reapply Axiom II-2 to
produce a fourth point, and then a fifth point, and so on. These ideas result
in the following theorem.

THEOREM.   Every line contains an infinity of points.

A similar argument can be used to establish theorems stating that lines
do not terminate at any point[9] and that the points of a line are serial in nature

---

[8] We haven't yet defined what is meant by "farther," but let's assume its usual meaning
for now.

[9] This is not quite the same as saying that lines are infinitely long, even though you can
always find a point "farther" from A than a previous one. This is a subtle distinction that can be
used to characterize two distinct branches of non-Euclidean geometry: hyperbolic and elliptic.

rather than cyclical. This assures us that the points of a line do not "wrap around," as would happen with a great circle on a sphere.

Some of the consequences of the axioms of order are self-evident. For example, we can prove the following obvious theorem.

THEOREM. If A and B are points, then there always exists a third point C, such that A-C-B.

This statement is not quite the same as Postulate II-2 (reread Postulate II-2 if this isn't clear) but is certainly a property we expect to be valid in the geometry. To show that it is, we begin by choosing an arbitrary line segment $\overline{AB}$ (Figure 2.4.1). Axiom I-3 assures us that there is a third point D that is not collinear with A and B. Axiom II-2 allows us to locate a point E so that A-D-E. Axioms I-1 and I-2 guarantee the unique line $\overleftrightarrow{EB}$, and a reapplication of Axiom II-2 allows us to place point F so that E-B-F. Now consider the three noncollinear points A, E, and B and the unique line that joins D and F. Axiom II-4 assures us that if $\overleftrightarrow{DF}$ contains a point of $\overline{AE}$ (in this case point D), then it must also contain a point on $\overline{EB}$ or $\overline{AB}$. Since $\overleftrightarrow{DF}$ intersects $\overleftrightarrow{EB}$ at point F, it cannot intersect at a second point on $\overline{EB}$ (which axiom excludes this possibility?) so that there is a point C on $\overline{AB}$ such that $\overleftrightarrow{DF} \cap \overline{AB} = C$. Since $\overline{AB}$ is comprised of the points that are between A and B, we have now located point C such that A-C-B.

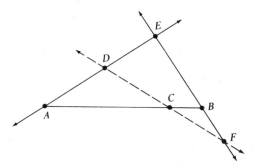

Figure 2.4.1

Axiom II-4, on which the preceding proof heavily relies, was actually stated first by the German mathematician Moritz Pasch in 1882 and is commonly referred to as *Pasch's axiom*. This axiom fills in a significant gap left by Euclid. Many of the proofs in the *Elements* rely on assumptions made from visual cues from diagrams. While the theorems Euclid proved in this way were true, it was found later that proofs of false statements could be constructed by using much the same type of reasoning.[10] So while Pasch's

[10] Remember the proof that all triangles are isosceles? This proof included hidden assumptions that resulted from placing too much emphasis on diagrams.

axiom may seem obvious (it is after all an assumption), it or a postulate equivalent to it is needed to provide valid means of proving several of Euclid's theorems.

Hilbert's third group of axioms, the axioms of congruence, listed here, are self-explanatory.

Group III   Axioms of Congruence

III-1. If A and B are two (distinct) points on line a, and if A' is a point on the same or another line a', then it is always possible to find a point B' on a given side of the line a' through A' such that segment $\overline{AB}$ is congruent to segment $\overline{A'B'}$.

III-2. If a segment $\overline{A'B'}$ and a segment $\overline{A''B''}$ are congruent to the same segment $\overline{AB}$, then the segment $\overline{A'B'}$ is also congruent to the segment $\overline{A''B''}$ or, briefly, if two segments are congruent to a third segment, they are congruent to each other.

III-3. On the line a, let $\overline{AB}$ and $\overline{BC}$ be two segments which except for B have no point in common. Furthermore, on the same or another line a', let $\overline{A'B'}$ and $\overline{B'C'}$ be two segments which except for B' have no point in common. In that case, if $\overline{AB} \cong \overline{A'B'}$ and $\overline{BC} \cong \overline{B'C'}$, then $\overline{AC} \cong \overline{A'C'}$. (This axiom expresses the additivity of segments.)

III-4. If $\angle ABC$ is an angle and if $\overrightarrow{B'C'}$ is a line, then there is exactly one point A' on each side of $\overrightarrow{B'C'}$ such that $\angle A'B'C' \cong \angle ABC$. Furthermore, every angle is congruent to itself. (This is often referred to as the *angle construction axiom.*)

III-5. If for two triangles $\triangle ABC$ and $\triangle A'B'C'$ the congruences $\overline{AB} \cong \overline{A'B'}$, $\overline{AC} \cong \overline{A'C'}$ and $\angle BAC \cong \angle B'A'C'$ are valid, then the congruence $\angle ABC \cong \angle A'B'C'$ is also satisfied.

Axiom III-4 says, in a somewhat indirect way, that any angle can be copied to a given ray, once on each side of the line that contains the ray. Axiom III-5 is in effect the SAS congruence condition (see Exercise Set 2.4, Problem 16). Euclid lists SAS as a theorem, however, the proof is flawed because he assumes the ability to move geometric figures around the plane without distortion. While this use of superposition is reasonable, any proof that makes use of this technique violates the rules of formal axiomatics unless a postulate is included to justify doing so. Eventually, a number of geometers legitimatized this technique by formally postulating the existence of geometric transformations that allow images of geometric figures to be used in the proof of a great many theorems. Hilbert chose not to travel the "transformation route," since an axiom set based on Euclidean transformations changes the nature of the development of Euclidean geometry, even though the results are clearly Euclidean. Since Euclidean transformations have become an important aspect of geometry, they will be discussed in detail in Chapter 5.

Hilbert, to his credit, recognized that at least a portion of the SAS congruence condition for triangles is truly independent of the other traditional Euclidean postulates. Consequently, he concluded that a postulate concerning the congruence of triangles was necessary. Since SAS is perhaps the most obvious of the Euclidean congruence conditions, Hilbert chose that one as his axiom of triangle congruence. As we shall see in Chapter 3, the other familiar congruence conditions can be derived as theorems.

By the time Hilbert presented his work in Euclidean geometry, it had already been established by Felix Klein that the parallel postulate is independent of the other, traditional, postulates for geometry. In order to help complete the axiom set, Hilbert needed to choose some form of a parallel postulate. He selected a form of Playfair's postulate (although in *Grundlagen der Geometrie* he calls it Euclid's axiom) as his axiom of parallels (Axiom IV-1), which for convenience is restated here.

IV-1. Let *a* be any line and *A* a point not on it. Then there is at most one line in the plane, determined by *a* and *A*, that passes through *A* and does not intersect *a*.

Note that this axiom postulates "at most one line" rather than "exactly one line parallel to *a*," even though in Euclidean geometry we expect to find exactly one parallel. The reason for this is, as we will prove in Chapter 3, that the other postulates imply the existence of at least one parallel so that to postulate exactly one would be somewhat redundant. In order to maintain the independence of the axioms, Hilbert opted for the weaker version of Playfair's postulate.

This left but one gaping hole to be filled in Euclid's geometry—the question of continuity of lines.[11] For convenience, the two Continuity postulates are stated here.

Group V   Axioms of Continuity

V-1. (*Axiom of Archimedes*) If $\overline{AB}$ and $\overline{CD}$ are any segments, then there exists a number *n* such that *n* copies of $\overline{CD}$ constructed contiguously from *A* along $\overline{AB}$ will pass beyond the point *B*.

V-2. (*Axiom of Line Completeness*) An extension of a set of points on a line with its order and congruence relations that would preserve the rela-

---

[11] The discussion of Euclid's proof for the construction of an equilateral triangle centered around the idea of continuity. Continuous lines or curves that cross must intersect in a Euclidean model. Hilbert's axioms of continuity address this issue.

tions existing among the original elements as well as the fundamental properties of line order and congruence that follow from Axioms I through III and V is impossible.

The idea behind Axiom V-1 is fairly straightforward, although the statement itself is confusing at first. The essence of the postulate is that line segments can be measured by units of an arbitrary size.

During the nineteenth century a number of mathematicians[12] studied geometries that involved spaces that were unbounded but finite. Distance in these and other geometries is measured in ways much different than in the Euclidean plane. Axiom V-1 guaranteed measurement of the usual (namely, Euclidean) type.

Axiom V-2 does not directly concern us in our discussion of Hilbert's geometry as an "upgrade" of Euclid's work. The completeness postulate is not necessary for the proof of any of the theorems in Euclidean geometry. By including it, Hilbert allowed himself to make the connection between his geometry and the formal systems of arithmetic that were being fine-tuned at the time. Hilbert's agenda involved establishing a one-to-one correspondence between the points on any line and the real numbers. This would allow him to declare the geometric system to be as consistent as the field of real numbers. Hilbert had no doubt that the consistency of the real numbers could be established, hence that his system would be also. This was a philosophical as well as a mathematical issue during that era which, alas, was not resolved to the satisfaction of Hilbert, and of many others.

Hilbert's geometry had mathematical significance beyond "patching up the holes" in Euclid's work. It was a classic example of the modern axiomatic method, and since it appeared at about the beginning of this century, it helped set the tone for much of twentieth century mathematical thought.

EXERCISE SET 2.4

Use Hilbert's axioms for Euclidean geometry to support informal proofs.

1. Have any of the theorems proved by Euclid in the *Elements* been shown to be false in Euclidean geometry? If so, which ones? If not, why did Hilbert feel the need to revamp Euclid's work?

2. Use Hilbert's axioms (you may assume that the full SAS condition has been established) to prove that the base angles of an isosceles triangle are congruent. (*Hint*: Prove that isosceles $\triangle ABC$ is congruent to $\triangle CBA$.)

3. Suppose that W, X, Y, and Z are points with X between W and Y and Y between W and Z. Prove, using Hilbert's axioms, that W, X, Y, and Z are distinct, collinear points.

---

[12] In particular, Bernhard Riemann and Felix Klein studied non-Euclidean geometries that denied the existence of parallels altogether. In doing so, they were forced to confront the notion of measure in a way that is vastly different from the Euclidean concept.

4. In geometry, if we choose a point on a line and consider that point along with all the points on one side of that point, we have what is usually called a *ray*. Give a more formal definition of the term "ray."

5. How many points are there on a line segment? Provide an argument in support of your response.

6. Explain what is meant by the following statement: Every line in a plane separates the plane into exactly two disjoint half-planes. Is this statement true? Would it be true if "line" were replaced by "line segment"? By "ray"? By "circle"?

7. Suppose you are given two distinct points X and Y and a line l that does not contain X or Y. Describe a criterion that could be used for determining whether the two points are in the same or different half-planes.

8. Is it possible for a line to intersect all three sides of a triangle? If so, does this contradict Axiom II-4? If not, explain why not.

9. Prove that if X, Y, and Z are points with Y between X and Z, then (a) $\overline{XY} \cap \overline{YZ} = Y$ and (b) $\overline{XY} \cup \overline{YZ} = \overline{XZ}$. (You may assume that line segments include their endpoints.)

10. Prove that two lines that are parallel to the same line are parallel to each other. (*Hint*: Proceed indirectly.)

11. What is meant by the term "interior of a triangle"? How can this term be defined formally? If a line contains a vertex of a triangle and interior points of the triangle, must it intersect the side opposite the vertex that it contains? Explain why or why not.

12. If a line intersects a side of a rectangle, must it intersect another side of the rectangle? Explain why or why not.

13. If a line intersects three sides of a rectangle, must it contain a vertex? Explain why or why not.

14. Hilbert did not define distance in his geometry. We can, however, allow for lengths to be compared by defining the relation "$<$" for segments as follows. $\overline{WX} < \overline{YZ}$ if and only if there is a point P between Y and Z such that $\overline{WX} \cong \overline{YP}$. Using this definition, show that for every set of points W, X, Y, and Z one of the following statements is true.

    (i) $\overline{WX} < \overline{YZ}$ (ii) $\overline{WX} \cong \overline{YZ}$ (iii) $\overline{YZ} < \overline{WX}$.

15. Show that the "$<$" relation defined in Problem 14 is transitive.

16. Congruence Axiom III-5 does not postulate, entirely, the SAS congruence con-

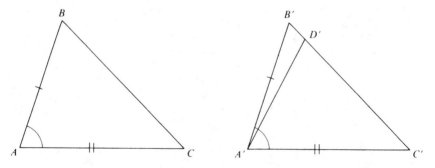

Figure 2.4.2   Given $\overline{AB} \cong \overline{A'B'}$, $\overline{AC} \cong \overline{A'C'}$, $\angle A \cong \angle A'$, show that $\overline{BC} \cong \overline{B'C'}$.

dition. By a change in notation it clearly implies that the remaining angles are congruent. We may not, however, conclude immediately that the remaining sides are congruent. Hilbert omitted postulating this final congruence because it can be proved as a theorem from the other axioms. The proof proceeds in the following manner. Suppose the remaining sides BC and $B'C'$ are _not_ congruent (Figure 2.4.2). (a) Under this hypothesis there is a unique point D' on $B'C'$ such that $D'C' = BC$. Which axiom guarantees this? (b) Next consider $\triangle ABC$ and $\triangle A'D'C'$. Verify that an application of Axiom III-5 to these triangles allows us to conclude that $\angle D'A'C' \cong \angle A$ and that this result, with the "given," contradicts Axiom III-4.

2.5 BIRKHOFF'S MODEL FOR EUCLIDEAN GEOMETRY

In the preceding section we saw that Hilbert was able to make Euclidean geometry rigorous by expanding Euclid's axiom set and by being meticulous in the logical development of the theorems. The price that Hilbert paid for the added rigor was an axiom set that was considerably larger than Euclid's and the need to provide rather difficult proofs for a number of "obvious" results, most notably in the area of betweenness. During Hilbert's lifetime a number of other geometers were also interested in this area of research, but Hilbert's work has survived as the effort that most nearly conforms to Euclid's approach.

An American mathematician, G. D. Birkhoff, provided a noteworthy exposition of Euclidean geometry in 1932.[13] Since Birkhoff's work, like Hilbert's, was constructed so as to be consistent with the work of Euclid, it provided no new results in terms of its theorems. Rather, its significance lies in the radically different set of axioms he chose to use in the development of the geometry. Birkhoff's axiom set is much smaller than Hilbert's (and even smaller in number than Euclid's). Birkhoff took as primitives the terms "point," "line," "distance," and "angle." His four postulates are as follows:

POSTULATE I.   The points A, B, . . . , of any line can be put into $1:1$ correspondence with the real numbers x so that $|x_b - x_a| = d(A,B)$[14] for all points A and B.

POSTULATE II.   One and only one line, l, contains any two distinct points P and Q.

POSTULATE III.   The half-lines (or rays) l, m, n, . . . , through any point O can be put into $1:1$ correspondence with the real numbers a (mod

[13] G. D. Birkhoff, "A Set of Postulates for Plane Geometry (Based on Scale and Protractor)," _Annals of Mathematics_, 33, 1932.

[14] Birkhoff used the notation $d(A,B)$ to represent the distance between A and B, and the notation x_a and x_b to represent the real numbers corresponding to points A and B, respectively.

2π) so that if A and B are points (other than O) of l and m, respectively, the difference $a_m - a_l$ (mod 2π) of the numbers associated with lines l and m is $m\angle AOB$.

POSTULATE IV.   If in two triangles $\triangle ABC$ and $\triangle A'B'C'$ and for some constant $k > 0$, $d(A',B') = kd(A,B)$, $d(A',C') = kd(A,C)$, and $m\angle B'A'C' = \pm m\angle BAC$, then also $d(B',C') = kd(B,C)$, $m\angle C'B'A' = m\angle CBA$, and $m\angle A'C'B' = m\angle ACB$.

And that's it. Whereas Hilbert needed 16 postulates[15] to derive the results of Euclid, Birkhoff was able to narrow the list down to 4. Of course, within Birkhoff's 4 stated postulates lies a wealth of other properties that occur as a consequence of postulating that (1) every line can be placed into a 1:1 correspondence with the real numbers, and (2) angles can be assigned unique measures in the range 0 to 2π (i.e., 0 to 360°).

For example, you will notice that no mention is made of the concept of betweenness in Birkhoff's postulate set. However, all the needed results concerning betweenness follow easily, since Birkhoff was able to define "between" in the following way.

DEFINITION.   If A, B, and C are distinct points, we say that B is *between* points A and C (A-B-C) if and only if $d(A,B) + d(B,C) = d(A,C)$.

From this definition and the field properties, all the needed results concerning betweenness follow immediately. It is, of course, the field properties that give this model the power needed to complete Euclid's geometry.

Postulate I is often referred to as the *postulate of linear measure* or the *ruler postulate*, since it allows measures (or lengths) to be assigned to each segment. From this postulate come the ideas needed in discussions of the congruence of line segments. Similarly, Postulate III has come to be known as the *protractor postulate*, since it provides a way to compare the size of angles. These two postulates, then, allow for complete development of the notions of congruence in general. Postulate IV does the same for the concept of similarity of geometric figures.

A more subtle consequence of these postulates involves the notion of parallelism as it applies to lines. Birkhoff proved the Euclidean fifth postulate in the following manner (Figure 2.5.1).

Suppose that A is a point not on line l and that B is the foot of a perpendicular drawn from A to l. Now let C be any point such that $\angle BAC$ is an acute angle. Clearly, the sum of the interior angles on the C side of $\overline{AB}$ is less than two right angles. In order to prove Euclid's fifth postulate, we

---

[15] Hilbert's postulates were revised over the course of years, and in *Grundlagen der Geometrie* he included some axioms for geometry in three dimensions. Thus the number of postulates may vary. The axiom set given here contains 16 postulates.

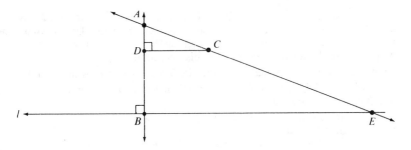

Figure 2.5.1

must show that $\overleftrightarrow{AC}$ meets l on the C side of $\overleftrightarrow{AB}$. To do this, construct $\overline{CD}$ perpendicular to $\overleftrightarrow{AB}$ as shown. Next locate E on l so that the following proportion is valid: $AD{:}DC = AB{:}BE$. Construct line $\overleftrightarrow{AE}$. What we now have is the SAS similarity condition (essentially Postulate IV), so that triangles $\triangle ADC$ and $\triangle ABE$ are similar. Consequently, $m\angle BAE = m\angle DAC$. The protractor postulate assures us that there is only one ray on the C side of $\overrightarrow{BA}$ that will do this. Consequently, $\overrightarrow{AC}$ must coincide with $\overrightarrow{AE}$, hence intersect line l at point E also. This shows that the postulate concerning triangle similarity can be used to establish Euclid's parallel postulate, so that any geometry that postulates similarity will be Euclidean in nature.

We see then that although Birkhoff opted for a significantly different set of postulates than Euclid or Hilbert did, the results are the same. While Hilbert was truer to the style of Euclid, the streamlined nature of Birkhoff's postulate set made it very attractive to many mathematicians. The ease with which one can address the issues of betweenness, congruence, and similarity (among other topics) made this approach pedagogically preferable to Hilbert's in many ways. In fact, the ruler and protractor postulates are standard in many current secondary school textbooks, since they allow for a rigorous, but not cumbersome, discussion of the related topics. In the following section we will study an axiom set offered by the School Mathematics Study Group (SMSG) in which the ruler and protractor postulates are assumed. The resulting axiom set is not independent, but the pedagogical advantages, it is felt, are worth the small mathematical sacrifice.

EXERCISE SET 2.5

1. Give a valid definition for the term "ray" that is consistent with Birkhoff's postulate set.
2. Birkhoff stated his protractor postulate in terms of radian measure. Restate it using degree measure.
3. Suppose that we define similarity of line segments in the following way: Two line

segments $\overline{AB}$ and $\overline{CD}$ are similar providing there exists a constant k such that $d(A,B) = kd(C,D)$. Are all line segments similar? Explain. Under what conditions will a pair of line segments be congruent? What restrictions, if any, should be placed on the legal values of k in this definition?

4. Define similarity for triangles. Define similarity for polygons in general.

5. Use your definition for similarity in Problem 4 to define congruence of triangles and polygons in general.

6. Prove, using Birkhoff's axioms, that
   (a) All right angles are equal in measure.
   (b) Similarity for triangles is a transitive relation.
   (c) Similarity for triangles is an equivalence relation.

7. Do you think that Birkhoff's axiom set is independent? Explain why or why not.

8. Do you think Birkhoff's geometry is isomorphic to Hilbert's? to Euclid's? Explain.

9. In Birkhoff's proof of Euclid's fifth postulate
   (a) What axiom assures us that a point E exists so that $AD:DC = AB:BE$?
   (b) Which axiom allows us to deduce that $\triangle ADC$ and $\triangle ABE$ are similar?

2.6 THE SMSG POSTULATES FOR EUCLIDEAN GEOMETRY

In the previous two sections we encountered two very different approaches to Euclidean geometry. Hilbert's axioms form the basis of what can be called a synthetic approach to geometry, since they provide the qualitative characteristics concerning points, lines, and planes needed to deduce all propositions from Euclid's *Elements* synthetically (i.e., constructively). In contrast, Birkhoff's axiom set is more analytical in nature, since the postulates given therein allow us to relate the terms "point," "line," and "angle" to numerical quantities by virtue of the one-to-one correspondences posited between them and the real numbers (see Postulates I and III). Remarkably, the two sets of axioms result in the same body of theorems, namely, those that constitute Euclid's geometry.

In the early 1960s, another set of axioms for Euclidean geometry was composed by a group of mathematicians and mathematics educators working as a component of a larger organization known as the School Mathematics Study Group (SMSG). This group was, in part, established to address the perceived failure of the United States to compete worldwide in the areas of science and mathematics.

During the late 1950s, when the Soviet Union began to show signs of technological superiority in comparison to the United States, Congress established the National Science Foundation through which it dedicated substantial funding to the goal of improving mathematics and science education in the United States. One facet of the overall program was curricular reform in school mathematics. SMSG was charged with the task of defining a "new math" for the nation's primary and secondary schools. While much of the new math was never successfully implemented, the axiom set for Euclidean geometry has survived, largely intact, to the present and will for the remainder of the text serve as a basis for the Euclidean and non-Euclidean geometries that follow. In this section we will discuss the relationships between the SMSG axioms and the axioms of Hilbert and Birkhoff.

The SMSG axioms (see Appendix D) can be separated into the following eight groups:

I. Axiom of incidence—Postulate 1.

II. Axioms concerning distance—Postulates 2 through 4.

III. Axioms concerning space relationships—Postulates 5 through 8.

IV. Axioms concerning separation—Postulates 9 and 10.

V. Axioms of angular measure—Postulates 11 through 14.

VI. Axiom of congruence—Postulate 15.

VII. Axiom of parallelism—Postulate 16.

VIII. Axioms concerning area and volume—Postulates 17 through 22.

The intent of the SMSG authors was to provide an axiom set that was (to the extent possible) complete and pedagogically sound, and, in addition, one that was accessible to students beginning their study of formal geometry. In order to fulfill the second of these objectives, the SMSG authors decided to sacrifice independence. The rationale behind this decision was that independent axiom sets require the proof of a large number of preliminary (and obviously true) theorems before the proofs of the major results.[16] The time and effort needed to get started in Hilbert's model, while a good exercise in the application of formal logic, does not give the student any new insight into geometry. Consequently, some of the axioms included in the SMSG set are redundant, since they can be proved using the others. This is a minor drawback and is counterbalanced by the comparative ease with which we will be able to move on to significant results using the SMSG set.

Of the axioms described above, group III and part of group VIII will be of little concern to us since they describe relationships in three-dimensional geometry which is not considered here. The remaining axioms can be related to axioms stated by Hilbert and/or Birkhoff.

The axiom of incidence, for example, is stated in much the same way by both Hilbert (among his axioms of connection) and Birkhoff (as Postulate II). The SMSG counterpart is as follows:

POSTULATE 1.   Given any two different points, there is exactly one line that contains them both.

---

[16] Recall the proof concerning betweenness given during the discussion of Hilbert's axioms. There are many other betweenness proofs of that type that must be proved in Hilbert's model before any substantial results can be established.

Compare this postulate to those given in the previous two sections. Although no explicit mention is made concerning the existence of points, the SMSG set ensures the existence of points and lines in much the same way that Birkhoff's axioms set does (see Exercise Set 2.6, Problem 1). Because of this, Hilbert's four axioms of connection are incorporated into the geometry although the SMSG set contains only one axiom of incidence.

The SMSG axioms concerning distance are based on Birkhoff's postulate of linear measure. The three postulates that comprise this group are listed and discussed here.

POSTULATE 2.   *The Distance Postulate:* To every pair of different points there corresponds a unique positive number.[17]

The unique positive number stipulated by Postulate 2 is, quite clearly, the distance between the points. This relationship is made more explicit in Postulate 3.

POSTULATE 3.   *The Ruler Postulate:* The points of a line can be put into one-to-one correspondence with the real numbers in such a way that (i) to every point there corresponds exactly one real number, (ii) to every real number there corresponds exactly one point of the line, and (iii) the distance between two points is the absolute value of the difference of the corresponding numbers.

This postulate is in fact a restatement of Birkhoff's postulate of linear measure. Items (i) and (ii) establish the one-to-one correspondence, and (iii) provides a means of assigning distance consistent with the intent of Birkhoff. By applying the field properties that apply to the coordinates of any points, A and B, we can show that $d(A,B)$ is in fact unique, so that SMSG Postulate 2 could be proved using Postulate 3. However, the proof would not be geometric in nature, which is why Postulate 2 is included in the SMSG set. In like fashion, Postulate 4 is not independent of the other axioms.

Another consequence of the ruler postulate is the Archimedian principle (see Hilbert's Axiom V-1). Since this property (or an equivalent one) must be postulated for real numbers, it comes "free" with the ruler postulate.

POSTULATE 4.   *The Ruler Placement Postulate:* Given two points P and Q of a line, the coordinate system (i.e., the one-to-one correspondence between the points of the line and the real numbers) can be chosen in such a way that the coordinate of P is zero and the coordinate of Q is positive.

---

[17] Some versions of the SMSG postulate set state Postulate 2 as follows: "To every pair of different points there corresponds a unique positive number called the *distance*."

Postulate 4 means that any line can be made into a number line with zero placed at any arbitrarily chosen point and with the positive and negative numbers placed according to choice. This is often convenient when coordinatizing a plane in analytic proofs such as those discussed in Chapter 5.

SMSG Postulates 5 through 8, the axioms concerning space relationships (which are listed in Appendix D), deal with the geometry of three dimensions which, while important, is not of central concern to us here. We therefore continue with a discussion of the axioms concerning separation.

POSTULATE 9.   *The Plane Separation Postulate:* Given a line *l* and a plane containing it, the points of the plane that do not lie on the line form two sets such that (i) each of the sets is convex,[18] and (ii) if point *P* is in one set and point *Q* is in the other, then segment $\overline{PQ} \cap l \neq \phi$.

POSTULATE 10.   *The Space Separation Postulate:* The points in space that do not lie in a given plane α form two sets such that (i) each of the sets is convex, and (ii) if point *P* is in one set and point *Q* is in the other, then $\overline{PQ} \cap \alpha \neq \phi$.

While Postulate 10 applies to three-dimensional geometry and therefore will not apply to the discussions that follow, Postulate 9 is essential to several results that will be derived in Chapter 3. The plane separation postulate plays the same role in the SMSG set that Pasch's axiom (Axiom II-4) plays in Hilbert's model. Assuming Postulate 9 will allow us, in Chapter 3, to prove Pasch's axiom as a theorem, which in turn will give rise to several other important theorems including the well-known crossbar theorem.

The next group of postulates deals with the measurement of angles.

POSTULATE 11.   *The Angle Measurement Postulate:* To every angle ∠*ABC* there corresponds a unique real number between 0 and 180.

POSTULATE 12.   *The Angle Construction Postulate:* Let $\overrightarrow{AB}$ be a ray on the edge of half-plane *H*. For every number *r* between 0 and 180 there is exactly one ray $\overrightarrow{AP}$, with *P* in *H* such that m∠*PAB* = *r*.

POSTULATE 13.   *The Angle Addition Postulate:* If *D* is a point in the interior of ∠*ABC*, then m∠*ABD* + m∠*DBC* = m∠*ABC*.

POSTULATE 14.   *The Supplement Postulate:* If two angles form a linear pair, then they are supplementary.

Postulates 11 and 12 together constitute the counterpart of Birkhoff's postulate of angle measure. A difference in the two approaches is evident,

---

[18] See Exercise Set 2.6, Problem 4, for a definition of the term "convex set."

since Birkhoff allows angle measures in the range $0 \leftrightarrow 2\pi$ ($0° \leftrightarrow 360°$) while the SMSG presentation restricts angle measures to between $0°$ and $180°$. The SMSG approach has become more standard in recent years, since with this restriction the interior of every angle is a convex set of points (see Exercise Set 2.6, Problem 4).

Postulate 13 can be proved using the others, but since the statement is intuitively obvious and the proof is not, it has been postulated for convenience. Similarly, Postulate 14, which is almost a definition, has also been included in order to eliminate the need for a tedious proof of an obvious result.

The axiom of congruence (Axiom 15) is the standard SAS congruence condition.

POSTULATE 15. *The SAS Postulate:* Given a correspondence between two triangles (or between a triangle and itself). If two sides and the included angle of the first triangle are congruent to the corresponding parts of the second triangle, then the correspondence is a congruence.

You may recall from the discussion of Hilbert's comparable Postulate III-5 that this statement provides more than is absolutely needed (see Exercise Set 2.4, Problem 16), but once again the SMSG authors chose convenience over independence.

The 15 postulates listed thus far form the basis of what is known as neutral geometry. SMSG Postulates 1 through 15 can be used in the development of both Euclidean and non-Euclidean geometry. Chapter 3 will investigate the implications of these postulates in detail.

The SMSG parallel postulate is a form of Playfair's postulate.

POSTULATE 16. *The Parallel Postulate:* Through a given external point there is at most one line parallel to a given line.

This is, as mentioned earlier, the postulate that characterizes Euclidean geometry. In Chapter 4 we will investigate many geometric statements that are true only if this statement (or one equivalent to it) is assumed.

None of the postulates concerning area and volume (Postulates 17 through 22) are absolutely necessary for the development of Euclidean or non-Euclidean geometry. Results concerning these quantities are addressed by both Euclid and Hilbert nonnumerically. Since, however, they can expedite the discussion of area and volume measurements, we will assume them in Chapter 4.

POSTULATE 17.   To every polygonal region there corresponds a unique positive number (called the *area* of the polygonal region).

POSTULATE 18.   If two triangles are congruent, then the triangular regions have the same area.

POSTULATE 19.   Suppose the region R is the union of two regions R_1 and R_2. Suppose also that R_1 and R_2 intersect at most in a finite number of segments and points. Then the area of R is the sum of the areas of R_1 and R_2.

POSTULATE 20.   The area of a rectangle is equal to the product of the length of its base and the length of its altitude.

POSTULATE 21.   The volume of a rectangular parallelpiped is equal to the product of the length of its altitude and the area of its base.

POSTULATE 22.   *Cavalieri's Principle:* Given two solids and a plane. If for every plane that intersects the solids and is parallel to the given plane the two intersections determine regions that have the same area, then the two solids have the same volume.

This completes the list of SMSG postulates. We have now seen four postulate sets intended to determine the same set of theorems: the theorems of Euclid, that is, Euclidean geometry.

1. Euclid's *Elements*. The early, but flawed, exposition that defined the discipline known as geometry for more than 2000 years.
2. Hilbert's *Grundlagen der Geometrie*. A modern (1899) treatment of Euclidean geometry that is true to the spirit of Euclid's work and uses a set of axioms that is acceptable under current standards.
3. Birkhoff's *A Set of Postulates for Plane Geometry (Based on Scale and Protractor)*. A second modern attempt to place Euclidean geometry on a firm foundational basis using an essentially different approach based on measurement.
4. School Mathematics Study Group's *Geometry*. A pedagogically oriented postulate set that combines (at the expense of independence) features from Hilbert and Birkhoff in a way that allows for an efficient development of Euclidean geometry.

Other axiom sets for Euclidean geometry have been offered over the years, but the ones listed above are perhaps the most historically significant. Other, inherently different, axiom sets for geometry are also possible. Some of these define non-Euclidean geometries. Two of these will be discussed briefly in the next section and then again in more depth in Chapter 6.

EXERCISE SET 2.6

1. Hilbert's axiom set for Euclidean geometry includes an axiom (Axiom I-3) that ensures the existence of at least three noncollinear points. The SMSG axioms contain no such axiom, at least not directly. Explain which of the SMSG axioms guarantee that the geometry is not vacuous.

2. Explain how SMSG Postulate 2 can be derived from Postulate 3.

3. Outline a proof describing how SMSG Postulate 4 can be deduced from the other SMSG postulates.

4. A set of points, S, is said to be convex if it is always true that $A \in S$ and $B \in S$ imply that AB is contained in S. If angle measures are restricted to values between 0° and 180°, is the interior of every angle a convex set? Explain. If, as Birkhoff did, we allow angle measures to have values between 0° and 360°, is the interior of every angle a convex set? Explain why or why not.

5. Which of Birkhoff's axioms implies SMSG Postulate 15?

6. What is meant by the term "adjacent angles"? Complete the following statement: Two angles are adjacent if and only if _____.

7. Reread SMSG Postulate 14. What do you think is meant by the term "linear pair"? How would you define the term "supplementary" with respect to angles? If a pair of adjacent angles form a linear pair, is their union an angle? Explain why or why not.

8. Find, in the library, a textbook that discusses the foundations of geometry. Make a list of the axioms that are posited.
   (a) Classify each axiom as being characteristic of Hilbert, Birkhoff, or neither.
   (b) Is the axiom set complete? Consistent? Independent?
   (c) Are there any axioms that are substantially different from those discussed so far in this chapter? If so, list them and explain why.
   (d) Do you think you can prove (using the axioms from the book) any theorems that cannot be proved using Hilbert's axioms? Birkhoff's axioms? SMSG axioms? Explain why or why not.

9. In Chapter 1 we discussed what is meant by the term "incidence geometry." Is the SMSG model an incidence geometry? Is Hilbert's model an incidence geometry? Do the results proved earlier for incidence geometries apply to the SMSG model? Explain why or why not.

2.7 NON-EUCLIDEAN GEOMETRIES

The works of Hilbert and Birkhoff successfully filled in the logical failures of Euclid's *Elements*. They did not, however, address another major issue that had troubled geometers for nearly 2000 years—what to do about the controversial fifth postulate. Although efforts to place development of the geometry on a more firm axiomatic basis were initiated during the 1800s (few mathematicians dared to question the perfection of the *Elements* before that time), a great many earlier mathematicians had attempted to resolve the issue of the parallel postulate. In fact, it is likely that Euclid himself was the first to be concerned with the question, since it was not until Proposition 29 that he first employed it, choosing to prove even some theorems concerning parallels without it. Proposition 16 gives further evidence that Euclid was inclined against the use of the parallel postulate. This proposition states that the measure of an exterior angle of a triangle is greater than that of either remote interior angle. Proposition 16 can, of course, be stated more specifically, since it is a Euclidean property that each exterior angle of a triangle is equal to *the sum* of the remote interior angles. This stronger statement is, as we will see in Chapters 3 and 4, a consequence of the parallel postulate. It seems then that Euclid was willing to delay a stronger version of the proposition rather than hasten the introduction of what he must have deemed to be the "weak link" in his postulate set.

Among the five postulates given by Euclid, there are actually two that stand out as substantially different from the others. Postulates 2 and 5 both require that we imagine the behavior of lines as they are extended "continuously" or "indefinitely." Postulate 2 has been less controversial, since it only postulates that this extension is possible. But from the very beginning Postulate 5 aroused suspicion because of its complexity. The statement of Postulate 5 has more the nature of a theorem than of an axiom. Consequently, a large number of mathematicians took on the task of proving Postulate 5 on the basis of the other four postulates.

Perhaps the first truly significant work in this direction was the volume *Euclides ab Omni Naevo Vindicatus (Euclid Freed From All Flaws)* published in 1733 by an Italian priest named Girolamo Saccheri. Saccheri committed himself to validating Euclid's geometry by pursuing the logical consequences of denying Postulate 5. Saccheri knew that the parallel postulate as stated by Euclid implied that there is a unique parallel to each line through any point off the line. To deny Postulate 5 meant to assume one of two things: either there were no parallels to a line through a point off the line, *or* there were multiple parallels. Since the first of these possibilities contradicted Proposition 27, which Euclid had established without the use of his parallel postulate,[19] Saccheri was easily able to dismiss this possibility. The other alternative was not as easy to remove, and most of *Euclides Vindicatus* involves deducing theorems that result from the hypothesis of multiple parallels. This hypothesis has come to be known as the hyperbolic parallel postulate. One form of this postulate is given here.

HYPERBOLIC PARALLEL POSTULATE.   There exists a line *l* and a point *P* such that at least two distinct lines pass through *P* that are parallel to *l*.

Saccheri displayed an impressive ability to apply deductive reasoning to the modified postulate set and deduced a remarkable number of theorems

---

[19] Proposition 27 states that if a transversal crosses two lines in such a way as to make the alternate interior angles congruent, then the lines are parallel. Thus, if we construct two perpendiculars to the same line, the alternate interior angles (both right angles) will sum to 180° and thus be parallel. From this we can conclude that parallel lines exist.

from this set, one of which still bears his name. (The Saccheri-Legendre theorem will be proved in Chapter 3.) Still, his goal was always to identify an inconsistency resulting from negating the fifth postulate. Saccheri pursued this goal religiously, but because some of his results were very far afield of Euclid's theorems (in Saccheri's words, they were "repugnant to the nature of a straight line"), he concluded erroneously that they must be inconsistent. Saccheri's story is really quite sad, since had he been willing to evaluate the consequences of his work more open-mindedly, he might well have taken a place among the great names in the history of mathematics.

As it happened, however, Saccheri published his work thinking that it was, in truth, a vindication of Euclid and the fifth Postulate. It was not. However, embedded within it were some radically different results that would, a century later, lead to a revolution in mathematical thought.

The first mathematician to recognize that the negation of Euclid's fifth postulate could exist within a consistent geometry may have been the eminent German mathematician Karl Friedrich Gauss. At about the beginning of the nineteenth century Gauss pursued the same line of reasoning as did Saccheri. He investigated the hypothesis of multiple parallels to a variety of logical consequences that were clearly in conflict with the results of the *Elements*. Gauss, however, unlike Saccheri, had the intellectual capacity to realize that the results he obtained were *not* inconsistent and that a geometry other than Euclid's could be valid in its own right.

Unfortunately, at that period in time it was often ill-advised to "rock the boat," as evidenced by the treatment Copernicus and Galileo had received in the preceding centuries. The most prominent philosopher of the eighteenth century was Immanuel Kant, who argued strongly that since one could not conceive of a geometry other than Euclid's, there could be no other geometry. It may very well be that Gauss was influenced by the strength of the Kantian position, for he chose not to make public his work in the area of non-Euclidean geometry.

A short time later (1831) a young Hungarian mathematician named Janos Bolyai published, as an appendix to a work by his father Wolfgang Bolyai, a short exposition of a non-Euclidean geometry based on the multiple-parallel alternative. Bolyai's work was forwarded to Gauss by the elder Boylai, and Gauss responded with the news that he had derived the same results independently. The younger Bolyai was devastated by this news and withdrew from mathematical research permanently.

Remarkably, at almost precisely the same time (1829), the Russian mathematician Nikolai Lobachevsky published an independent development of the same non-Euclidean geometry. Unfortunately, because the work was published in Russian and was not well received in Russia, Lobachevsky's work attracted only minimal attention. By 1840 Gauss had become aware of the work of both J. Bolyai, and Lobachevsky, and even though these works replicated his own results, he still failed to give open support to the existence of non-Euclidean geometries.

To understand Gauss's temerity, it will help to have an idea of the results that he, J. Bolyai, and Lobachevsky had derived. Earlier in this chapter you encountered some of the consequences of the Euclidean parallel postulate. If one uses the postulate of multiple parallels (i.e., the hyperbolic parallel postulate), the difference in results is striking. For example (compare these to the Euclidean results listed earlier),

1. Through a given point can be drawn an infinite number of parallels to a given line.
2. There exist no triangles in which the sum of the angle measures is 180°.
3. There exist no triangles that are similar but not congruent.
4. There are no lines that are everywhere equidistant.
5. There exist triangles that cannot be circumscribed.
6. The sum of the measures of the interior angles of triangles varies among triangles and is always less than 180°.

In addition, some other strange results in non-Euclidean geometry are

7. No rectangles exist.
8. There is an upper limit to the area of a triangle.
9. As triangles become larger in area, their angle sums become smaller.
10. The distance between certain pairs of parallel lines goes to zero in one direction and becomes infinite in the other direction.
11. If two parallel lines are crossed by a transversal, the alternate interior angles may not be equal.

Clearly, this is a strange geometry. However, even though the results are not reinforced by our senses, they are consistent within the axiomatic system that produced them. Somehow, we have the feeling that these results are nonsense and that they have no practical place in the world in which we live. Gauss himself is said to have tried to address this issue using a physical experiment in which he placed measuring devices on the peaks of three mountains and measured the angles between the peaks as the interior angles of a triangle. An angle sum significantly smaller (more than could be attributed to measurement error) than 180° would have indicated that physical space is indeed non-Euclidean in nature. Needless to say, the results were inconclusive. However, even today this remains an open question. Newtonian physics was derived under the hypothesis that the universe in which we live is Euclidean. Einstein's theory of relativity assumes the existence of a

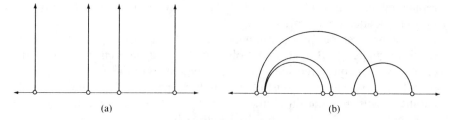

(a)   (b)

Figure 2.7.1

type of non-Euclidean geometry and serves to explain some phenomena in physics that had troubled physicists and mathematicians for centuries. Does this mean that we live in a non-Euclidean universe? Possibly. But even if we do, all the geometries yield very similar results for relatively small regions (e.g., regions close to earth).

The geometry developed by Gauss, J. Bolyai, and Lobachevsky is usually called *hyperbolic geometry*. A model that is often used to illustrate hyperbolic geometry is the Poincaré[20] half-plane (Figure 2.7.1). Within this model the undefined term "point" is interpreted as the points contained in a Euclidean half-plane. (*Note:* Any line in a plane separates the plane into two disjoint half-planes.) Hyperbolic lines in this model appear in two ways. First, any set of hyperbolic points contained in a Euclidean line perpendicular to the separation line constitutes a hyperbolic line in this model (Figure 2.7.1a). Second, any set of hyperbolic points contained in a Euclidean circle centered on the separation line constitutes a hyperbolic line. While these interpretations seem odd, in order to be valid they need only satisfy the axioms of hyperbolic geometry, namely, the neutral postulates (SMSG Postulates 1 through 15) and the hyperbolic parallel postulate.

In Chapter 6 we will justify the validity of hyperbolic axioms for a modification of the half-plane model, a model called the Poincaré disk. Assume for now, however, that the hyperbolic axioms are valid in the half-plane model. Figure 2.7.2a shows a hyperbolic line *l* and a hyperbolic point *P* through which several lines parallel to *l* pass, a phenomenon that is characteristically hyperbolic. Figure 2.7.2b shows a hyperbolic triangle $\triangle ABC$. The measures of the angles at points *A*, *B*, and *C* in the model are the measures of the angles formed by the tangents to the semicircles at those points. Because of this, the sum of the measures of the interior angles of a hyperbolic triangle is apparently less than 180°, another non-Euclidean property.

---

[20] The Frenchman Henri Poincaré was one of the world's leading mathematicians during the last part of the nineteenth century and the first part of the twentieth. He was among the first to show the consistency of hyperbolic geometry by producing a valid model of it (see Section 6.6).

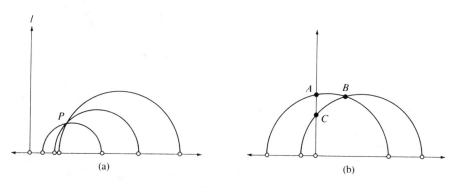

Figure 2.7.2

The preceding discussion is informal and proves nothing. It should, however, provide some evidence that models can be created in which all the hyperbolic postulates are valid. A formal discussion is included in Chapter 6.

While the term "hyperbolic geometry" refers to the non-Euclidean geometry in which multiple parallels exist, the term "elliptic geometry" refers to a third possibility in which the existence of parallels is denied. One form of the elliptic parallel postulate is as follows:

ELLIPTIC PARALLEL POSTULATE.   Given any line *l* and any point *P* not on *l*, there is no line through *P* that is parallel to *l*.

Saccheri was able to eliminate this alternative from consideration by noting that it is possible, using only Euclidean Postulates 1 through 4 (and some tacit Euclidean assumptions), to prove that parallels exist. In particular, Proposition 27 of the *Elements* says that if two lines are crossed by a transversal in such a way as to make the alternate interior angles equal, the lines will be parallel. In addition, Proposition 12 allows us to drop perpendiculars to a line from points not on the line. So if we take a line *l* and construct two perpendiculars to it, *m* and *n*, we obtain two lines (*m* and *n*) crossed by a transversal (*l*) in which the alternate interior angles are equal (both are right angles). Consequently, the lines are parallel. So it seems that the elliptic postulate is not a valid alternative. Yes and no. To see the discrepancy, consider geometry on the surface of a sphere where points are surface points and lines are great circles drawn on the surface of the sphere (Figure 2.7.3).

If we choose a line (the equator is a convenient choice) and construct two perpendicular lines (great circles) to it, we see that the perpendiculars are not parallel but rather are intersecting lines with the north pole as a point of intersection. So we have a model in which a proposition derived from Euclidean Postulates 1 through 4 is not valid. How is this possible?

As mentioned earlier, there were really two "objectionable" postulates in Euclid's set. In addition to the fifth postulate, Postulate 2 can be ques-

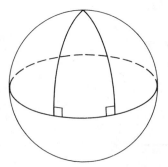

Figure 2.7.3

tioned because it requires us to imagine the behavior of lines as they approach infinity.[21] This is an assumption that we can never verify empirically. Geometry on a sphere doesn't allow us to extend lines infinitely, even though lines have no end. In elliptic geometry, we alter our notion of the undefined term "line" to allow for lines that are endless but bounded or finite. Much of the work in this area of geometry was done during the middle of the nineteenth century by the German mathematician Bernhard Riemann.

One other problem that needs to be addressed using this model of elliptic geometry arises from the fact that lines (i.e., great circles) meet in two distinct points (e.g., any two perpendiculars to the equator meet at both poles). The problems that arise from this phenomenon can be overcome and will be addressed in detail in Section 6.8, where various theorems will be proved using the elliptic parallel postulate.

As a preview of the consequences of the elliptic parallel postulate, consider the question of angle sum in a triangle. To begin, we will consider the equator and build a triangle on it. To do so, we'll construct two perpendiculars to the equator one-quarter of a great circle apart. The perpendiculars meet at the pole at an angle of 90°. Since each of the base angles is also 90°, the angle sum for this triangle is 270°, a result that is clearly contrary to the results of Euclidean and hyperbolic geometries. Other results from elliptic geometry are discussed in Chapter 6, but with no more than this brief introduction we may characterize the three distinct geometries in terms of the angle sum for triangles:

> Hyperbolic geometry: angle sum < 180°
>
> Euclidean geometry: angle sum = 180°
>
> Elliptic geometry: angle sum > 180°

Of course, there are a great many other differences that distinguish the three geometries mentioned above. These will be discussed when we develop the geometries in greater depth in Chapter 6.

[21] Postulate 2: To produce a finite straight line continuously in a straight line.

EXERCISE SET 2.7

1. Which geometries (Euclidean, hyperbolic, or elliptic) do you think would exhibit each of the following properties?
   (a) Pairs of lines enclose an area.
   (b) Similar triangles are congruent.
   (c) Rectangles exist.
   (d) The sum of the measures of the interior angles of a pentagon is less than 540°.
   (e) All lines cross at least once.
   (f) Lines parallel to the same line are parallel to each other.
   (g) Two lines cannot have more than one common perpendicular.
   (h) The sum of the angles of a quadrilateral is less than two straight angles.
   (i) If A, B, and C are points on a line, then one must be between the other two.
   (j) The distance between parallel lines changes if we choose different points on the lines.

2. In Euclidean geometry the measure of an exterior angle of a triangle is equal to the sum of the measures of the remote interior angles.
   (a) Outline a proof of this fact from the Euclidean results you recall from earlier studies of geometry.
   (b) Do you think this theorem is true in hyperbolic geometry? Explain why or why not.
   (c) Do you think this theorem is true in elliptic geometry? Explain why or why not.

3. Neutral Postulate 1 states that there is a unique line that contains any two points. Is this postulate valid in the Poincaré half-plane model of hyperbolic geometry? Explain why or why not. (*Note*: Since the half-plane model is embedded in a Euclidean plane, the justification can make use of Euclidean results.)

4. Imagine that the line that defines the half-plane in the Poincaré model is wrapped around into a circle. The resulting geometry might look like Figure 2.7.4.

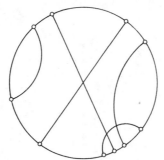

Figure 2.7.4

   (a) What role, with respect to the circle, do the lines that were previously perpendicular to the separation line now play?
   (b) Consider the points where the semicircles intersect the circle in Figure 2.7.4. If you were to draw tangents to the circle and to the semicircle at a point of intersection, what relationship do you think would exist between the lines?
   (c) The geometry shown in Figure 2.7.4 is called the Poincaré disk model of hyperbolic geometry. Sketch a triangle within this model. Does it appear

that the triangle has an angle sum greater than, less than, or equal to 180°? Justify your response.

(d) If the Poincaré disk is a model for hyperbolic geometry, there should exist multiple parallels to any line from an external point. Choose any diameter of the circle in Figure 2.7.4 and any external point. Sketch two lines parallel to the diameter that contain the external point. How many such parallels exist? Explain.

3

TRAVELING
TOGETHER

Neutral Geometry

3.1 INTRODUCTION

In the previous chapter we were briefly introduced to three different models
of infinite incidence geometry. Each of these models was established using
the same collection of undefined terms and essentially the same set of axioms, with one notable exception, the parallel postulate. The Euclidean
models assumed the Euclidean parallel postulate.[1] The non-Euclidean
models assumed some form of a negation of the Euclidean parallel postulate.
If we were to compare the theorems of these distinctly different geometries,
certainly we would find many that are strikingly different, but we would also
find a large collection of theorems common to all three. This body of common theorems would be those whose proofs do not depend on that geometry's parallel postulate or its logical consequences. Since the geometry in
which these theorems are true is neither Euclidean nor non-Euclidean, we
shall call it *neutral geometry*.[2] As mentioned in Chapter 2, Euclid proved

---

[1] For the purposes of this development, when the term "Euclidean parallel postulate" is
used, it will refer to Playfair's postulate. Later in this development Euclid's fifth postulate,
Playfair's postulate (Hilbert's Axiom IV-1), and Birkhoff's Postulate IV (postulate of similarity)
will be shown to be equivalent statements.

[2] Historically, this geometry has been called *absolute geometry*, a name that reflected the
feeling that the first four postulates of Euclid were intuitively acceptable as necessary but that
the fifth postulate is not independent and thus should be proven from the first four.

Propositions 1 through 28 in Book I of the *Elements* without using his parallel postulate, therefore these propositions could be regarded as theorems in
neutral geometry.

In this chapter we will investigate neutral geometry and a variety of its
theorems. The reasons for studying neutral geometry are twofold. First, by
studying the body of theorems that do not depend on a parallel postulate, we
are better able to clarify the roles that the various parallel postulates play in
the development of Euclidean and non-Euclidean geometries. Second, since
they do not require a parallel postulate for their proof, theorems proven in
the context of neutral geometry are valid theorems in either Euclidean or
non-Euclidean geometry.

For the purposes of our development of neutral geometry, we will use
the undefined terms "point," "line," and "plane" and the relation "on." In
addition, we will make as our fundamental assumptions the first 15 axioms
from the set developed by SMSG[3] in the early 1960s. As previously indicated, these axioms, while not independent, combine aspects of both
Hilbert's and Birkhoff's axiom sets yet provide a pedagogical compromise
between high rigor and student accessibility. This lack of independence, as
discussed in Chapter 1, should in no way detract from our development of
neutral geometry.

3.2 PRELIMINARY NOTIONS

We will begin our discussion of neutral geometry by observing that many of
the notions we have become familiar with in earlier experiences in geometry
are neutral in character. For example, the ideas associated with angle measure (such as right, acute, and obtuse angles), the ideas associated with angle
addition (such as adjacent, complementary, and supplementary angles), and
many ideas associated with polygons (such as triangles and quadrilaterals)
are all elements of neutral geometry. A large number of theorems can be
proven in neutral geometry.[4] For example, since all the axioms of incidence
geometry are included in some form in our axiom set (see Exercise Set 3.2,
Problem 2), their logical consequences are theorems in neutral geometry.
For example, Incidence Theorem 1, which states, if two distinct lines intersect, then that intersection is exactly one point—is a theorem in neutral
geometry. Furthermore, any theorems from the previous chapter, either in
the text or the exercises, that are consequences of Hilbert's, Birkhoff's, or
the SMSG postulate sets and whose proofs do not require the use of a
parallel axiom are also theorems in neutral geometry. In this chapter we will

---

[3] A complete listing of these axioms can be found in Appendix D.

[4] See the end of the chapter for a listing of some theorems from neutral geometry.

prove a variety of theorems, many of which will be familiar but a few of which may prove to be quite strange looking.

Before we continue, a short comment on notation is appropriate. Since the SMSG postulate set provides axioms dealing with the measure of line segments and angles, we will use the notation $\overline{AB}$ to represent line segment AB, and AB to represent the measure or length of $\overline{AB}$. Likewise, we will use the notation $\angle ABC$ to represent angle ABC, and $m\angle ABC$ to represent its measure. Employing this notation, we offer the following definitions.

DEFINITION.   Two line segments are congruent if and only if their measures are equal. For example, $\overline{AB} \cong \overline{CD}$ if and only if $AB = CD$.

DEFINITION.   Two angles are congruent if and only if their measures are equal. That is, $\angle ABC \cong \angle DEF$ if and only if $m\angle ABC = m\angle DEF$.

Furthermore, polygons will be defined to be congruent in the following way.

DEFINITION.   Two polygons are congruent if and only if there exists a one-to-one correspondence between their vertices such that all their corresponding sides (line segments) are congruent and all their corresponding angles are congruent. For example, $\triangle ABC \cong \triangle DEF$ if and only if $\overline{AB} \cong \overline{DE}$, $\overline{BC} \cong \overline{EF}$, $\overline{AC} \cong \overline{DF}$, $\angle ABC \cong \angle DEF$, $\angle BCA \cong \angle EFD$, and $\angle CAB \cong \angle FDE$.[5]

The following theorems deal with the previous definitions of congruence and should appear familiar to anyone acquainted with elementary geometry. For that reason, the proofs will be left as exercises (see Exercise Set 3.2).

THEOREM 3.2.1.   The relations of line segment congruence, angle congruence, and polygonal congruence are equivalence relations.

THEOREM 3.2.2.   (i) Every line segment has exactly one midpoint, and (ii) every angle has exactly one bisector.

THEOREM 3.2.3.   Supplements (and complements) of the same or congruent angles are congruent.

THEOREM 3.2.4.   Any pair of vertical angles is congruent.

THEOREM 3.2.5.   *Pasch's Axiom.* If a line *l* intersects $\triangle PQR$ at a point S such that $P\text{-}S\text{-}Q$, then *l* intersects $\overline{PR}$ or $\overline{RQ}$ (Figure 3.2.1).

---

[5] It is important to note that we will use the relation "equa,s" in conjunction with line segments, angles, and polygons only when they represent identical figures.

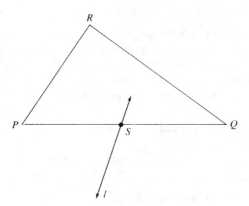

Figure 3.2.1

Proof. Since l intersects $\overline{PQ}$ at S, we know that P and Q are on differ-ent sides[6] of l. Employing an indirect proof, we shall assume that l does not intersect either $\overline{PR}$ or $\overline{RQ}$. Since P and R are on the same side of l and R and Q are on the same side of l, then P and Q are on the same side of l. This contradiction implies that l must intersect at least one of $\overline{PR}$ or $\overline{RQ}$ and completes the proof.

The following theorem is an important consequence of the plane sepa-ration axiom (SMSG Postulate 9) and Theorem 3.2.5. However, since it is not the authors' intent to complete a rigorous development of separation and convexity, it is simply stated here without proof.[7]

THEOREM 3.2.6.   *The Crossbar Theorem.* If X is a point in the interior of $\triangle UVW$, then $\overrightarrow{UX}$ intersects $\overline{WV}$ at a point Y such that W-Y-V (Figure 3.2.2).

THEOREM 3.2.7.   *Isosceles Triangle Theorem.* If two sides of a trian-gle are congruent, then the angles opposite those sides are congruent (Figure 3.2.3).

Proof. Suppose that we are given $\triangle PQR$ with $\overline{PQ} \cong \overline{RQ}$. First draw the bisector of $\angle PQR$, which as a result of the crossbar theorem must intersect $\overline{PR}$ at a point T such that P-T-R. Now $\angle PQT \cong \angle RQT$ (definition of an angle bisector) and $\overline{QT} \cong \overline{QT}$. Therefore $\triangle PQT \cong \triangle RQT$ by the SAS Postulate (SMSG Postulate 15) and $\angle QPT \cong \angle QRT$.

___

[6] We say that points A and B are on different sides of l if and only if l intersects $\overline{AB}$ at a point C such that A-C-B, and we say that points A and B are on the same side of l if l does not intersect $\overline{AB}$.

[7] For a complete development of separation and convexity, see E. E. Moise, *Elementary Geometry From an Advanced Standpoint,* 2nd ed. (Reading, Mass.: Addison-Wesley Publish-ing Co., 1974). pp 61–73.

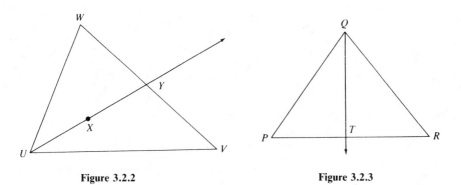

Figure 3.2.2   Figure 3.2.3

THEOREM 3.2.8.   A point is on the perpendicular bisector of a line segment if and only if it is equidistant from the endpoints of the line segment.

Proof. Note that the logical structure of an "if and only if" theorem requires that we prove both a statement and its converse. We will first prove the following statement: *If* a point is equidistant from the endpoints of a line segment, *then* it is on the perpendicular bisector of that line segment. Consider the line segment $\overline{AB}$ and the point P which is equidistant from both A and B (i.e., $\overline{PA} \cong \overline{PB}$) (Figure 3.2.4). In a fashion similar to the proof of the previous theorem, we will construct $\overrightarrow{PQ}$ as the bisector of $\angle APB$, and as a result, $\triangle APQ \cong \triangle BPQ$. By the definition of congruent triangles, we have $\overline{AQ} \cong \overline{BQ}$, which makes $\overrightarrow{PQ}$ a bisector of $\overline{AB}$. Also, since $\angle AQP$ and $\angle BQP$ are both supplementary and congruent, then PQ is perpendicular to $\overline{AB}$. Hence $\overrightarrow{PQ}$ is the perpendicular bisector of $\overline{AB}$. Proof of the converse of this statement is left as an exercise.

The next theorem will prove to be very useful in the further development of neutral geometry, but before we prove it, we need an additional definition.

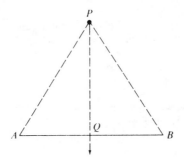

Figure 3.2.4

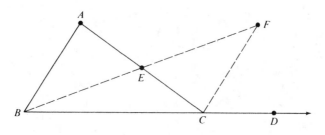

Figure 3.2.5

DEFINITION.   Any angle that is both supplementary and adjacent to an angle of a triangle is called an *exterior angle of the triangle*.

THEOREM 3.2.9.   *The Exterior Angle Theorem.* The exterior angle of a triangle is greater in measure than either of the nonadjacent interior angles of the triangle.

Proof. Consider $\triangle ABC$ with D on $\overrightarrow{BC}$ such that B-C-D[8] (Figure 3.2.5). We must show that $m\angle ACD$ is greater than both $m\angle BAC$ and $m\angle ABC$. Let E be the midpoint of $\overline{AC}$ and locate point F on $\overrightarrow{BE}$ such that B-E-F and $\overline{EF} \cong \overline{BE}$. Now if we draw $\overline{FC}$, it is easy to show, by the SAS postulate, that $\triangle AEB \cong \triangle CEF$, and consequently $\angle BAE \cong \angle FCE$. Since F is interior to $\angle ACD$,[9] we have $m\angle ACD > m\angle FCE$, and by substitution, $m\angle ACD > m\angle BAE$. Similarly, $m\angle ACD > m\angle ABC$.

EXERCISE SET 3.2

1. In this discussion of neutral geometry, many familiar elementary geometric terms have not been formally defined. Provide a definition for each of the following terms:
   (a) Point B is between A and C (denoted A-B-C)
   (b) Line segment
   (c) Angle
   (d) Right, acute, and obtuse angles
   (e) Adjacent angles
   (f) Vertical angles
   (g) Supplementary angles (and complementary angles)
   (h) Midpoint of a line segment
   (i) Bisector of an angle
   (j) Perpendicular lines
   (k) Triangle

---

[8] Such a point exists as a consequence of the ruler postulate (SMSG Postulate 3) and the ruler placement postulate (SMSG Postulate 4).

[9] The fact that F is interior to $\angle ACD$ is a consequence of the plane separation postulate (see Exercise 3.2, Problem 10).

(l) Quadrilateral (polygon)

(m) Interior of an angle (a triangle and a polygon)

2. Show that the axioms of incidence geometry stated in Chapter 1 are implied by the SMSG postulates.

3. Compare the axiom sets of Hilbert and Birkhoff with the SMSG axiom set. Show that each of Hilbert's axioms is equivalent to or implied by a postulate in the SMSG set. Do the same for Birkhoff's axioms.

4. Prove Theorem 3.2.1.

5. Prove Theorem 3.2.2.

6. Prove Theorem 3.2.3.

7. Prove Theorem 3.2.4.

8. An *equilateral triangle* has three congruent sides. Prove that an equilateral triangle is equiangular.

9. Complete the proof of Theorem 3.2.8 by proving the converse of the proven statement.

10. In the proof of Theorem 3.2.9 we indicated that a point F was in the interior of $\angle ACD$. Provide a proof for this fact.

11. Complete the proof of Theorem 3.2.9 by showing that m$\angle ACD$ > m$\angle ABC$.

12. Prove that a perpendicular drawn to a point on a line is unique.

13. Prove that a perpendicular drawn from a point P to a line l is unique.

3.3 CONGRUENCE CONDITIONS

In this section we will investigate several conditions under which polygons are congruent. In addition, we will develop several results that will be of importance to the proofs of several later theorems. Many of these theorems will be familiar to the reader, however, the proofs may prove to be less familiar.

THEOREM 3.3.1. *Angle-Side-Angle (ASA) Triangle Congruence Theorem.* If the vertices of two triangles are in one-to-one correspondence such that two angles and the included side of one triangle are congruent, respectively, to two angles and the included side of the second triangle, then the triangles are congruent.

Proof. Suppose we are given $\triangle ABC$ and $\triangle DEF$ with $\angle CAB \cong \angle FDE$, $\angle CBA \cong \angle FED$, and $\overline{AB} \cong \overline{DE}$ (Figure 3.3.1). Now if either $\overline{AC} \cong \overline{DF}$ or $\overline{CB} \cong \overline{EF}$, we would be finished by using the SAS postulate; therefore without loss of generality we will assume that $\overline{AC}$ is not congruent to $\overline{DF}$ and in particular that $AC > DF$. Since $AC > DF$, there exists a point C' (not equal to C) on $\overline{AC}$ such that $\overline{AC'} \cong \overline{DF}$ and, by SAS, $\triangle ABC' \cong \triangle DEF$, and thus $\angle ABC' \cong \angle DEF$. Now by the transitive property of angle congruence, $\angle ABC' \cong \angle ABC$, which contradicts the angle construction postulate (SMSG Postulate 12). Therefore $\overline{AC} \cong \overline{DF}$ and $\triangle ABC \cong \triangle DEF$ by SAS.

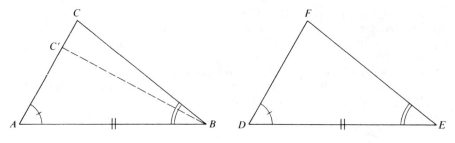

Figure 3.3.1

THEOREM 3.3.2. *Converse of the Isosceles Triangle Theorem*. If two angles of a triangle are congruent, then the sides opposite those angles are congruent. The proof is left as an exercise.

THEOREM 3.3.3. *Angle-Angle-Side (AAS) Triangle Congruence Theorem*. If the vertices of two triangles are in one-to-one correspondence such that two angles and the side opposite one of them in one triangle are congruent to the corresponding parts of the second triangle, then the triangles are congruent.

Proof. Consider $\triangle ABC$ and $\triangle DEF$ with $\angle CAB \cong \angle FDE$, $\angle CBA \cong \angle FED$ and $\overline{AC} \cong \overline{DF}$ (Figure 3.3.2). We must show that $\triangle ABC \cong \triangle DEF$. Now if $\overline{AB} \cong \overline{DE}$, the triangles would be congruent by SAS; therefore we will employ an indirect proof by assuming that $\overline{AB}$ is not congruent to $\overline{DE}$. Without losing generality we may assume that $AB > DE$, and therefore we can find a point B' on $\overline{AB}$ such that $\overline{AB'} \cong \overline{DE}$. By our construction, $\triangle AB'C \cong \triangle DEF$. Now by the definition of congruent triangles, $\angle AB'C \cong \angle DEF$, and consequently $\angle AB'C \cong \angle ABC$. But $\angle AB'C$ is an exterior angle to $\triangle CB'B$, hence $m\angle AB'C > m\angle ABC$, which is a contradiction. Therefore $\overline{AB}$ is congruent to $\overline{DE}$ and $\triangle ABC \cong \triangle DEF$.

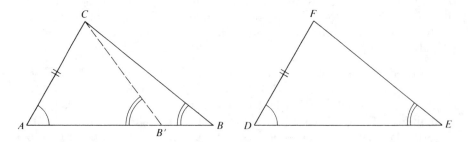

Figure 3.3.2

THEOREM 3.3.4. *Side-Angle-Side-Angle-Side (SASAS) Congruence Theorem for Quadrilaterals*. If the vertices of two quadrilaterals are in one-

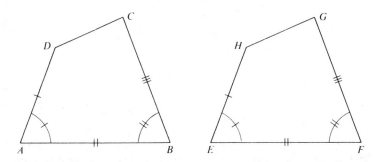

Figure 3.3.3 If $\overline{AD} \cong \overline{HE}$, $\overline{AB} \cong \overline{EF}$, $\overline{BC} \cong \overline{FG}$, $\angle A \cong \angle E$, and $\angle B \cong \angle F$, then $\square ABCD \cong \square EFGH$.

one correspondence such that three sides and the two included angles of one quadrilateral are congruent to the corresponding parts of a second quadrilateral, then the quadrilaterals are congruent (Figure 3.3.3). The proof is left as an exercise.

The next theorem will allow us to make some conclusions with regard to the relative sizes of the sides and angles in a triangle.

THEOREM 3.3.5.   *Inverse of the Isosceles Triangle Theorem*. If two sides of a triangle are not congruent, then the angles opposite those sides are not congruent and the larger angle is opposite the larger side.

Proof. Consider $\triangle ABC$ in which $AC > BC$ (Figure 3.3.4). Since $AC > BC$, there exists a point D on $\overrightarrow{CB}$ such that C-B-D and $\overline{CD} \cong \overline{AC}$. By the isosceles triangle theorem, $\angle CAD \cong \angle CDA$. By definition, $\angle CBA$ is an exterior angle of $\triangle ABD$; therefore $m\angle CBA > m\angle CDA$. Now since B is interior to $\angle CAD$, we have $m\angle CAD > m\angle CAB$, and by substitution $m\angle CBA > m\angle CAB$.

It is important to note that a repeated application of this theorem produces the fact that in any triangle the largest angle is opposite the largest side.

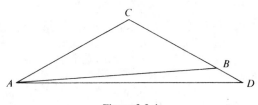

Figure 3.3.4

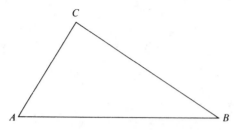

Figure 3.3.5 If m∠*A* > m∠*B*, then *BC* > *AC*.

THEOREM 3.3.6.   If two angles of a triangle are not congruent, then the sides opposite them are not congruent and the larger side is opposite the larger angle (Figure 3.3.5). The proof is left as an exercise.

The stage is now set to prove two very familiar and useful theorems.

THEOREM 3.3.7.   *Triangle Inequality.* The sum of the measures of any two sides of a triangle is greater than the measure of the third side.

Proof. Consider △*PQR*. We must show that *PQ* + *QR* > *PR*. Suppose that there exists point *S* on $\overrightarrow{PQ}$ such that *P-Q-S* and $\overline{QS} \cong \overline{QR}$ (Figure 3.3.6). By construction, △*QRS* is isosceles, and therefore ∠*QRS* ≅ ∠*QSR*. Also, since *P-Q-S*, we see that *PQ* + *QS* = *PS*, and by substitution, *PQ* + *QR* = *PS*. Now since *Q* is interior to ∠*PRS*, we have m∠*PRS* > m∠*QRS*, and substituting, we have m∠*PRS* > m∠*QSR*. Finally, applying Theorem 3.3.6 to △*PRS*, we observe that *PS* > *PR* and therefore *PQ* + *QR* > *PR*. The proof is similar for any two other sides.

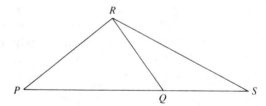

Figure 3.3.6

THEOREM 3.3.8.   *The Hinge Theorem.* If two sides of one triangle are congruent to two sides of a second triangle and the included angle of the first triangle is larger in measure than the included angle of the second triangle, then the measure of the third side of the first triangle is larger than the measure of the third side of the second triangle.

Proof. If we are given △*ABC* and △*DEF* such that $\overline{AB} \cong \overline{DE}$, $\overline{AC} \cong \overline{DF}$, and m∠*CAB* > m∠*FDE*, we must show that *BC* > *EF* (Figure 3.3.7). Since m∠*CAB* > m∠*FDE*, there exists a point *P* in the interior of ∠*CAB* such that △*ABP* ≅ △*DEF*. Now let $\overrightarrow{AR}$ bisect ∠*CAP* and intersect $\overline{BC}$ at *S*. By our construction, △*ACS* ≅ △*APS* and $\overline{CS} \cong \overline{SP}$. Using the previous theorem, we have *BS* + *SP* > *BP*. Since $\overline{SP} \cong \overline{SC}$, we have *BS* + *SC* > *BP*.

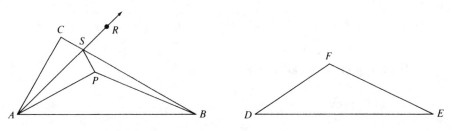

Figure 3.3.7

Now $BS + SC = BC$, hence $BC > BP$. But $\overline{BP} \cong \overline{EF}$, and therefore $BC > EF$.

Finally we introduce the last of the most common triangle congruence theorems, the side-side-side congruence condition. Its proof is a consequence of the previous theorems and is left as an exercise.

THEOREM 3.3.9.   *Side-Side-Side (SSS) Triangle Congruence Theorem*. If the vertices of two triangles are in one-to-one correspondence such that all three sides of one triangle are congruent, respectively, to all three sides of the second triangle, then the triangles are congruent.

EXERCISE SET 3.3

1. Prove Theorem 3.3.2.
2. Prove Theorem 3.3.4.
3. State and prove the ASASA congruence theorem for quadrilaterals.
4. In a *scalene triangle*, all three sides have different lengths. Show that the largest angle of a scalene triangle is opposite the largest side.
5. Prove Theorem 3.3.6.
6. Prove Theorem 3.3.9. (*Hint*: Use the hinge theorem).
7. Given $\triangle ABC$ and $\triangle A'B'C'$ where A-B-D, A'-B'-D', $\overline{AB} \cong \overline{A'B'}$, $\overline{BD} \cong \overline{B'D'}$, $\overline{AC} \cong \overline{A'C'}$ and $\overline{BC} \cong \overline{B'C'}$. Prove $\overline{CD} \cong \overline{C'D'}$.
8. Is there an analog to the SSS triangle congruence theorem for quadrilaterals? Explain your answer.
9. What would be a sufficient congruence condition for pentagons to be congruent? Explain your answer.

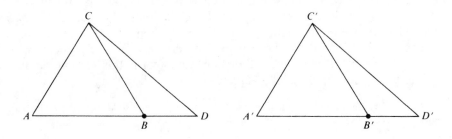

10. Prove that two right triangles are congruent if the hypotenuse and leg of one are congruent, respectively, to the hypotenuse and leg of the other.

3.4 THE PLACE OF PARALLELS

As it did in our previous discussion of incidence geometry, the question of the existence of parallel lines arises. But in contrast to the discussion of incidence geometry, we find, as the following series of theorems shows, that we can actually prove that in neutral geometry at least one line can be drawn parallel to a given line through a point not on that line. We begin with the following theorem.

THEOREM 3.4.1.   *Alternate Interior Angle Theorem.* If two lines are intersected by a transversal such that a pair of alternate interior angles formed are congruent, then the lines are parallel (Figure 3.4.1).

Proof. Suppose that lines *l* and *m* are intersected by a transversal at distinct points *P* and *Q*. Furthermore, assume that a pair of alternate interior angles ∠1 and ∠2 are congruent. We must show that *l* and *m* are parallel. Using an indirect proof, we will assume that *l* and *m* intersect. By a theorem from incidence geometry, that intersection is exactly one point *R*. Now ∠2 is an exterior angle of △*PQR*, hence m∠2 > m∠1, which contradicts our hypothesis. Therefore *l* is parallel to *m*.

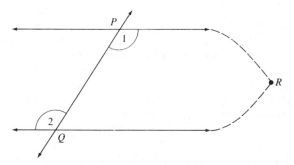

Figure 3.4.1

Important Note: In past geometrical experiences you may recall using the converse of the alternate interior angle theorem which states, If two parallel lines are intersected by a transversal, then the pairs of alternate interior angles formed are congruent. The converse is *not* a theorem in neutral geometry. It is, as we will demonstrate later in this section, equivalent to the Euclidean parallel postulate.[10]

---

[10] The reader is reminded that by the Euclidean parallel postulate we mean Playfair's postulate, as opposed to Euclid's fifth postulate.

As a consequence of the alternate interior angle theorem, we may now introduce several related corollaries.

COROLLARY 3.4.2.   Two lines perpendicular to the same line are parallel. The proof is left as an exercise.

An important consequence of Theorem 3.4.1 and Corollary 3.4.2 is the fact that we may now ensure the existence of at least one line parallel to a given line through a point not on that line.

COROLLARY 3.4.3.   If two lines are intersected by a transversal such that a pair of corresponding angles formed are congruent, then the lines are parallel. The proof is left as an exercise.

COROLLARY 3.4.4.   If two lines are intersected by a transversal such that a pair of interior angles on the same side of the transversal are supplementary, then the lines are parallel. The proof is left as an exercise.

As you recall from Chapter 2, the essence of the Euclidean parallel postulate may take on a variety of forms. In particular, earlier in this section, a claim was made that the converse of the alternate interior angle theorem is equivalent to the Euclidean parallel postulate. The remainder of this section is devoted to demonstrating the logical equivalence of a variety of statements related to the Euclidean parallel postulate. The reader is reminded that by proving the equivalence of two statements, we have not proven either statement as a theorem but have simply demonstrated that we could prove one *only if* we were willing to assume the other.

THEOREM 3.4.5.   Euclid's fifth postulate is equivalent to the Euclidean parallel postulate.[11]

Proof. First, we will assume Euclid's fifth postulate and then use it to prove that through a given point A not on a given line m there passes at most one line that does not intersect m (i.e., is parallel to m) (Figure 3.4.2a.) Let t be a line passing through A perpendicular to m and intersecting m at B, and let l be a line passing through A perpendicular to t. By Corollary 3.4.2, l is parallel to m. We must now show that there cannot be another line through A parallel to m. Let n be any *other* line through A. Since $n \neq l$, one of the

[11] The reader should recall that the Euclidean parallel postulate is the same as Hilbert's parallel postulate (Axiom IV-1), Playfair's postulate, and the SMSG parallel postulate (Postulate 16).

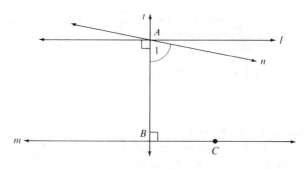

Figure 3.4.2a

angles formed by *n* and *t* is acute. Call this angle ∠1. Now ∠1 and ∠*ABC* are together less than two right angles; therefore by Euclid's fifth postulate, *n* and *m* must intersect. Thus we have at most one line through *A* parallel to *m*.

To prove the converse, we will assume the Euclidean parallel postulate and use it to prove Euclid's fifth postulate as a theorem. Let *t* be a line intersecting lines *l* and *m* in distinct points such that ∠1 and ∠2 are together less than two right angles; that is, m∠1 + m∠2 < 180 (Figure 3.4.2b). We must now show that *l* and *m* intersect and that they meet on the same side of *t* as point *C*. Since m∠3 + m∠2 = 180, we can conclude that m∠1 < m∠3, and by the angle construction postulate (SMSG Postulate 12), there exists a unique ray $\overrightarrow{AD}$ such that ∠*BAD* ≅ ∠3. Therefore by Theorem 3.4.1, $\overrightarrow{AD} \parallel m$, and since *l* ≠ $\overrightarrow{AD}$, the Euclidean parallel postulate guarantees that *l* must intersect *m*. In order to show that *l* and *m* meet on the same side of *t* as *C*, we will use an indirect proof and assume that they intersect at *E* which is on the side of *t* opposite *C* (Figure 3.4.2c). Now ∠1 is an exterior angle of △*ABE*, and since m∠1 < m∠3, we have a contradiction to the exterior angle theorem, which completes the proof of Euclid's fifth postulate.

Now, as promised earlier, we will demonstrate the equivalence of the Euclidean parallel postulate and the converse of the alternate interior angle theorem.

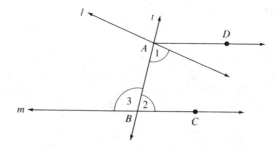

Figure 3.4.2b

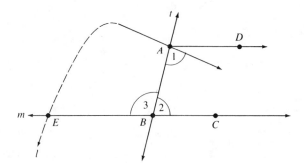

Figure 3.4.2c

THEOREM 3.4.6.   The Euclidean parallel postulate is equivalent to the converse of the alternate interior angle theorem.

Proof. We will begin by assuming the Euclidean parallel postulate and use it to prove the following statement: If two parallel lines are intersected by a transversal, then the alternate interior angles formed are congruent. Let $l \parallel m$, and let t be a transversal intersecting l and m and forming alternate interior angles $\angle 1$ and $\angle 2$. We now must show that $\angle 1 \cong \angle 2$ (Figure 3.4.3). Assume that $\angle 1$ and $\angle 2$ are not congruent and that $m\angle 1 > m\angle 2$. Using the angle construction postulate, there exists a ray $\overrightarrow{BC}$ such that $\angle ABC \cong \angle 2$. Now by the alternate interior angle theorem (Theorem 3.4.1), $\overleftrightarrow{BC} \parallel m$, and since $\overleftrightarrow{BC} \neq l$, we have two lines through B parallel to m, which contradicts the Euclidean parallel postulate. The second part of this proof is to assume the converse of the alternate interior angle theorem and use it to prove the Euclidean parallel postulate. That proof is left as an exercise.

The next three theorems establish the equivalence of various statements and the Euclidean parallel postulate. While one of the following statements may appear to be intuitively equivalent, the others may not.

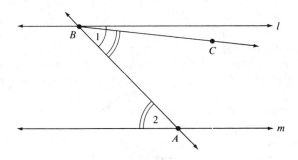

Figure 3.4.3

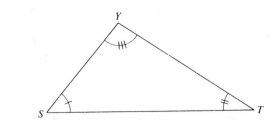

Figure 3.4.4 $\triangle PQR \sim \triangle STU \Leftrightarrow \angle P \cong \angle S, \angle Q \cong \angle T, \angle R \cong \angle U$, and $PQ/ST = QR/TU = RP/US$.

THEOREM 3.4.7.   The Euclidean parallel postulate is equivalent to the following statement: If a line intersects one of two parallel lines, then it intersects the other. The proof is left as an exercise.

THEOREM 3.4.8.   The Euclidean parallel postulate is equivalent to the following statement: If a line is perpendicular to one of two parallel lines, then it is perpendicular to the other. The proof is left as an exercise.

If we define *similar triangles* as two triangles whose vertices can be placed in one-to-one correspondence such that their corresponding angles are congruent and the ratios of each pair of corresponding sides are equal[12] (Figure 3.4.4), we can produce another statement that is equivalent to the Euclidean parallel postulate.

THEOREM 3.4.9.   The Euclidean parallel postulate is equivalent to the following statement: Given any $\triangle PQR$ and any line segment $\overline{AB}$, there exists a triangle having a side congruent to $\overline{AB}$ that is similar to $\triangle PQR$.[13]

Proof. First we will assume the Euclidean parallel postulate and then use it to prove Wallis' postulate. Given $\triangle PQR$ and line segment $\overline{AB}$, we will construct a triangle that is similar to $\triangle PQR$ and that has a side congruent to $\overline{AB}$. First, we will assume that $\overline{AB}$ is not congruent to any of the sides of $\triangle PQR$, for if it were, a triangle congruent to $\triangle PQR$ and having $\overline{AB}$ as a side would complete the proof. In particular, we will assume that $AB < PQ$ (if $AB > PQ$, the proof would be analogous) (Figure 3.4.5a). Since $AB < PQ$, there exists a point P' on $\overline{PQ}$ such that $P\text{-}P'\text{-}Q$ and $\overline{P'Q} \cong \overline{AB}$. Now by the Euclidean parallel postulate, which we have assumed, there exists a unique line m through P' parallel to $\overline{PR}$. It is left as an exercise to show that line m must intersect QR at R' ($R' \neq R$ or Q) (see Exercise Set 3.4, Problem 9). As a

---

[12] When the ratios of each pair of corresponding sides of two triangles are equal, we say that the corresponding sides are proportional.

[13] This statement was posed as an alternative for Euclid's fifth postulate by John Wallis, a British mathematician who lived during the seventeenth century. This statement is also equivalent to Birkhoff's postulate of similarity.

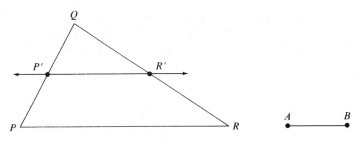

Figure 3.4.5a

result, $\angle QP'R' \cong \angle QPR$ and $\angle QR'P' \cong \angle QRP$ (see Exercise Set 3.4, Problem 10). Now clearly, $\angle Q \cong \angle Q$ and, furthermore, given the Euclidean parallel postulate, the proportionality of the sides is a consequence of the congruence of the angles (see Theorem 4.4.5). Therefore $\triangle P'R'Q$ is similar to $\triangle PRQ$.

Conversely, we will assume Wallis' postulate and use it to prove the Euclidean parallel postulate. Let m be a line and A be a point not on m. We must show that there is at most one line through A parallel to m. Let l be a line through A that shares a common perpendicular with m. Corollary 3.4.2 assures us that l is parallel to m (Figure 3.4.5b). Now let n be any other line through A, and we will show that n intersects m. Since $n \neq l$, let $\overrightarrow{AD}$ (which is a subset of n) be between $\overrightarrow{AC}$ and $\overrightarrow{AB}$ and drop $\overline{DE}$ perpendicular to $\overline{AB}$. Now consider $\triangle AED$ and line segment $\overline{AB}$. By our hypothesis there exists a point F such that $\triangle ABF$ is similar to $\triangle AED$. Since $\triangle AED$ is similar to $\triangle ABF$, we can conclude that $\angle BAF \cong \angle EAD$, and by the angle construction postulate, F must lie on $\overrightarrow{AD}$, hence on n. Similarly, since $\angle ABF \cong \angle AED$, $\overline{FB}$ must be perpendicular to $\overline{AB}$ at B. Therefore since m is the unique perpendicular to $\overline{AB}$ through B, F must also lie on m. Our conclusion is then that n intersects m, and our proof is finished.

As the previous theorems show, the road through neutral geometry can be a treacherous one. Each of the statements we have proven equivalent to

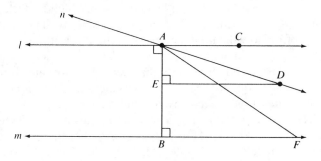

Figure 3.4.5b

the Euclidean parallel postulate provides a quick detour into Euclidean geometry. The reader is cautioned to be alert for these detours as we venture further into neutral geometry. Some other potential detours will be found in the exercises.

EXERCISE SET 3.4

1. Prove Corollary 3.4.2.
2. Prove Corollary 3.4.3.
3. Prove: If two lines are intersected by a transversal such that a pair of alternate interior angles are congruent, then the lines have a common perpendicular.
4. Prove Corollary 3.4.4.
5. Complete the proof of Theorem 3.4.6 by proving that the converse of the alternate interior angle theorem implies the Euclidean parallel postulate.
6. Prove Theorem 3.4.7.
7. Prove Theorem 3.4.8.
8. Prove that the Euclidean parallel postulate is equivalent to the following statement: If two parallel lines are intersected by a transversal, then the pairs of interior angles on the same side of the transversal are supplementary.
9. In the proof of Theorem 3.4.9 we indicated that line m, which passes through P' parallel to $\overleftrightarrow{PR}$, must intersect $\overleftrightarrow{QR}$ at a point R' which is distinct from both Q and R. Explain why this must be true.
10. Prove that the Euclidean parallel postulate is equivalent to the converse of Corollary 3.4.3.
11. If we define the *angle sum* of $\triangle ABC$ as $m\angle A + m\angle B + m\angle C$, prove that the Euclidean parallel postulate implies that every triangle has an angle sum of $180°$.

3.5 THE SACCHERI-LEGENDRE THEOREM

A large portion of the body of theorems known as neutral geometry can be found in the work of the Jesuit monk, Girolamo Saccheri (1667–1733.) As previously indicated, early mathematicians devoted considerable effort to investigating the question of the independence of Euclid's fifth postulate. Saccheri, as indicated in Chapter 2, attempted to employ the indirect method of proof in an effort to validate the parallel postulate. Without assuming any parallel postulate (i.e., in neutral geometry), Saccheri undertook an exhaustive study of a class of quadrilaterals that he called isosceles birectangular quadrilaterals. These quadrilaterals, known today as *Saccheri quadrilaterals,* have two sides that are congruent and perpendicular to a third side (Figure 3.5.1). After proving that $\angle R \cong \angle S$, Saccheri considered three possibilities for $\angle R$ and $\angle S$: (1) $m\angle R = m\angle S > 90°$, the hypothesis of the obtuse angle; (2) $m\angle R = m\angle S = 90°$, the hypothesis of the right angle; (3) $m\angle R = m\angle S < 90°$, the hypothesis of the acute angle. His intent was to

Figure 3.5.1 $\overline{SP} \perp \overline{PQ}$, $\overline{RQ} \perp \overline{PQ}$, and $\overline{SP} \cong \overline{RQ}$.

show that possibilities 1 and 3 lead to contradictions and that possibility 2, as we will prove in a subsequent section, was equivalent to Euclid's fifth postulate. The following theorem provides the basis for the contradiction to the obtuse angle hypothesis that Saccheri found. The contradiction to the acute angle hypothesis, as we know today, was never to be found.

THEOREM 3.5.1.   *Saccheri-Legendre Theorem*.[14] The angle sum of any triangle is less than or equal to 180°.

Before we begin our proof several preliminary results are necessary.

LEMMA 3.5.2.   The sum of the measures of any two angles of a triangle is less than 180°.

Proof. Consider $\triangle ABC$ and let D be on $\overrightarrow{BC}$ such that B-C-D (Figure 3.5.2). By definition, $\angle 4$ is an exterior angle of $\triangle ABC$, and therefore $m\angle 4 > m\angle 1$. Since $m\angle 4 + m\angle 2 = 180°$, we know that $m\angle 4 = 180° - m\angle 2$. Therefore by substitution, $m\angle 1 < 180° - m\angle 2$ and $m\angle 1 + m\angle 2 < 180°$. In similar fashion, it can be shown that $m\angle 2 + m\angle 3 < 180°$ and $m\angle 1 + m\angle 3 < 180°$.

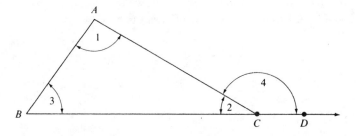

Figure 3.5.2

[14] An independent proof of this theorem was given by Adrien-Marie Legendre (1752–1833.)

LEMMA 3.5.3.   For any $\triangle ABC$ there exists $\triangle A_1B_1C_1$ having the same angle sum as $\triangle ABC$ but where $m\angle A_1 \leq \frac{1}{2}(m\angle A)$.

Proof. Consider $\triangle ABC$ where E is the midpoint of $\overline{BC}$. Locate F on $\overline{AE}$ such that $A\text{-}E\text{-}F$ and $\overline{AE} \cong \overline{EF}$ (Figure 3.5.3). Now if we draw $\overline{FC}$, it is easy to show that $\triangle BEA \cong \triangle CEF$ and, as a result, $\angle 2 \cong \angle 5$ and $\angle 3 \cong \angle 6$. Now if we denote the angle sum of $\triangle ABC$ by $S(\triangle ABC)$, then $S(\triangle ABC) = m\angle A + m\angle B + m\angle C = m\angle 1 + m\angle 2 + m\angle 3 + m\angle 4$. Substituting, we find $S(\triangle ABC) = m\angle 1 + m\angle 5 + m\angle 6 + m\angle 4 = m\angle CAF + m\angle AFC + m\angle FCA$, which is the angle sum of $\triangle AFC$. Therefore $\triangle AFC$ has the same angle sum as $\triangle ABC$. Since $m\angle A = m\angle 1 + m\angle 2 = m\angle 1 + m\angle 5$, either $m\angle 1$ or $m\angle 5$ is less than or equal to $\frac{1}{2}(m\angle A)$. If $m\angle 1 \leq \frac{1}{2}(m\angle A)$, let $A = A_1$, $F = B_1$, and $C = C_1$. If $m\angle 5 \leq \frac{1}{2}(m\angle A)$, let $F = A_1$, $C = C_1$, and $A = B_1$, and $\triangle A_1B_1C_1$ is the required triangle.

We are now in a position to prove the Saccheri-Legendre theorem, which is restated here.

THEOREM 3.5.1.   The angle sum of any triangle is less than or equal to 180°.

Proof. We will employ an indirect proof and assume that there exists a $\triangle ABC$ with an angle sum of $180° + p$, where p is any positive number. Using Lemma 3.5.3, we can produce $\triangle A_1B_1C_1$ having the same angle sum ($180° + p$) as $\triangle ABC$ but where $m\angle A_1 \leq \frac{1}{2}(m\angle A)$. Now we may apply Lemma 3.5.3 to $\triangle A_1B_1C_1$ to produce $\triangle A_2B_2C_2$ with the same angle sum as $\triangle A_1B_1C_1$, hence the same angle sum as $\triangle ABC$, with $m\angle A_2 \leq \frac{1}{2}(m\angle A_1) \leq \frac{1}{4}(m\angle A)$. If we repeat this process, we can construct a sequence of triangles, $\triangle A_1B_1C_1$, $\triangle A_2B_2C_2$, $\triangle A_3B_3C_3$, . . . , $\triangle A_nB_nC_n$, each with an angle sum of $180° + p$, such that for any $n > 0$, $m\angle A_n \leq (\frac{1}{2^n})(m\angle A)$. Now the Archimedean property of real numbers allows us to select an n sufficiently large so that $m\angle A_n$ is as small as we choose, and in particular such that $m\angle A_n \leq p$. Now since $m\angle A_n + m\angle B_n + m\angle C_n = 180° + p$, it follows that $m\angle B_n + m\angle C_n > 180°$, which contradicts Lemma 3.5.2 and proves our result.

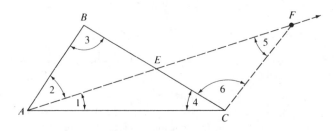

Figure 3.5.3

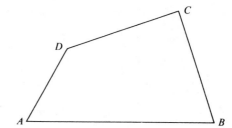

Figure 3.5.4 m∠ *A* + m∠ *B* + m∠ *C* + m∠ *D* ≤ 360.

The Saccheri-Legendre theorem leads to the proof of the following corollary which clearly implies the falsity of Saccheri's hypothesis of the obtuse angle.

COROLLARY 3.5.4.   The angle sum of any convex quadrilateral[15] is less than or equal to 360° (Figure 3.5.4). The proof is left as an exercise.

Additional properties of Saccheri quadrilaterals and their consequences will be investigated in the next section.

EXERCISES 3.5

1. Prove Corollary 3.5.4.
2. Prove: The measure of an exterior angle of a triangle is greater than or equal to the sum of the measures of the two opposite interior angles.
3. State and prove the converse of Euclid's fifth postulate.

3.6 THE SEARCH FOR A RECTANGLE

In the last section we briefly mentioned Saccheri's investigations of a class of quadrilaterals that we refer to today as Saccheri quadrilaterals. Recall that a Saccheri quadrilateral has two sides that are congruent and perpendicular to a third side (Figure 3.6.1). The congruent segments $\overline{AD}$ and $\overline{BC}$ are called the *sides* or legs, $\overline{AB}$ is the *base*, and $\overline{DC}$ is called the *summit*. ∠*C* and ∠*D* are called the *summit angles*, and they provided the focal point for Saccheri's investigations. His goal was to show that the summit angles were both right angles—essentially proving Euclid's fifth postulate and thereby validating Euclid's geometry.

---

[15] □*ABCD* is convex if it has a pair of opposite sides $\overline{AB}$ and $\overline{CD}$ such that $\overline{CD}$ is contained in one of the half-planes bounded by $\overleftrightarrow{AB}$, and $\overline{AB}$ is one of the half-planes bounded by $\overleftrightarrow{CD}$.

Figure 3.6.1

Now if a *rectangle* is defined as a quadrilateral with four right angles, Saccheri's investigations are equivalent to asking the question, Do rectangles exist in neutral geometry? As we will see, this is by no means a trivial question, and its answer is essential to understanding the role of the parallel postulate in Euclid's geometry. At this time, the reader is asked to consider various ways that a rectangle might be constructed in neutral geometry.

The following theorems represent some of Saccheri's results. Proofs for the first five theorems require only elementary applications of the triangle congruence theorems; therefore they are left as exercises.

THEOREM 3.6.1.   The diagonals of a Saccheri quadrilateral are congruent.

THEOREM 3.6.2.   The summit angles of a Saccheri quadrilateral are congruent.

THEOREM 3.6.3.   The summit angles of a Saccheri quadrilateral are not obtuse and thus are both acute or both right.[16]

THEOREM 3.6.4.   The line joining the midpoints of both the summit and the base of a Saccheri quadrilateral is perpendicular to both (Figure 3.6.2).

COROLLARY 3.6.5.   The summit and the base of a Saccheri quadrilateral are parallel.

THEOREM 3.6.6.   In any Saccheri quadrilateral the length of the summit is greater than or equal to the length of the base.

Proof. Let $\square ABCD$ be a Saccheri quadrilateral where $\overline{DA} \perp \overline{AB}$, $\overline{CB} \perp \overline{AB}$, and $\overline{AD} \cong \overline{CB}$ (Figure 3.6.3). We must now show that $AB \le CD$. First, draw $\overline{BD}$ and note that $\angle 1 \cong \angle 2$ or $m\angle 1 < m\angle 2$ or $m\angle 1 > m\angle 2$. If $\angle 1 \cong \angle 2$, then $\triangle ADB \cong \triangle CBD$ and $\overline{AB} \cong \overline{CD}$. Thus $AB = CD$, and the proof is completed. If $m\angle 1 < m\angle 2$ then, by the hinge theorem, $AB < CD$, and the proof is completed. Therefore we will assume that $m\angle 1 > m\angle 2$. Now $m\angle 2 + m\angle 3 = 90°$, which implies that $m\angle 1 + m\angle 3 > 90°$, and therefore

---

[16] If they are both right angles, we have found our rectangle.

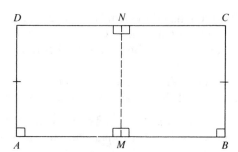

Figure 3.6.2 If *ABCD* is a Saccheri quadrilateral with *M* the midpoint of the base $\overline{AB}$ and *N* the midpoint of summit $\overline{CD}$, then $\overline{MN} \perp \overline{CD}$ and $\overline{MN} \perp \overline{AB}$.

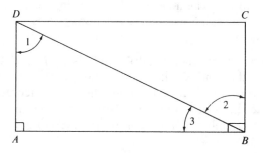

Figure 3.6.3

$m\angle 1 + m\angle 3 + m\angle DAB > 180°$, which contradicts the Saccheri-Legendre theorem. Hence $m\angle 1 \leq m\angle 2$, and our proof is completed.

In his book *Die Theorie der Parallellinien*, Johann Heinrich Lambert (1728–1777), a Swiss mathematician, attempted to prove the Euclidean parallel postulate by studying the properties of a different quadrilateral; one also motivated by an attempt to "construct" a rectangle. The *Lambert quadrilateral*, as it is called today, is a quadrilateral with at least three right angles (Figure 3.6.4). Lambert's investigation paralleled Saccheri's by creating acute, right, and obtuse angle hypotheses with regard to the fourth angle. Lambert's work went beyond that of Saccheri in producing theorems based on the acute angle and obtuse angle hypotheses. In fact, it was Lambert who

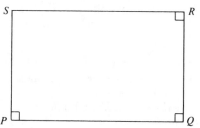

Figure 3.6.4 If $\angle P$, $\angle Q$, and $\angle R$ are right angles, then *PQRS* is a Lambert quadrilateral.

first recognized the similarities between the obtuse angle hypothesis and the geometry of a sphere (see Section 6.8). Several of Lambert's results are included among the following theorems and also in the exercises.

THEOREM 3.6.7.   The fourth angle of a Lambert quadrilateral is not obtuse and thus is acute or right.[17] The proof is left as an exercise.

THEOREM 3.6.8.   The measure of the side included between two right angles of a Lambert quadrilateral is less than or equal to the measure of the side opposite it. The proof is left as an exercise.

THEOREM 3.6.9.   The measure of the line joining the midpoints of the base and the summit of a Saccheri quadrilateral is less than or equal to the measure of its sides.

Proof. Consider Saccheri quadrilateral, $\square ABCD$ with $\overline{AD} \cong \overline{CB}$, $\overline{DA} \perp \overline{AB}$, $\overline{CB} \perp \overline{AB}$, and M and N midpoints of the base $\overline{AB}$ and the summit $\overline{DC}$, respectively (Figure 3.6.2). By a previous theorem (Theorem 3.6.4), $\overline{MN}$ is perpendicular to both $\overline{AB}$ and $\overline{CD}$. Now by definition, $\square AMND$ and $\square CBMN$ are both Lambert quadrilaterals, and therefore, by Theorem 3.6.8, $MN \leq AD$ and $MN \leq CB$.

As we have seen, Saccheri and Lambert were both successful in eliminating the obtuse angle hypothesis, but both were equally unsuccessful in eliminating the acute angle hypothesis and therefore were unable to prove the existence of a rectangle in neutral geometry.

As we now know, the contradiction that both Saccheri and Lambert sought regarding the acute angle hypothesis does not exist, and the assumption of that hypothesis gives rise to a non-Euclidean geometry mentioned in the last chapter, hyperbolic geometry. We will investigate hyperbolic geometry in more detail in a later chapter. The remainder of this section follows a path previously traveled by Prenowitz and Jordan[18] and is devoted to the consequences of the assumption of the existence of a rectangle and its equivalence to the Euclidean parallel postulate.

THEOREM 3.6.10.   If one rectangle exists,[19] then there exists a rectangle with an arbitrarily large side.

Proof. Suppose that $\square ABCD$ is a rectangle and that $\overline{PQ}$ is any segment. We must show that we can construct a rectangle with one side whose

---

[17] If it is a right angle, we have found our rectangle.

[18] Prenowitz, Walter, and Meyer Jordan, *Basic Concepts in Geometry*, (New York: Ardsley House, Inc., 1989) pp. 40–46.

[19] The reader is reminded that this theorem does not imply the existence of a rectangle. More importantly, once proven, this theorem is true whether a rectangle exists or not.

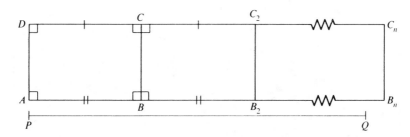

Figure 3.6.5

length is greater than PQ. By the segment construction, there exists point C_2 on $\overrightarrow{DC}$ such that $D\text{-}C\text{-}C_2$ and $\overline{CC_2} \cong \overline{DC}$ (Figure 3.6.5). Similarly there exists a point B_2 on $\overrightarrow{AB}$ such that $A\text{-}B\text{-}B_2$ and $\overline{BB_2} \cong \overline{AB}$. Now by Theorem 3.3.5 (SASAS), $\square BB_2C_2C$ is congruent to rectangle $\square ABCD$, and therefore $\angle BB_2C_2$ and $\angle CC_2B_2$ are both right angles. By definition, $\square AB_2C_2D$ is a rectangle and $AB_2 = 2(AB)$. Repeating this construction, we can find a rectangle $\square AB_nC_nD$ such that $AB_n = n(AB)$. Now the Archimedean property of real numbers allows us to choose n large enough so that $n(AB) > PQ$, and our proof is complete.

COROLLARY 3.6.11.   If one rectangle exists, then there exists a rectangle with two arbitrarily large sides. The proof is left as an exercise.

THEOREM 3.6.12.   If one rectangle exists, then there exists a rectangle with two adjacent sides congruent to the given line segments $\overline{PQ}$ and $\overline{RS}$.

Proof. By the previous corollary, we can construct rectangle $\square WXYZ$ such that $WX > PQ$ and $WZ > RS$ (Figure 3.6.6). If we locate X' on $\overline{WX}$ such that $\overline{WX'} \cong \overline{PQ}$ and drop a perpendicular from X' intersecting $\overline{ZY}$ at Y', we create $\square WX'Y'Z$, which we will show is a rectangle. The angles at Y', Z, and W are right angles; therefore we only have to show that $\angle WX'Y'$ is a right angle. Since $\square WX'Y'Z$ is a Lambert quadrilateral, $m\angle WX'Y' \leq 90°$. But if

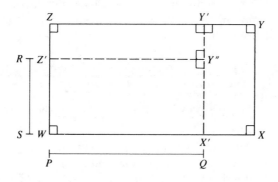

Figure 3.6.6

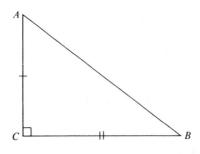
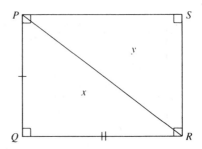

Figure 3.6.7

we assume that m∠$WX'Y'$ < 90°, then m∠$XX'Y'$ > 90°, which is a contradiction since □$X'XYY'$ is also a Lambert quadrilateral and its fourth angle cannot be obtuse. Therefore ∠$WX'Y'$ is a right angle and □$WX'Y'Z$ is a rectangle. In a similar fashion, we can locate Z' on $\overline{WZ}$, and Y'' on $\overline{X'Y'}$, such that □$WX'Y''Z'$ is our required rectangle.

THEOREM 3.6.13.   If one rectangle exists, then every right triangle has an angle sum of 180°.

Proof. Let △ABC be any right triangle with its right angle at vertex C (Figure 3.6.7). By the previous theorem, we can construct rectangle □$PQRS$ such that $\overline{PQ} \cong \overline{AC}$ and $\overline{QR} \cong \overline{CB}$. If we draw diagonal $\overline{PR}$, we can easily show that △PRQ is congruent to △ABC and therefore that they have the same angle sum. Now let x be the angle sum of △PRQ and y be the angle sum of △RPS. Since □$PQRS$ is a rectangle, $x + y = 360°$. We know by the Saccheri-Legendre theorem that both x and y are less than or equal to 180°. Therefore we will assume that $x < 180°$, but this implies that $y > 180°$, which is a contradiction. Hence the angle sum of right triangle △ABC, which is also x, must be 180°.

The next two theorems give additional insight into the relationship between the assumption of the existence of even a single rectangle and the Euclidean parallel postulate.

THEOREM 3.6.14.   If one rectangle exists, then every triangle has an angle sum of 180°. The proof is left as an exercise.

THEOREM 3.6.15.   If one triangle having an angle sum of 180° exists, then a rectangle exists.

Proof. Consider △ABC such that m∠A + m∠B + m∠C = 180° (Figure 3.6.8). Since at least two angles of a triangle must be acute, say ∠A and ∠B, we let $\overline{CD}$ be the altitude to side $\overline{AB}$ such that A-D-B (see Exercise Set 3.6, Problem 15). Now let x be the angle sum of △ADC, and let y be the angle

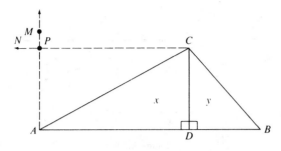

Figure 3.6.8

sum of $\triangle BDC$. Since $m\angle A + m\angle B + m\angle C + m\angle ADC + m\angle BDC = 360°$, we know that $x + y = 360°$, and using the same method employed in the last proof, $x = 180°$. We now have right triangle $\triangle ADC$ whose angle sum is $180°$. Now by the angle construction postulate, we can find $\overrightarrow{CN}$ on the side of $\overrightarrow{AC}$ opposite D such that $\angle ACN \cong \angle CAD$. If we let P be a point on $\overrightarrow{CN}$ such that $\overline{CP} \cong \overline{AD}$, we claim that $\square ADCP$ is our required rectangle. Why? (The proof is left as an exercise.)

CoROLLARY 3.6.16.   If there exists one triangle with an angle sum of 180°, then all triangles have angle sums of 180°. The proof is left as an exercise.

CoROLLARY 3.6.17.   If one triangle exists with an angle sum of less than 180°, then every triangle has an angle sum of less than 180°. The proof is left as an exercise.

The two previous corollaries provide that our geometry has the "all-or-none" property, since either *all* of our triangles have an angle sum of 180° or *none* of them do. And, as we see with the next theorem, if all the triangles have an angle sum of 180°, our geometry is Euclidean.

THEOREM 3.6.18.   The Euclidean parallel postulate is equivalent to the following statement: The angle sum of every triangle is 180°.

Proof. The first implication is left as an exercise (see Exercise Set 3.6, Problem 13). Next we will assume that every triangle has an angle sum of 180° and use this fact to prove the Euclidean parallel postulate. Consider line l and point P which is not on line l, and let m be a parallel line through P that shares a common perpendicular with l. (Such a line exists as a result of Corollary 3.4.2.) Now if we let n be any *other* line through P, we must show that n intersects l (Figure 3.6.9). Using an indirect approach, we will assume that n is parallel to l. Since n is different than m, consider $\overrightarrow{PS}$ which lies between $\overrightarrow{PR}$ and $\overrightarrow{PQ}$ (i.e., $m\angle RPS > 0$). By the Archimedean property of

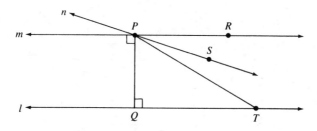

Figure 3.6.9

real numbers, choose point T on l (on the same side of $\overleftrightarrow{PQ}$ as R and S) such that $m\angle QTP < m\angle SPR$. Since $\overrightarrow{PT}$ is in the interior of $\angle QPS$ (otherwise $\overrightarrow{PS}$ would be interior to $\angle QPT$, hence by the crossbar theorem would intersect $\overline{QT}$), $m\angle TPQ < m\angle SPQ$. Now $m\angle TPQ + m\angle QTP < m\angle SPQ + m\angle SPR = 90°$, and thus the angle sum of $\triangle PQT$ is less than 180°, which contradicts our hypothesis. Therefore n must intersect l, hence there is only one parallel through P.

EXERCISE SET 3.6

1. Determine which of the following statements are true in neutral geometry.
   (a)   The alternate interior angle theorem.
   (b)   The converse of the alternate interior angle theorem.
   (c)   If the Euclidean parallel postulate is true, then when two parallel lines are intersected by a transversal, the pairs of interior angles on the same side of the transversal are supplementary.
   (d)   A Saccheri quadrilateral exists.
   (e)   A rectangle exists.
   (f)   If a rectangle exists, then every Saccheri quadrilateral is a rectangle.
   (g)   The summit and base of a Saccheri quadrilateral are parallel.
   (h)   If two lines are intersected by a transversal such that the alternate interior angles formed are congruent, then the lines must share a common perpendicular.
   (i)   If the alternate interior angle theorem is true, then the converse of the alternate interior angle theorem is true.
   (j)   If there exists a triangle having an angle sum of less than 180°, then no rectangle exists.
   (k)   If no rectangle exists, then a triangle having an angle sum of less than 180° exists.
   (l)   If a line is perpendicular to one of two parallel lines, then it is perpendicular to the other.
   (m) A quadrilateral that is both Saccheri and Lambert is a rectangle.
   (n)   Parallelograms have both pairs of opposite sides congruent.
   (o)   The diagonals of a rectangle are congruent.
   (p)   The measure of the side included between two right angles in a Lambert quadrilateral is less than or equal to the measure of the side opposite it.
   (q)   Similar but noncongruent triangles exist.

2. Prove Theorem 3.6.1.
3. Prove Theorem 3.6.2.
4. Prove Theorem 3.6.3.
5. Prove Theorem 3.6.4.
6. Prove Corollary 3.6.5.
7. Prove: If □*ABCD* is a quadrilateral where ∠*A* and ∠*B* are right angles and ∠*C* ≅ ∠*D*, then □*ABCD* is a Saccheri quadrilateral.
8. Prove: If two Saccheri quadrilaterals have congruent bases and congruent legs, then they are congruent.
9. Given any △*ABC*, let *M* and *N* be the midpoints of sides $\overline{AC}$ and $\overline{BC}$, respectively. If the lines drawn perpendicular to $\overleftrightarrow{MN}$ from *A* and *B* intersect $\overleftrightarrow{MN}$ at *E* and *D*, respectively, prove that □*ABDE* is a Saccheri quadrilateral. □*ABDE* is called the *Saccheri quadrilateral associated with* △*ABC*.

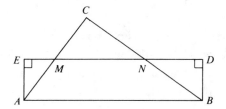

10. If a Saccheri quadrilateral is associated with △*ABC*, then the sum of the measures of its summit angles is equal to the angle sum of △*ABC*.
11. Prove Theorem 3.6.7.
12. Prove Theorem 3.6.8.
13. Prove Corollary 3.6.11.
14. In the proof of Theorem 3.6.15 we let $\overline{CD}$ be the altitude to $\overline{AB}$ such that *A-D-B*. Justify this step by proving that if ∠*A* and ∠*B* of △*ABC* are acute, then the altitude from ∠*C* must intersect $\overline{AB}$ at *D* such that *A-D-B*.
15. Complete the proof of Theorem 3.6.15 by showing that □*ADCP* is a rectangle.
16. Prove Corollary 3.6.16.
17. Prove Corollary 3.6.17.
18. Complete the proof of Theorem 3.6.18 by proving that the Euclidean parallel postulate implies that every triangle has an angle sum of 180°.
19. Prove: If the sum of the angles of a triangle is a constant *n*, then *n* = 180° and thus the geometry is Euclidean.
20. Prove: If there exists a triangle such that the segment joining the midpoints of any two sides is equal in measure to one-half the third side, then the geometry is Euclidean.
21. Prove: If a quadrilateral is both Saccheri and Lambert, then it is a rectangle.
22. Prove: If one rectangle exists then the geometry is Euclidean.
23. Prove: If there exists a quadrilateral that has an angle sum of 360° then the geometry is Euclidean.
24. Prove: If a rectangle exists then its opposite sides are congruent.
25. Prove: If a rectangle exists then its diagonals bisect each other.

CHAPTER 3 SUMMARY

3.2 Preliminary Notions

Definition. Two segments are congruent if and only if their measures are equal. For example, $\overline{AB} \cong \overline{CD}$ if and only if $AB = CD$.

Definition. Two angles are congruent if and only if their measures are equal. That is, $\angle ABC \cong \angle DEF$ if and only if $m\angle ABC = m\angle DEF$.

Definition. Two polygons are congruent if and only if there exists a one-to-one correspondence between their vertices such that all their corresponding sides (line segments) are congruent and all of their corresponding angles are congruent. For example, $\triangle ABC \cong \triangle DEF$ if and only if $\overline{AB} \cong \overline{DE}$, $\overline{BC} \cong \overline{EF}$, $\overline{AC} \cong \overline{DF}$, $\angle ABC \cong \angle DEF$, $\angle BCA \cong \angle EFD$, and $\angle CAB \cong \angle FDE$.

Theorem 3.2.1. The relations of line segment congruence, angle congruence, and polygonal congruence are equivalence relations.

Theorem 3.2.2. (i) Every line segment has exactly one midpoint, (ii) every angle has exactly one bisector.

Theorem 3.2.3. Supplements (and complements) of the same or congruent angles are congruent.

Theorem 3.2.4. Any pair of vertical angles are congruent.

Theorem 3.2.5 *Pasch's Axiom.* If a line l intersects $\triangle PQR$ at a point S such that P-S-Q, then l intersects $\overline{PR}$ or $\overline{RQ}$.

Theorem 3.2.6 *The Crossbar Theorem.* If X is a point in the interior of $\triangle UVW$, then $\overrightarrow{UX}$ intersects $\overline{WV}$ at a point Y such that W-Y-V.

Theorem 3.2.7 *Isosceles Triangle Theorem.* If two sides of a triangle are congruent, then the angles opposite those sides are congruent.

Theorem 3.2.8. A point is on the perpendicular bisector of a line segment if and only if it is equidistant from the endpoints of the line segment.

Theorem 3.2.9 *The Exterior Angle Theorem.* The exterior angle of a triangle is greater in measure than either of the nonadjacent interior angles of the triangle.

3.3 Congruence Conditions

Theorem 3.3.1 *Angle-Side-Angle (ASA) Triangle Congruence Theorem.* If the vertices of two triangles are in one-to-one correspondence such that two angles and the included side of one triangle are congruent, respectively, to two angles and the included side of the second triangle, then the triangles are congruent.

Theorem 3.3.2 *Converse of the Isosceles Triangle Theorem.* If two angles of a triangle are congruent, then the sides opposite those angles are congruent.

Theorem 3.3.3 *Angle-Angle-Side (AAS) Triangle Congruence Theorem.* If the vertices of two triangles are in one-to-one correspondence such that two angles and the side opposite one of them in one triangle are congruent to the corresponding parts of the second triangle, then the triangles are congruent.

Theorem 3.3.4 *Side-Angle-Side-Angle-Side (SASAS) Congruence Theorem for Quadrilaterals.* If the vertices of two quadrilaterals are in one-to-one correspondence such that three sides and the two included angles of one quadrilateral are congruent to the corresponding parts of a second quadrilateral, then the quadrilaterals are congruent.

Theorem 3.3.5 *Inverse of the Isosceles Triangle Theorem.* If two sides of a triangle are not congruent, then the angles opposite those sides are not congruent and the larger angle is opposite the larger side.

Theorem 3.3.6 If two angles of a triangle are not congruent, then the sides opposite them are not congruent and the larger side is opposite the larger angle.

Theorem 3.3.7 *Triangle Inequality.* The sum of the measures of any two sides of a triangle is greater than the measure of the third side.

Theorem 3.3.8 *The Hinge Theorem.* If two sides of one triangle are congruent to two sides of a second triangle, and the included angle of the first triangle is larger in measure than the included angle of the second triangle, then the measure of the third side of the first triangle is larger than the measure of the third side of the second triangle.

Theorem 3.3.9 *Side-Side-Side (SSS) Triangle Congruence Theorem.* If the vertices of two triangles are in one-to-one correspondence such that all three sides of one triangle are congruent, respectively, to all three sides of the second triangle, then the triangles are congruent.

3.4 The Place of Parallels

Theorem 3.4.1 *Alternate Interior Angle Theorem.* If two lines are intersected by a transversal such that a pair of alternate interior angles formed are congruent then the lines are parallel.

Corollary 3.4.2. Two lines perpendicular to the same line are parallel.

Corollary 3.4.3. If two lines are intersected by a transversal such that a pair of corresponding angles formed are congruent, then the lines are parallel.

Corollary 3.4.4.   If two lines are intersected by a transversal such that a pair of interior angles on the same side of the transversal are supplementary, then the lines are parallel.

Theorem 3.4.5.   Euclid's fifth postulate is equivalent to the Euclidean parallel postulate.

Theorem 3.4.6.   The Euclidean parallel postulate is equivalent to the converse of the alternate interior angle theorem.

Theorem 3.4.7.   The Euclidean parallel postulate is equivalent to the following statement: If a line intersects one of two parallel lines, then it intersects the other.

Theorem 3.4.8.   The Euclidean parallel postulate is equivalent to the following statement: If a line is perpendicular to one of two parallel lines, then it is perpendicular to the other.

Theorem 3.4.9.   The Euclidean parallel postulate is equivalent to the following statement. Given any $\triangle PQR$ and any line segment $\overline{AB}$, there exists a triangle having a side congruent to $\overline{AB}$ that is similar to $\triangle PQR$.

3.5 The Saccheri-Legendre Theorem

Theorem 3.5.1   *Saccheri-Legendre Theorem.*   The angle sum of any triangle is less than or equal to 180°.

Lemma 3.5.2.   The sum of the measures of any two angles of a triangle is less than 180°.

Lemma 3.5.3.   For any $\triangle ABC$, there exists $\triangle A_1B_1C_1$ having the same angle sum as $\triangle ABC$ but where $m\angle A_1 \leq \frac{1}{2}(m\angle A)$.

Corollary 3.5.4.   The angle sum of any convex quadrilateral is less than or equal to 360°.

3.6 The Search for a Rectangle

Theorem 3.6.1.   The diagonals of a Saccheri quadrilateral are congruent.

Theorem 3.6.2.   The summit angles of a Saccheri quadrilateral are congruent.

Theorem 3.6.3.   The summit angles of a Saccheri quadrilateral are not obtuse and thus are both acute or both right.

Theorem 3.6.4.   The line joining the midpoints of both the summit and the base of a Saccheri quadrilateral is perpendicular to both.

Corollary 3.6.5.   The summit and the base of a Saccheri quadrilateral are parallel.

Theorem 3.6.6.   In any Saccheri quadrilateral, the length of the summit is greater than or equal to the length of the base.

Theorem 3.6.7.   The fourth angle of a Lambert quadrilateral is not obtuse and thus is acute or right.

Theorem 3.6.8.   The measure of the side included between two right angles of a Lambert quadrilateral is less than or equal to the measure of the side opposite it.

Theorem 3.6.9.   The measure of the line joining the midpoints of the base and the summit of a Saccheri quadrilateral is less than or equal to the measure of its sides.

Theorem 3.6.10.   If one rectangle exists, then there exists a rectangle with an arbitrarily large side.

Corollary 3.6.11.   If one rectangle exists, then there exists a rectangle with two arbitrarily large sides.

Theorem 3.6.12.   If one rectangle exists, then there exists a rectangle with two adjacent sides congruent to the given line segments $\overline{PQ}$ and $\overline{RS}$.

Theorem 3.6.13.   If one rectangle exists, then every right triangle has an angle sum of 180°.

Theorem 3.6.14.   If one rectangle exists, then every triangle has an angle sum of 180°.

Theorem 3.6.15.   If one triangle having an angle sum of 180° exists, then a rectangle exists.

Corollary 3.6.16.   If there exists one triangle that has a angle sum of 180°, then all triangles have angle sums of 180°.

Corollary 3.6.17.   If one triangle exists with an angle sum of less than 180°, then every triangle has an angle sum of less than 180°.

Theorem 3.6.18.   The Euclidean parallel postulate is equivalent to the following statement: The angle sum of every triangle is 180°.

4

ONE WAY TO GO

Euclidean Geometry of the Plane

4.1 INTRODUCTION

All the theorems of neutral geometry were derived without making any assumptions about the number of parallels that can be drawn to a line through a point not on the line.[1] Geometers, such as Saccheri, working prior to the nineteenth century, were convinced that neutral geometry could be extended to include all the theorems of Euclidean geometry. As great as their efforts were, none succeeded—and for good reason. During the 1800s several mathematicians were able to show that Euclid's fifth postulate was in fact independent of the others by creating consistent geometric models in which Postulates 1 through 4 were valid, but in which Postulate 5 was negated. These works led to the development of non-Euclidean geometries.

With this issue settled, geometers could in effect "choose their geometry" and proceed to develop it without concern for the developments in other, independent, geometries. In this chapter and the two that follow it, we will consider some of the main results from both Euclidean and non-Euclidean geometries. We begin with a survey of some elementary results from Euclidean geometry.[2]

---

[1] While it is true that we have not made any assumptions about the number of parallels, we have proved that there is at least one such line using the postulates from neutral geometry.

[2] By *Euclidean geometry* we mean the body of theorems that can be derived using the neutral postulates and the Euclidean parallel postulate. This means that all the neutral theorems proved in Chapter 3 are Euclidean theorems as well. Some of the theorems derived in this chapter are neutral as well.

4.2 THE PARALLEL POSTULATE (AND SOME IMPLICATIONS)

We saw earlier that Euclid's fifth postulate[3] and the Euclidean parallel postulate (or Playfair's postulate[4]) are equivalent in the sense that each can be derived as a theorem given the other and the remaining neutral postulates. Less obvious are the many other equivalent statements that could have been postulated instead of Euclid's fifth postulate. For example, in the discussion of neutral geometry Wallis' postulate (that similar but noncongruent triangles exist) was shown to imply the Euclidean parallel postulate, and conversely. This equivalence is somewhat surprising, since at first glance there does not seem to be a direct connection between the existence of similar triangles and the existence of unique parallels. In fact, there are numerous statements which, if postulated, would allow us to "prove" Euclid's fifth postulate.

One of the earliest of these proofs was given by Proclus in which he claimed to prove that if a line intersects one of two parallel lines, it must intersect the other. A successful proof of the "transitivity of parallelism" would subsequently lead to a proof of Euclid's fifth postulate. A proof of this property, as given by Proclus in his *Commentary,* is outlined below. Read it with a critical eye, since there is a subtle flaw included. Problem 1 in Exercise Set 4.2 addresses this question.

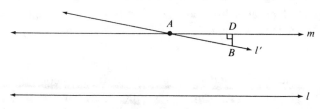

Figure 4.2.1

In Figure 4.2.1 lines *l* and *m* are parallel, and line *l'* intersects line *m* at point *A*. Proclus's proof attempts to show that lines *l* and *l'* must therefore intersect.

To show this, choose any point *D* on *m* and construct a perpendicular to *m* at that point. Label the intersection of this perpendicular and line *l'* point *B*. If we imagine that point *D* moves away from *A*, we will see that the distance *BD* continually increases. However, since the distance between any two parallel lines *l* and *m* is constant, say *k*, there will eventually be a point *D* such that *DB = k*. Then *l' ∩ l = B*, so that *l'* and *l* are not parallel.

[3] If two lines are crossed by a transversal, the lines will meet on the side where the sum of the measures of the interior angles is less than two right angles.

[4] Given a line and a point not on the line, there is at most one line through the point that is parallel to the line.

The statement proved above (that a line which intersects one of two parallel lines must intersect the other) is equivalent to Euclid's fifth postulate and Playfair's postulate, hence cannot be derived as a consequence of the other Euclidean postulates, but this wasn't recognized until long after Proclus' time. Because of this, all efforts in this direction were destined for failure.

In Chapter 2 a list of results from hyperbolic geometry is given. Each of these results has a Euclidean counterpart, some of which are the following:

1. Through a point P there can be drawn at most one line m that is parallel to any line l that does not contain P.
2. There exists a triangle in which the sum of the measures of the interior angles is 180°.
3. There exist triangles that are similar but not congruent.
4. Parallel lines are everywhere equidistant.
5. Every triangle can be circumscribed.
6. The sum of the measures of the interior angles of a triangle is constant for all triangles.
7. A rectangle exists.
8. There is no upper limit to the area that a triangle can enclose.
9. If a pair of parallel lines are crossed by a transversal, the alternate interior angles are congruent.

None of these statements was proved in neutral geometry, since each requires the use of some form of the Euclidean parallel postulate in its derivation. Theorems of this sort will be called *strictly Euclidean*. In fact, most of the theorems from the *Elements* require some form (or consequence) of the Euclidean parallel postulate. Euclid's *Elements* contains more than 470 propositions. The first 28 of these postulates are derived before Postulate 5 is used for the first time. Of the remaining, nearly all are strictly Euclidean in the sense that they cannot be proved in neutral geometry.[5] In Chapter 6 we will discuss a method that can be used to distinguish neutral theorems from those that are strictly Euclidean.

In order to ensure that we are in fact traveling the Euclidean road, we must commit ourselves to a parallel postulate in one form or another. Therefore the following postulate, along with all the neutral postulates, will be available in the proofs of Euclidean theorems:

---

[5] Some propositions after Book I can be derived without the use of the parallel postulate. For example, in Book III several results pertaining to circles are derived independently of Postulate 5.

THE EUCLIDEAN PARALLEL POSTULATE.   Given a line and a point not
on the line there is *at most* one line through the point that is parallel to the
line.

With this added postulate, we now begin the task of proving theorems
from Euclidean geometry.

One result from Euclidean geometry that is often thought of as characteristic of the geometry concerns the sum of the measures of the interior
angles of a triangle. You will probably recall that in neutral geometry the
Saccheri-Legendre theorem established the result that the sum of these angles is at most 180°. In Euclidean geometry we can prove a related but more
specific theorem. To begin, recall a result proved in Chapter 3.

THEOREM 3.4.5.   The Euclidean parallel postulate is equivalent to the
converse of the alternate interior angle theorem.

Since our geometry is now Euclidean, Theorem 3.4.5 allows us to
conclude that if two parallel lines are crossed by a transversal, the alternate
interior angles are congruent, that is, equal in measure. The traditional proof
of the Euclidean angle sum theorem for triangles proceeds in the following
way:

THEOREM 4.2.1.   The sum of the measures of the interior angles of a
triangle is 180°.

Proof. Consider the general $\triangle ABC$ shown in Figure 4.2.2. According
to the Euclidean parallel postulate there is a unique line m through B that is
parallel to line $\overleftrightarrow{AC}$. Since $\angle 1$, $\angle 2$, and $\angle 3$ form a linear triple, their sum is
180°. Applying the converse of the alternate interior angle theorem, we see
that $m\angle 1 = m\angle 4$ and $m\angle 3 = m\angle 5$. Since $m\angle 1 + m\angle 2 + m\angle 3 = 180°$, we
have, by substitution, $m\angle 4 + m\angle 2 + m\angle 5 = 180°$, completing the proof.

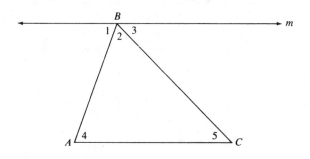

Figure 4.2.2

Clearly, Theorem 4.2.1 is a strictly Euclidean version of the Saccheri-Legendre theorem proved in Chapter 3. Without the use of some form of the Euclidean parallel postulate, the best we can say about the sum of the measures of the interior angles of a triangle is that it is *less than or equal* to 180°. A related corollary also corresponds to a neutral theorem from Chapter 3:

COROLLARY 4.2.2.   The measure of a exterior angle of a triangle is *equal to* the sum of the measures of the two remote interior angles.

This is a direct consequence of Theorem 4.2.1, and its proof is left as an exercise.

Another result that is characteristic of Euclidean geometry involves parallelograms, defined below as follows:

DEFINITION.   *Parallelogram:* A quadrilateral is a parallelogram if and only if both pairs of opposite sides are parallel.

Note that the definition says nothing about the lengths of the opposite sides, even though it is intuitively clear that they are (at least in Euclidean geometry) equal in length. We formalize this idea in Theorem 4.2.3 which, like the previous two theorems, depends on the converse of the alternate interior angle theorem.

THEOREM 4.2.3.   The opposite sides of a parallelogram are congruent.

Proof. Let $\square ABCD$ be a parallelogram with diagonal $\overline{AC}$ (Figure 4.2.3). Since $\overrightarrow{BC} \parallel \overrightarrow{AD}$ the alternate interior angles $\angle 1$ and $\angle 3$ are congruent. Similarly, since $\overrightarrow{AB} \parallel \overrightarrow{CD}$, it follows that $\angle 2$ and $\angle 4$ are congruent. As a result, $\triangle ABC$ and $\triangle CDA$ are congruent [Theorem 3.3.1 (ASA congruence)]. From this it follows that $\overline{AB} \cong \overline{CD}$ and $\overline{BC} \cong \overline{DA}$.

For our next Euclidean theorem we will consider a set of three parallel lines and a transversal that intersects them all. (Is it possible for a transversal to intersect one but not the other two?) In particular, suppose that the transversal intersects the three parallels in such a way as to make the segments between these parallel lines congruent (Figure 4.2.4). Will other trans-

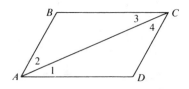

Figure 4.2.3

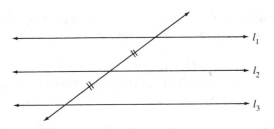

Figure 4.2.4

versals exhibit this property? In other words, if one transversal contains congruent segments between the parallels will all other transversals do the same? Theorem 4.2.4 addresses this question.

THEOREM 4.2.4.   If a transversal intersects three parallel lines in such a way as to make congruent segments between the parallels, then every transversal intersecting these parallel lines will do likewise.

Proof. Let l, m, and n be the three parallel lines, and let t_1 be a transversal intersecting l, m, and n at points A, B, and C so that $\overline{AB} \cong \overline{BC}$ (Figure 4.2.5). In addition, let t_2 be any other transversal with points of intersection A', B', and C'. We wish to show that $\overline{A'B'} \cong \overline{B'C'}$.

To do this, we will use the Euclidean parallel postulate to construct t_p, the transversal through B' that is parallel to t_1 and intersects l at E and n at D. Then, since $n \parallel m$ and $t_1 \parallel t_p$, we may conclude that $\square ABB'D$ is a parallelogram. In a similar fashion, we can show that $\square BCEB'$ is a parallelogram. As a result, we may apply Theorem 4.2.3 to conclude that $\overline{AB} \cong \overline{DB'}$ and $\overline{BC} \cong \overline{B'E}$. Since, by hypothesis, $\overline{AB} \cong \overline{BC}$, it follows that $\overline{DB'} \cong \overline{B'E}$.

Next, since m$\angle$1 = m$\angle$2 (why?) and m$\angle$3 = m$\angle$4 (why?) we may conclude that $\triangle DB'A'$ and $\triangle EB'C'$ are congruent (Theorem 3.3.1) so that $\overline{A'B'} \cong \overline{B'C'}$, completing the proof.

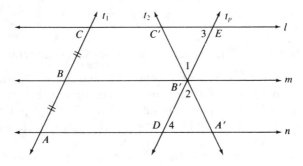

Figure 4.2.5

Theorem 4.2.4 can be generalized so that it applies to more than three parallel lines, which results in the following corollary:

COROLLARY 4.2.5. If a transversal crosses three *or more* parallel lines in such a way as to result in congruent segments between the parallels, then every transversal will do likewise.

The proof of this corollary proceeds by induction on $n \geq 3$ and is included as an exercise at the end of this section.

The next theorem we will consider involves line segments within triangles. In particular, we will examine the segments known as the medians of a triangle.

DEFINITION. A *median* of a triangle is a line segment that has as its endpoints a vertex of the triangle and the midpoint of the opposite side of the triangle.

Every triangle has three medians, and if in a particular triangle we take a pair of medians, it is obvious (although the proof is not entirely trivial) that they will meet at an interior point of the triangle (Figure 4.2.6). What is not as obvious is whether the third median will contain the point where medians 1 and 2 intersect. Three (or more) lines that intersect in a single point are said to be *concurrent*. The next theorem is generally called the *median concurrence theorem*.

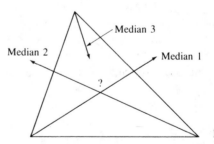

Figure 4.2.6

THEOREM 4.2.6. The three medians of a triangle are concurrent.

Proof. Consider $\triangle ABC$ with medians $\overline{BD}$ and $\overline{AE}$ as shown in Figure 4.2.7. Locate points F and G as the midpoints of $\overline{BE}$ and $\overline{EC}$, respectively. (*Note:* F, E, and G are the "quarter points" of $\overline{BC}$.) Through points B, F, G, and C construct lines l_1, l_2, l_3, and l_4, each parallel to line l, the line containing median $\overline{AE}$.

We now have a transversal, line $\overleftrightarrow{BC}$, that intersects a set of five parallel lines (l and l_1 through l_4) with congruent segments ($\overline{BF}$, $\overline{FE}$, $\overline{EG}$, and $\overline{GC}$)

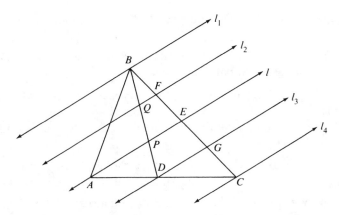

Figure 4.2.7

between the parallels. Corollary 4.2.5 tells us that any other transversal crossing these parallels will also produce congruent segments between the parallels. In particular, consider transversal $\overleftrightarrow{AC}$. Since D is the midpoint of $\overline{AC}$, we may conclude (by virtue of Corollary 4.2.5) that l_3 intersects $\overline{AC}$ at point D. In addition, $\overline{BQ} \cong \overline{QP} \cong \overline{PD}$ (why?). From this we may deduce that medians $\overline{AE}$ and $\overline{BD}$ interseet at P, a point that is two-thirds of the way from B to D, so that $BP = \frac{2}{3}BD$.

To complete this proof we need only repeat the process by drawing parallels at the quarter-points of $\overline{AB}$ and showing that the median from C to $\overline{AB}$ also intersects $\overline{BD}$ at a point two-thirds the way from B to D (Exercise Set 4.2, Problem 4). As a result, all three of the medians contain point P and are, by definition, concurrent.

Along the way, we have established the following corollary.

COROLLARY 4.2.7.   Any two medians of a triangle intersect at a point that is two-thirds of the distance from any vertex to the midpoint of the opposite side.

The following theorems are also direct consequences of the Euclidean parallel postulate. Since the proofs of these theorems are elementary, they have been left as exercises.

THEOREM 4.2.8.   Two lines parallel to the same line are parallel to each other.

THEOREM 4.2.9.   If a line intersects one of two parallel lines, then it intersects the other.

THEOREM 4.2.10.   Each diagonal of a parallelogram partitions the parallelogram into a pair of congruent triangles.

THEOREM 4.2.11.   The diagonals of a parallelogram bisect each other.[6]

THEOREM 4.2.12.   If the diagonals of a quadrilateral bisect each other, then the quadrilateral is a parallelogram.

THEOREM 4.2.13.   If a line segment has as its endpoints the midpoints of two sides of a triangle, then the segment is contained in a line that is parallel to the line containing the third side and the segment is one-half the length of the third side.

THEOREM 4.2.14.   The diagonals of a rhombus are perpendicular. (A *rhombus* is a quadrilateral in which all four sides are congruent.)

THEOREM 4.2.15.   If the diagonals of a quadrilateral bisect each other and are perpendicular, then the quadrilateral is a rhombus.

THEOREM 4.2.16.   The median to the hypotenuse of a right triangle is one-half the length of the hypotenuse.

THEOREM 4.2.17.   If, in a right triangle, one of the angles measures 30°, then the side opposite this angle is one-half the length of the hypotenuse.

THEOREM 4.2.18.   If one leg of a right triangle is half the length of the hypotenuse, then the angle opposite that leg has a measure of 30°.

THEOREM 4.2.19.   The sum of the measures of the interior angles of a convex n-gon is $(n - 2)(180°)$.

THEOREM 4.2.20.   The sum of the exterior angles (one at each vertex) of a convex n-gon is 360°.

EXERCISE SET 4.2

1. Identify the errors in the proof of the parallel postulate given by Proclus.
2. Do you think that the median concurrence theorem could have been proved as a theorem in neutral geometry? Explain why or why not.
3. Prove Corollary 4.2.5.
4. Complete the proof of the median concurrence theorem (Theorem 4.2.6).
5. Prove, using the axioms from neutral geometry, that each pair of medians of a triangle meet at an interior point of the triangle.
6. Prove Theorem 4.2.8.   13. Prove Theorem 4.2.15.
7. Prove Theorem 4.2.9.   14. Prove Theorem 4.2.16.
8. Prove Theorem 4.2.10.   15. Prove Theorem 4.2.17.
9. Prove Theorem 4.2.11.   16. Prove Theorem 4.2.18.
10. Prove Theorem 4.2.12.   17. Prove Theorem 4.2.19.
11. Prove Theorem 4.2.13.   18. Prove Theorem 4.2.20.
12. Prove Theorem 4.2.14.

[6] You may assume that the diagonals intersect at a point in the interior of the parallelogram.

19. Which, if any, of Theorems 4.2.1 through 4.2.19 do you think could have been proved in Chapter 3? Explain.

4.3 CONGRUENCE AND AREA

In Chapter 3 we proved that the ASA, SSS, and AAS congruence conditions for triangles are theorems in neutral geometry. Consequently, those results may be used as we proceed in the development of Euclidean theorems. In this section we will apply our ideas concerning congruence in an investigation of a related concept—area.

One advantage we gain by using SMSG postulates rather than Hilbert's (or Euclid's) axioms, is that the SMSG set includes several postulates that relate directly to the concept of area. Specifically, Postulates 17 through 20 provide us with assurances that (1) every polygon has a unique area, (2) congruent triangles (and by implication congruent polygons) have the same area, (3) area is additive in nature, and (4) we can find the area of a rectangle in the usual way.

Euclid derived many results concerning area without stating any postulates concerning area. He did, however, implicitly assume that congruent figures are "equal" (presumably meaning equal in area) and that polygonal regions are equal in area if they are comprised of parts that are congruent. For example, consider the following proof which was given by Euclid in the *Elements*.

THEOREM 4.3.1.   *Proposition 35, Book I:* Parallelograms that share a common base and which have sides opposite this base contained in the same (parallel) line are equal in area.

Proof. $\overline{BC}$ is shared by $\square ABCD$ and $\square EBCF$. The sides opposite $\overline{BC}$ ($\overline{AD}$ and $\overline{EF}$, respectively) are both contained in line $\overleftrightarrow{AF}$ as shown in Figure 4.3.1. We are to show that the two parallelograms are equal in area.

Note first that $\triangle ABE$ and $\triangle DCF$ are congruent, since $\overline{AB} \cong \overline{DC}$, $\overline{BE} \cong \overline{CF}$, and m$\angle ABE$ = m$\angle DCF$. (Verify each of these equalities.) Note also that $\square ABCD$ and $\square EBCF$ can be partitioned so that the areas can be represented as follows:

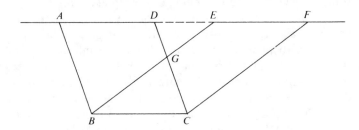

Figure 4.3.1

Area($\square ABCD$) = area($\triangle ABE$) + area($\triangle BCG$) − area($\triangle DGE$)

Area($\square EBCF$) = area($\triangle DCF$) + area($\triangle BCG$) − area($\triangle DGE$)

Since $\triangle ABE$ and $\triangle DCF$ are congruent, and the remainders of the two area expressions are identical, we may conclude that area($\square ABCD$) = area($\square EBCF$).

The proof is flawed in several ways. First, the configuration of points used is only one of several possible arrangements. For example, the proof needs to be altered somewhat if point D lies between points E and F (Figure 4.3.2). However, valid proofs are possible for this and other configurations of points (see Exercise Set 4.2, Problems 21 and 22), and the theorem is universally valid. Second, Euclid assumes throughout that congruent figures bound areas that are equal and that areas can be added or subtracted as if they were numbers without ever postulating the ideas. These are the same types of flaws mentioned in Chapter 2 when we discussed shortcomings of the *Elements*.

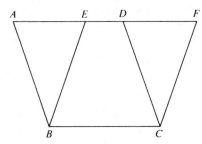

Figure 4.3.2

This proof is significant, however, because it is typical of the way in which Euclid approached the idea of area. No mention of a number to represent the area is made. No use of an area formula is needed. Euclid can be faulted because he has made some unstated assumptions about the way areas are related. But these and other problems were disposed of by Hilbert in his more rigorous development of Euclidean geometry, illustrating that it is not absolutely necessary to use numbers to represent areas.

Why then do we routinely use numerical values to represent areas? The answer is straightforward—convenience and practicality. It is more convenient to refer to an area as a number than as the interior of a polygonal region $P_1P_2P_3 \ldots P_n$, and it is practical to be able to order carpet, for example, in terms of the number of square yards needed with the assurance that that quantity will be sufficient to cover the floor regardless of its shape.

Postulate 20 from the SMSG set allows us to determine the area of a rectangle as the product of the lengths of its base and its height. This postulate is of course strictly Euclidean, since it presupposes the existence of rectangles that exist only in the Euclidean plane. Area in the hyperbolic and elliptic planes is a more difficult issue, as we will see in Chapter 6.

The formulas for areas of nonrectangular polygonal regions follow as theorems from Postulate 20 and the others. For example, if we define the height of a parallelogram as the perpendicular distance between its bases, we can state the following theorem.

THEOREM 4.3.2.   The area of a parallelogram is the product of the measure (length) of its base and the measure of its height.

Proof. Consider □*ABCD* as shown in Figure 4.3.3. Extend $\overline{BC}$ sufficiently so that perpendiculars constructed at *A* and *D* intersect it at *E* and *F*. Now $\overleftrightarrow{AE}$ and $\overleftrightarrow{FD}$ are both perpendicular to $\overleftrightarrow{AD}$ and so are parallel. (Which neutral theorem guarantees this?) Also, $\overleftrightarrow{AD}$ and $\overleftrightarrow{EF}$ are contained in parallel lines and so are parallel. Consequently, □*AEFD* is a parallelogram and so, by Theorem 4.3.1, has the same area as □*ABCD*. But □*AEFD* is also a rectangle, so that by Postulate 20 its area is given by the product *AD* × *FD*. If we note that $\overline{AD}$ is a base of the parallelogram and that the measure of $\overline{FD}$ is the parallelogram's height, we see that the area of the parallelogram is equal to the product of the lengths of its base and height.

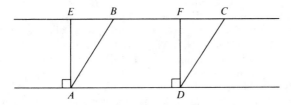

Figure 4.3.3

The elementary theorems listed below are derived in a similar fashion, and the proofs are left as exercises. Some theorems involve the use of terms that are familiar but have not yet been defined (e.g., the height of a triangle). As you read the theorems, try to formulate valid definitions for the terms. Several of the exercises involve writing these definitions.

THEOREM 4.3.3.   The area of a right triangle is one-half the product of the lengths of its legs.

THEOREM 4.3.4.   The area of a triangle is one-half the product of the length of any base and the corresponding height.

THEOREM 4.3.5.   The area of a trapezoid is the product of its height and the arithmetic mean of the lengths of its bases.

THEOREM 4.3.6.   The area of a rhombus is one-half the product of the lengths of the diagonals.

In addition, several other standard area formulas concerning polygonal regions are explored as exercises in the following set.

EXERCISE SET 4.3

1. Prove Theorem 4.3.3.
2. Define what is meant by the height of a triangle and then prove Theorem 4.3.4.
3. Define a trapezoid and height for a trapezoid and then prove Theorem 4.3.5.
4. Define a rhombus and prove Theorem 4.3.6.
5. Prove that if two triangles have congruent bases, then their areas are in the same ratio as their heights.
6. Suppose that $\triangle ABC$ is a right triangle with the right angle at C. Let M be the midpoint of $\overline{AB}$. Prove that the area of $\triangle AMC$ is the same as the area of $\triangle MCB$.
7. Suppose that $\triangle ABC$ is a right triangle with the right angle at C. Let $\overline{CD}$ be the perpendicular from point C to the hypotenuse $\overline{AB}$. Show that the ratio of the areas of $\triangle ADC$ and $\triangle DCB$ is the same as the ratio $AD:DB$.
8. In Euclid's proof of Theorem 4.3.1 the areas of $\square ABCD$ and $\square EBCF$ were represented as the sums of the areas of two triangles less the area of a third [i.e., the area of $\square ABCD$ was represented as $A(\triangle ABE) + A(\triangle BGC) - A(\triangle DGE)$]. The available postulate involves area addition (SMSG Postulate 19) and says nothing about differences in areas. Is Euclid's proof valid? Is the theorem true? How might a valid proof of Theorem 4.3.1 proceed?
9. Suppose that $\square ABCD$ is a square and that line m contains the point where the diagonals of the square intersect. Prove that m partitions the square into regions of equal area.
10. State and prove the converse to the statement proved in Problem 9.
11. Do the statements in Problems 9 and 10 generalize to the case where $\square ABCD$ is a rectangle? A parallelogram? A trapezoid? Justify your answer.
12. Prove that the area of a triangle is equal to one-half the product of its perimeter and the length of the radius of a circle inscribed within the triangle. (Assume for now that every triangle has such a circle and that the radii of the inscribed circle are perpendicular to the sides of the triangle at the points of contact.)
13. Suppose that $\triangle ABC$ is an equilateral triangle and that P is a point in the interior of this triangle. Prove that the sum of the perpendicular distances from P to each of the sides of the triangle is equal to the height of the triangle.
14. Do you think that the notion of a "square unit" can be used to measure area in hyperbolic geometry? Explain why or why not.
15. A regular n-gon is a polygon with n sides in which all sides are the same length and all interior angles have the same measure. Prove that there is a point in the interior of every regular n-gon that is equidistant from each vertex of the n-gon. (This point serves as the center of the circle that circumscribes the n-gon.)
16. Prove that the point that serves as the center of the circle that circumscribes a regular n-gon also serves as the center of a circle that contains the midpoint of each side of the n-gon. (This is the inscribed circle.)

17. The length of the radius of the inscribed circle of a regular n-gon is called the *apothem* of the n-gon. Prove that the area A of a regular n-gon can be computed using the formula

$$A = \frac{aP}{2}$$

where a represents the apothem and P the perimeter of the regular n-gon.

18. Consider regular n-gons with the following numbers of sides and in which the radius of the circumscribed circle is 1 unit:

      (a) $n = 3$ (b) $n = 4$ (c) $n = 6$ (d) $n = 100$

What is the length of the apothem in each polygon? What is the length of each side? What is the perimeter? What is the area? (You may use trigonometry to solve this problem.)

19. Using the results from Problem 18, determine the ratio of the perimeter to the apothem for a regular 100-gon.

20. Prove Theorem 4.3.1 for the case where point D lies between points E and F (see Figure 4.3.2).

21. Redraw Figure 4.3.2 so that A is between E and D and then prove Theorem 4.3.1.

4.4 SIMILARITY

Informally, we may describe congruence as a property relating geometric figures that have "the same size and same shape." A natural extension of this description involves similarity, a property relating figures that have the same shape but which are not necessarily the same size. In particular, if we examine a pair of similar polygons, we expect one to be an expanded version of the other. While the terms "same shape" and "expanded" are intuitive, they are not precise. In order to study the similarity of polygons more rigorously, a better working definition for the term is needed.[7]

    Since the shape of a polygon is in some ways determined by the measure of its interior angles, it might seem reasonable to attempt to define similar polygons as polygons whose interior angles are congruent. Unfortunately, this definition fails since, for example, it leads us to believe that all rectangles are similar to all squares (Figure 4.4.1), an unacceptable result.

    A second approach might involve the idea of proportionality of lengths. Using this approach, we could define similar polygons as polygons whose sides have measures that are proportional. (We'll be more precise about what is meant by this shortly.) But this too fails, since then the rectangle and parallelogram in Figure 4.4.2 would be similar.

    We see then that neither of the preceding conditions taken alone is sufficient to define similarity. Taken together, however, they are sufficient, so that we may state the following definition:

---

[7] This was not a problem with "congruence," since that term was taken as a primitive.

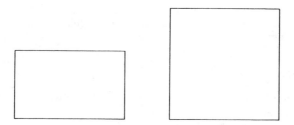

Figure 4.4.1

 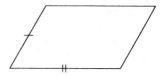

Figure 4.4.2

DEFINITION.   *Similar Polygons:* Two polygons $P_1P_2P_3 \ldots P_n$ and $P_1'P_2'P_3' \ldots P_n'$ are similar if and only if:

(i) $m\angle P_i = m\angle P_i'$ for all $i = 1$ to n, and

(ii)
$$\frac{P_1P_2}{P_1'P_2'} = \frac{P_2P_3}{P_2'P_3'} = \ldots = \frac{P_nP_1}{P_n'P_1'}$$

Condition (i) implies that to be similar, the corresponding angles of the polygons must be congruent. Condition (ii) implies that the ratios of the measures of corresponding sides must be constant for all pairs. These are naturally the two intuitive conditions discussed above. The definition may seem confusing at first, but if $n = 3$, it simply states that to be similar two triangles must have three pairs of congruent angles and a constant ratio for the lengths of the corresponding sides.

If we let s_i represent the length P_iP_{i+1} and s_i' represent the length $P_i'P_{i+1}'$, condition (ii), may be stated in the following, equivalent, form:

(ii) $\dfrac{s_1}{s_1'} = \dfrac{s_2}{s_2'} = \dfrac{s_3}{s_3'} = \ldots = \dfrac{s_n}{s_n'}$

Two sequences of non-zero numbers, $a_1, a_2, a_3, \ldots, a_n$ and $b_1, b_2, b_3, \ldots, b_n$, are said to be proportional providing

$$b_1 = ka_1 \quad \text{or} \quad a_1/b_1 = 1/k$$

$$b_2 = ka_2 \quad \text{or} \quad a_2/b_2 = 1/k$$
$$b_3 = ka_3 \quad \text{or} \quad a_3/b_3 = 1/k$$

$$
\begin{array}{cc}
\cdot & \cdot \\
\cdot & \cdot \\
\cdot & \cdot
\end{array}
$$

$$b_n = ka_n \quad \text{or} \quad a_n/b_n = 1/k$$

so that $a_1/b_1 = a_2/b_2 = a_3/b_3 = \ldots = a_n/b_n$. From this we may conclude that if two polygons are similar, the lengths of their sides form proportional sequences. By treating the lengths as proportional sequences, we can use number relationships to help establish properties regarding similarity, such as the following theorem, the proof of which is left as an exercise.

THEOREM 4.4.1.   Similarity of polygons is an equivalence relation.

The preceding discussion addresses the idea of similarity in a general sense, since it pertains to polygons with n vertices. We now look at some theorems pertaining to the simplest of polygons—triangles. To begin, let's consider an important result that will underlie our discussion of similarity.

THEOREM 4.4.2.   *The Basic Proportionality Theorem:* If a line parallel to one side of a triangle intersects the other two sides in two different points, then it divides these sides into segments that are proportional.

As an illustration of this theorem, suppose a line l is drawn through point D, the midpoint of side $\overline{AB}$, and is parallel to side $\overline{AC}$ (Figure 4.4.3). Since D is the midpoint of $\overline{AB}$ we have $\overline{AD} = \overline{DB}$ implying that $AD/DB = 1$. Since E must divide $\overline{BC}$ proportionally, we have $CE/EB = 1$, so that $CE = EB$. This means that E is the midpoint of $\overline{BC}$. In this case, then, we conclude that a line that is parallel to a side of a triangle and which contains the midpoint of a second side must also contain the midpoint of the third side.

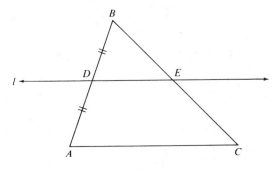

Figure 4.4.3

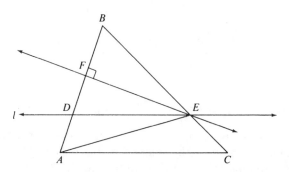

Figure 4.4.4

Theorem 4.4.2 does not, however, restrict our choice of D to the mid-point of $\overline{AB}$. If D had been chosen two-thirds of the way from A to B, E would then be two-thirds of the way from C to B. The proof of this depends largely on some results involving area developed in the preceding section. In particular, we will need to recall the way in which we can compute the area of a triangle (Theorem 4.3.4).

Proof. Begin with $\triangle ABC$ and line l, parallel to $\overleftrightarrow{AC}$, which intersects $\overline{AB}$ and $\overline{BC}$ in points D and E, respectively. We wish to show that $BD/DA = BE/EC$. To do this, draw segments $\overline{AE}$ and $\overline{EF}$, with $\overleftrightarrow{EF}$ constructed so that it is perpendicular to $\overleftrightarrow{AB}$ as shown in Figure 4.4.4.

Now consider the ratio of the areas of $\triangle DEB$ and $\triangle AED$.

$$\frac{\text{Area}(\triangle DEB)}{\text{Area}(\triangle AED)} = \frac{\frac{1}{2}(BD)(EF)}{\frac{1}{2}(DA)(EF)} = \frac{BD}{DA} \tag{1}$$

Next, we partition $\triangle ABC$ differently (as shown in Figure 4.4.5) and construct the second ratio involving the area of $\triangle DEB$, this time comparing it to the area of $\triangle CDE$.

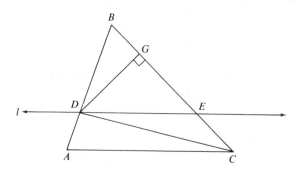

Figure 4.4.5

$$\frac{\text{Area}(\triangle DEB)}{\text{Area}(\triangle CDE)} = \frac{\frac{1}{2}(BE)(GD)}{\frac{1}{2}(EC)(GD)} = \frac{BE}{EC} \tag{2}$$

Ratios (1) and (2) are closely related. To see this, compare the areas of $\triangle AED$ and $\triangle CDE$. By choosing $\overline{DE}$ as the base, we see that $\triangle AED$ and $\triangle CDE$ share a common base and also have heights that are equal (each is the distance between the parallel lines $\overline{DE}$ and $\overleftrightarrow{AC}$; see Exercise Set 4.4 Problem 2). As a result of this we may conclude that area($\triangle AED$) = area($\triangle CDE$), and we have

$$\frac{\text{Area}(\triangle DEB)}{\text{Area}(\triangle AED)} = \frac{\text{area}(\triangle DEB)}{\text{area}(\triangle CDE)}$$

Substituting from Equations (1) and (2), we see that

$$\frac{BD}{DA} = \frac{BE}{EC}$$

which completes the proof.

An immediate consequence of the basic proportionality theorem is the following corollary:

COROLLARY 4.4.3.   If a line parallel to one side of a triangle intersects the other two sides in different points, then it cuts off segments that are proportional to the sides of the triangle.

Using $\triangle ABC$ in Figure 4.4.4, Corollary 4.4.3 implies the following proportion.

$$\frac{BA}{BD} = \frac{BC}{BE}$$

The proof of this corollary involves expressing the ratio

$$\frac{BD}{DA} = \frac{BE}{EC}$$

in its reciprocal form

$$\frac{DA}{BD} = \frac{EC}{BE}$$

and by then adding an *appropriate form* of *1* to each side of this proportion (see Exercise Set 4.4, Problem 3, for a clue concerning what is appropriate).

Next consider the converse to the corollary of the basic proportionality theorem.

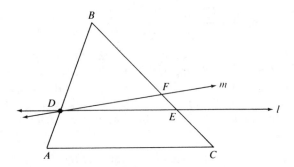

Figure 4.4.6

THEOREM 4.4.4.   If a line *l* intersects two sides of a triangle in different points so that it cuts off segments that are proportional to the sides, then
the line is parallel to the third side.   *that the lines intersect*

Proof. (Refer to Figure 4.4.6.) By hypothesis, we know that

$$\frac{BD}{BA} = \frac{BE}{BC} \qquad (3)$$

The objective is to show that line l is parallel to side $\overline{AC}$.

The Euclidean parallel postulate assures us that there is a unique line *m*
through point *D* that is parallel to $\overleftrightarrow{AC}$. This line must intersect side $\overline{BC}$
(why?) at a point we will call *F*. Applying Corollary 4.4.3 to line *m*, we see
that

$$\frac{BD}{BA} = \frac{BF}{BC} \qquad (4)$$

Combining equations (3) and (4), we can deduce that

$$\frac{BE}{BC} = \frac{BF}{BC}$$

from which it follows that $BE = BF$. The ruler postulate then allows us to
conclude that points *E* and *F* are the same point, meaning *l* and *m* are the
same line (Which neutral postulate assures this?), so that *l* is parallel to $\overleftrightarrow{AC}$.

As mentioned earlier, similarity of polygons involves two related
ideas—angle measure and proportionality. Figures 4.4.1 and 4.4.2 show
that, in general, similarity cannot be deduced based on the presence of just
one of these properties. There is, however, one classification of polygons,
triangles, for which congruence of angles *or* proportionality of sides is sufficient to demonstrate similarity. The following theorem justifies the first half
of the preceding statement.

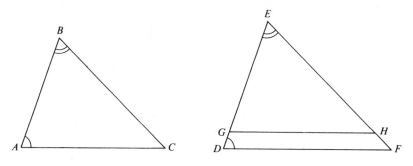

Figure 4.4.7

THEOREM 4.4.5.   *The AAA Similarity Theorem:* If the three interior angles of one triangle are congruent to the three interior angles of a second triangle, then the triangles are similar.

Proof. Let $\triangle ABC$ and $\triangle DEF$ be triangles whose corresponding angles are congruent. If segments $\overline{AB}$ and $\overline{DE}$ are congruent, then $\triangle ABC$ and $\triangle DEF$ are congruent (ASA) and therefore similar. Assume, then that $AB \neq DE$. Without loss of generality, we may assume that $DE > AB$. We may then locate point G between D and E so that $GE = AB$. Likewise, locate H on $\overrightarrow{EF}$ so that $HE = CB$ (Figure 4.4.7). Since m$\angle B$ = m$\angle E$, we know that $\triangle ABC \cong \triangle GEH$ (SAS), so that m$\angle EGH$ = m$\angle BAC$ = $\angle EDF$. This allows us to conclude (see Exercise Set 4.4, Problem 18) that line $\overleftrightarrow{GH}$ is parallel to line $\overleftrightarrow{DF}$. Now, if we apply corollary 4.4.3, we have the proportion

$$\frac{EG}{ED} = \frac{EH}{EF}$$

but since $EG = BA$ and $EH = BC$, we have the proportion

$$\frac{BA}{ED} = \frac{BC}{EF}$$

This shows that two of the pairs of sides are proportional. To complete the proof we must show also that the ratio AC/DF has the same value. To do this, we make another copy of $\triangle ABC$ within $\triangle DEF$ (Figure 4.4.8) and proceed in a similar fashion. The remainder of the proof is left as an exercise.

The AAA similarity theorem can of course be shortened to the AA similarity theorem, since the sum of the measures of the interior angles is fixed in Euclidean geometry.

Two other theorems that define similarity conditions and can be proved using the results established in this section are the following:

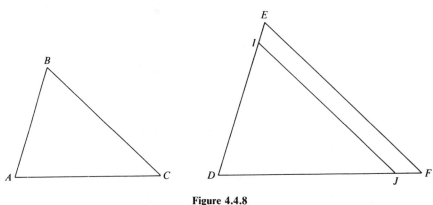

Figure 4.4.8

THEOREM 4.4.6.   *SAS Similarity Theorem:* If an angle of one triangle is congruent to an angle from another triangle, and if the sides that surround this angle are proportional, then the triangles are similar.

THEOREM 4.4.7.   *SSS Similarity Theorem:* If the three sides of one triangle are proportional to the three sides of a second triangle, then the triangles are similar.

A line segment from a vertex of a triangle that is perpendicular to the side of the triangle opposite that vertex is called an *altitude* of the triangle. (See Figure 4.4.9 where $\overline{CD}$ is the altitude from C to $\overline{AB}$.) The altitude to the hypotenuse of a right triangle has the interesting property that it partitions the triangle into two smaller triangles, each of which is similar to the right triangle itself (see Figure 4.4.9).

In particular, $\triangle ADC \approx \triangle ACB$ by Theorem 4.4.5, since they share $\angle A$ and each contains a right angle. From this similarity relation we can construct the proportion

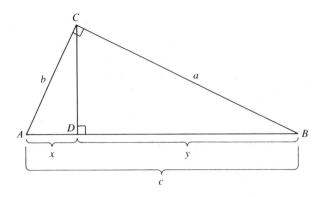

Figure 4.4.9

$$\frac{b}{c} = \frac{x}{b}$$

which is equivalent to

$$b^2 = xc \tag{5}$$

Similarly, we can show that $\triangle ACB \approx \triangle CDB$ (AAA), which gives the proportion

$$\frac{a}{c} = \frac{y}{a}$$

$$a^2 = yc \tag{6}$$

Adding equations (5) and (6), we find

$$a^2 + b^2 = xc + yc$$

or

$$a^2 + b^2 = (x + y)c$$

If we note that $x + y = c$, we have the well-known Pythagorean relation

$$a^2 + b^2 = c^2$$

THEOREM 4.4.8.   *The Pythagorean Theorem:* If a and b are the measures of the legs of a right triangle, and if c is the measure of the hypotenuse, then $a^2 + b^2 = c^2$.

Some other consequences of similarity are explored in the following exercise set.

EXERCISE SET 4.4

1. Prove that the similarity of polygons is an equivalence relation (Theorem 4.4.1).
2. In the proof of the basic proportionality theorem (Theorem 4.4.2), we showed that the areas of $\triangle AED$ and $\triangle CDE$ were equal since they share a common base and have heights that are the same length. Complete the proof by verifying that the altitudes of these triangles are in fact of equal length.
3. Prove Corollary 4.4.3. (*Hint*: Refer to Figure 4.4.4 and note that $BD/BD = BE/BE = 1$.)
4. Similarity of noncongruent triangles is not possible in hyperbolic geometry. Where is the Euclidean parallel assumption used in the proof of the basic proportionality theorem?
5. Complete the proof of the AAA similarity theorem (Theorem 4.4.5).
6. Prove Theorem 4.4.6.
7. Prove Theorem 4.4.7.

8. Prove that any angle bisector of a triangle separates the opposite side into segments whose lengths have the same ratio as the ratio of the lengths of the remaining two sides.

9. Prove that if two triangles are similar, the lengths of corresponding angle bisectors are in the same ratio as the lengths of corresponding sides.

10. Prove or disprove: In any triangle the ratio of any two sides is equal to a ratio of the corresponding altitudes.

11. If $\triangle ABC$ is the right triangle shown in Figure 4.4.9, prove that $b^2 - a^2 = x^2 - y^2$.

12. Prove that if the ratio of corresponding sides of two triangles is k, then the ratio of the areas is k^2.

13. Determine the height of an equilateral triangle whose sides are s units in length. Determine the area of the same triangle.

14. Suppose that $\triangle ABC$ is a 30°-60°-90° triangle with the right angle at C and the 30° angle at A. Determine the ratios AB/BC and BC/CA.

15. The geometric mean of two numbers a and b is, by definition, $\sqrt{ab}$. Prove that the length of the altitude to the hypotenuse of a right triangle is the geometric mean of the lengths of the segments into which the altitude partitions the hypotenuse.

16. Prove that each leg of a right triangle is the geometric mean between the length of the hypotenuse and the length of that leg's projection on the hypotenuse. (In the accompanying diagram $\overline{DB}$ is the projection of side $\overline{CB}$ upon the hypotenuse $\overline{AB}$.)

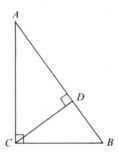

17. Let $\triangle ABC$ be a right triangle with the right angle at C. Suppose that the bisector of the interior angle at A intersects $\overline{BC}$ at D. Suppose also that the bisector of the exterior angle at A intersects $\overline{BC}$ (extended) at E. Prove that $BD/BE = CD/CE$. (*Hint*: Use the result from Problem 8.)

18. In the proof of the AAA similarity theorem, we showed that $\angle EGH \cong \angle EDF$. Show that if we extend $\overline{GH}$ we may conclude that $\overleftrightarrow{GH} \parallel \overline{DF}$ by applying the alternate interior angle theorem.

4.5 SOME EUCLIDEAN RESULTS CONCERNING CIRCLES

One of the most familiar of all geometric shapes is the circle. Euclid dedicated Book III of the *Elements* to the study of the properties of circles, and geometers since that time have elaborated on those results. In this section and the next, we will investigate some of the important results concerning circles in Euclidean geometry. We begin with a formal definition of the term "circle."

DEFINITION.   *Circle:* A circle is a set of points each of which is equidistant from a given point. The given point is called the *center* of the circle, and the common distance is called the *radius*[8] of the circle.

You probably recall that in neutral geometry it was postulated that two distinct points on a line uniquely determine the line. It should be nearly as clear that two of the points on a circle are not sufficient to determine the circle, as illustrated in Figure 4.5.1 where points A and B are contained by two distinct circles.

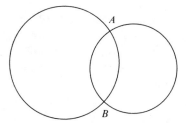

Figure 4.5.1

We may, however, prove the following theorem which is in a sense analogous to Postulate 1 in the SMSG set.

THEOREM 4.5.1.   In Euclidean geometry three distinct, noncollinear points determine a unique circle.

Proof. Suppose that A, B and C are three distinct, noncollinear points. We must show, first, the existence of a circle containing these three points and, second, that this circle is unique. We begin by showing existence.

Let l_1 and l_2 be the perpendicular bisectors of $\overline{AB}$ and $\overline{BC}$, respectively (Figure 4.5.2). We claim (1) that there is a point P such that $l_1 \cap l_2 = P$ (i.e., the perpendicular bisectors intersect), (2) that P is the center of a circle containing A, B, and C and (3) that $PA = PB = PC = r$, the radius of a circle containing A, B, and C.

For part 1 suppose that $l_1 \cap l_2 = \varnothing$, so that l_1 and l_2 are parallel (Figure 4.5.3). Since line $\overleftrightarrow{AB}$ intersects l_1 (at M_1) and, by hypothesis, $l_1 \parallel l_2$, we may conclude that $\overleftrightarrow{AB}$ also intersects l_2 (Theorem 4.2.9) at some point which we will call D. We also know that $D \neq M_2$, since, otherwise, A, B, and C would be collinear, contrary to the hypothesis. If we apply the converse of the

---

[8] The term "radius" is used in two senses with respect to circles. "The radius" of a circle is the (unique) distance from the center to any point of the circle. "A radius" refers to any line segment that connects the center of the circle to any point of the circle. The meaning is in general clear from the context.

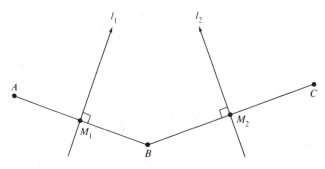

Figure 4.5.2

alternate interior angle theorem, we will see that $\angle BDM_2$ is a right angle. However, this results in a triangle ($\triangle BDM_2$) with an interior angle ($\angle BDM_2$) congruent to an exterior angle ($\angle CM_2E$), which contradicts the exterior angle theorem (Theorem 3.2.6). We may therefore conclude that l_1 and l_2 are *not* parallel, so that there is a point $P = l_1 \cap l_2$.

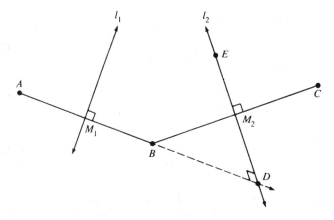

Figure 4.5.3

Items 2 and 3 involve showing that each of A, B, and C is the same distance from P as the others, so that there is a circle containing all three. This follows directly from the fact that every point on the perpendicular bisector of a segment is equidistant from the endpoints of the segment (Theorem 3.2.5) and the definition of a circle. The details of the proof are left as an exercise.

Thus far, we have shown that there exists a circle C_1 containing any three noncollinear points. Left to be shown is that this circle is unique. Once again we will proceed indirectly by assuming that a second such circle C_2 exists. If P_2 is the center of a circle containing A and B, then P_2 is equidistant from A and B and, by Theorem 3.2.5, is on l_1, the perpendicular bisector of $\overline{AB}$. In a similar fashion, we know that $P_2 \in l_2$. But since $l_1 \cap l_2 = P_1$, the center of C_1, $P_2 = P_1$, so that C_1 and C_2 have the same center. Next consider r_2, the radius of C_2. We know that $r_2 = P_2A = P_1A = r_1$, the radius of C_1. As a result we see that C_1 and C_2 are circles with the same center and the same radius so that $C_1 = C_2$. Thus the circle containing A, B, and C is unique.

Since any three noncollinear points can be thought of as the vertices of a triangle, an alternate form of Theorem 4.5.1 can be stated as follows.

THEOREM 4.5.1.   (*Alternate*). In Euclidean geometry, every triangle can be circumscribed, and the center of the circumscribing circle is the point where the perpendicular bisectors of the sides meet.

You may recall that this theorem was listed earlier in this chapter among the equivalences to the Euclidean parallel postulate. By implication, there must exist triangles in hyperbolic geometry that cannot be circumscribed. This phenomenon will be explored in greater detail in Chapter 6. We mention it here, however, to point out that the consequences of the Euclidean parallel postulate go beyond statements about lines and angles. Theorem 4.5.1 states a characteristically Euclidean property that deals with points (or triangles) and circles.

Several results concerning circles involve the terms "chord" and "diameter." Working definitions for these terms are given here.

DEFINITION.   *Chord:* A chord of a circle is a line segment joining two of the points of the circle.

DEFINITION.   *Diameter:* A diameter of a circle is a chord that contains the center of the circle.

We often think of a diameter of a circle as being a chord with the greatest possible length. This notion is accurate and is established by the following theorem:

THEOREM 4.5.2.   If $\overline{AB}$ is a diameter of a circle and if $\overline{CD}$ is another chord of the same circle that is not a diameter, then $AB > CD$.

Proof. Let O be the center of the circle that contains A, B, C, and D (Figure 4.5.4). Since $\overline{CD}$ is not a diameter, C, D, and O are not collinear, so that we may consider $\triangle ODC$. By the triangle inequality theorem (Theorem 3.3.7), we know that $CO + OD > CD$. Since $\overline{CO}$, $\overline{OD}$, $\overline{AO}$ and $\overline{OB}$ are all radii of the circle centered at O (and are therefore equal in length), we may say that

$$AB = AO + OB = CO + OD > CD$$

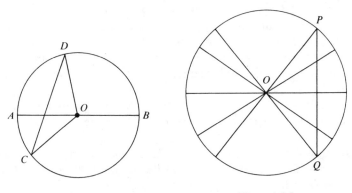

Figure 4.5.4   Figure 4.5.5

so that

$$AB > CD$$

which completes the proof of the theorem.

Figure 4.5.5 shows a circle $C(O,OP)$[9] with a chord $\overline{PQ}$ (which is not a diameter) and several diameters that intersect $\overline{PQ}$. The next sequence of theorems will focus on (1) the diameter that is perpendicular to $\overleftrightarrow{PQ}$, (2) the diameter that bisects $\overline{PQ}$, and (3) the perpendicular bisector of $\overline{PQ}$. We will show that the three sets of points that result from these descriptions are closely related.

THEOREM 4.5.3.   If a diameter of a circle is perpendicular to a chord of the circle, then the diameter bisects the chord.

Proof. In Figure 4.5.6 let $\overline{AB}$ be a diameter of $C(O,OA)$. Let $\overline{CD}$ be a chord of $C(O,OA)$ such that $\overleftrightarrow{CD} \perp \overleftrightarrow{AB}$. If $\overline{CD}$ is a diameter of $C(O,OA)$, then $O \in \overline{CD}$, so that $\overline{AB} \cap \overline{CD} = O$, hence $\overline{AB}$ bisects $\overline{CD}$. Assume then that $\overline{CD}$ is not a diameter, and so that $\overline{AB} \cap \overline{CD} = P$. We may then conclude that $\triangle DPO$ and $\triangle CPO$ are congruent (Hypotenuse-leg congruence condition; see Exercise Set 3.3 Problem 10), so that $\overline{CP} \cong \overline{PD}$ and $\overline{AB}$ bisects $\overline{CD}$.

Items 2 and 3 above are stated as Theorems 4.5.4 and 4.5.5. The proofs are similar to the proof of Theorem 4.5.3 and are left as exercises.

THEOREM 4.5.4.   If a diameter of a circle bisects a chord of the circle (which is not a diameter), then the diameter is perpendicular to the chord.

[9] The notation $C(O,OP)$ denotes the circle with center O and radius OP. In Figure 4.5.5 it should be clear that $C(O,OP) = C(O,OQ)$.

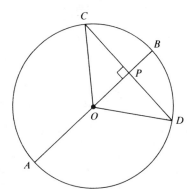

Figure 4.5.6

THEOREM 4.5.5.   The perpendicular bisector of a chord of a circle contains a diameter of the circle.

Having discussed diameters and chords, we may now extend our investigation to other related geometric sets. First, consider a line that contains a chord. (Remember, a chord is a line *segment*). Such a line is called a secant. Next imagine that the secant is moved toward the edge of the circle until it touches the circle at just one point. This line is called a tangent. Formally, the terms are defined as follows:

DEFINITION.   *Secant:* A secant to a circle is a line that contains exactly two points of the circle.

DEFINITION.   *Tangent:* A tangent to a circle is a line that contains exactly one point of the circle.

One important result concerning tangent lines can be visualized in the following way. Imagine a tangent t drawn to a circle O. If we call the point of tangency P and construct the radius $\overline{OP}$, what relationship is there between line $\overleftrightarrow{OP}$ and tangent t? Intuitively, we think that these lines should be perpendicular. Theorem 4.5.6 provides the justification for this hunch.

THEOREM 4.5.6.   If a line is tangent to a circle, then it is perpendicular to the radius drawn to the point of tangency.

Proof. We proceed indirectly and assume otherwise. Suppose that line t is tangent to $C(O,OP)$ at point P, but that $\overleftrightarrow{OP}$ is not perpendicular to line t (Figure 4.5.7). We then draw $\overline{OQ}$, the unique perpendicular from O to t. Next, we locate on line t the point R with the following properties: R is on the opposite side of Q from P and $\overline{QR} \cong \overline{QP}$. We then construct $\overline{OR}$.

Next, we note that $\triangle QOP \cong \triangle QOR$ (SAS), so that $\overline{OR} \cong \overline{OP}$. How-

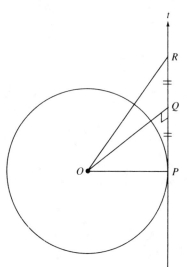

Figure 4.5.7

ever, if $\overline{OR} \cong \overline{OP}$, then R should be a point of $C(O,OP)$. But P is the only point common to both $C(O,OP)$ and tangent t. Because of this contradiction, we may conclude that $OP \perp t$.

Several other results concerning circles are addressed in the following exercise set.

EXERCISE SET 4.5

1. Reread the statement of Theorem 4.5.1. Note that the theorem applies only to Euclidean geometry. Do you think that this statement is a theorem in neutral geometry? In hyperbolic geometry? Explain why or why not.
2. Provide the missing details for the proof of Theorem 4.5.1 by showing that if A, B, and C are non-collinear points and if P is the intersection of the perpendicular bisectors of $\overline{AB}$ and $\overline{BC}$, then P will serve as the center of a circle containing A, B, and C and that if the radius of this circle is r, then $r = PA$ (or PB or PC).
3. Prove Theorem 4.5.4.
4. Prove Theorem 4.5.5.
5. Which of Theorems 4.5.2 through 4.5.5 do you think are neutral and which do you think are strictly Euclidean?
6. In the proof of Theorem 4.5.2 a diagram was used that implied that the chord in question crosses the diameter in question. Is the theorem true if the chord fails to cross the diameter? Does the proof need to be amended if the chord fails to cross the diameter? Explain.
7. Consider a point P in the exterior of $C(O,OA)$. Imagine that two tangent segments are drawn to $C(O,OA)$ from point P. Prove that these two tangent segments are congruent.

8. Suppose that line t is tangent to $C(O,OP)$ at point P and that secant l, which is parallel to t, intersects $C(O,OP)$ at the two points A and B. Prove that chords $\overline{AP}$ and $\overline{BP}$ are congruent.

9. In Theorem 4.5.6 we proved that a tangent is perpendicular to a radius drawn to the point of tangency. Prove the related theorem that says that a line perpendicular to a diameter at an endpoint is tangent to the circle at the endpoint of the diameter.

10. Suppose that lines l and n are secants to $C(O,OA)$ and are parallel. Suppose also that secant l intersects $C(O,OA)$ at points A and B and that secant n intersects $C(O,OA)$ at C and D. Prove that chord $\overline{AC}$ is congruent to chord $\overline{BD}$. (*Note*: There are two possible ways that this can be configured depending on the placement of A, B, C, and D; the theorem is true regardless, so consider both cases.)

11. Prove that if two chords are equidistant from the center of a circle, then the chords are congruent.

12. In a fixed circle a chord of a given length is chosen. The midpoint of this chord and all other chords of the same length in that circle are located. Describe the resulting locus of points.

13. Describe a method for locating the center of a circle using only a compass and a straightedge.

14. Define the following terms with respect to spheres: tangent lines, tangent planes, secant lines, secant planes.

15. Theorem 4.5.1 essentially guarantees that every triangle can be circumscribed. Which of the following types of polygons can always be circumscribed: (a) squares? (b) rectangles? (c) trapezoids? (d) parallelograms? (e) regular n-gons? Formulate a rule that will allow you to decide whether or not a quadrilateral can be circumscribed.

4.6 MORE EUCLIDEAN RESULTS CONCERNING CIRCLES

Our discussion of circles in Euclidean geometry continues in this section with the development of some quantitative aspects of circles, namely, measures of angles, arcs, and line segments. To begin, we will define several terms that are pertinent to the discussions that follow.

DEFINITION.   *Central Angle:* Any angle whose vertex is the center of a circle is called a central angle for the circle (Figure 4.6.1).

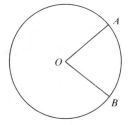

Figure 4.6.1

You may recall that the protractor postulate (SMSG Postulate 11) restricts angle measures to values between 0° and 180°. As a result, each central angle contains in its interior a portion of its circle that is in some sense less than one-half of the circle. Portions of the circle such as this are called minor arcs, while portions greater than one-half the circle are called major arcs. As you might expect, arcs comprising exactly one-half the circle are called semicircles.

Since it will be necessary for us to distinguish among these three types of arcs, we now state the following formal definitions:

DEFINITION.   *Semicircle:* If line $\overleftrightarrow{AB}$ contains a diameter of $C(O,OA)$, and if H represents either half plane defined by $\overleftrightarrow{AB}$ (see SMSG Postulate 9), then the union of points A, B and all points of $C(O,OA)$ that lie in H is called a semicircle of $C(O,OA)$ (Figure 4.6.2a).

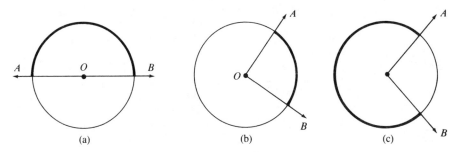

(a)   (b)   (c)

Figure 4.6.2

DEFINITION.   *Minor Arc:* If A and B are points of $C(O,OA)$ that are not endpoints of the same diameter, then the union of A, B, and all points of $C(O,OA)$ that are in the interior of $\angle AOB$ is called a minor arc of $C(O,OA)$ (Figure 4.6.2b).

DEFINITION.   *Major Arc:* If A and B are points of $C(O,OA)$ that are not endpoints of the same diameter, then the union of A, B and all points of $C(O,OA)$ that are in the exterior of $\angle AOB$ is called a major arc of $C(O,OA)$ (Figure 4.6.2c).

We see then that when we discuss a circle, for example, $C(O,OA)$, and two of its points A and B, there are actually two arcs connecting A and B: one proceeding clockwise from A to B, and the other proceeding counterclockwise. Because of the possible ambiguity, arcs are often named using three points, for example, $\overparen{ACB}$, where C represents a point contained in the arc. In instances where the intention is not obvious, the notation involving three points will be used.

It seems reasonable that there should be a relationship between the measure of an arc and the measure of a central angle associated with it (namely, the larger the central angle, the larger the associated arc, and conversely). A problem arises, however, since we normally think that arcs can have measures between 0° and 360° (360° being the entire circle), while angle measures are restricted to values between 0° and 180°. This discrepancy motivates the following definition for measures of arcs:

DEFINITION.   *Measure of Arc $\overset{\frown}{AB}$:* The degree measure of arc $\overset{\frown}{AB}$ of a circle $C(O,OA)$ is: (i) the degree measure of the angle $\angle AOB$ if $\overset{\frown}{AB}$ is a minor arc, (ii) 180° if $\overset{\frown}{AB}$ is a semicircle, or (iii) 360° minus the measure of $\angle AOB$ if arc $\overset{\frown}{AB}$ is a major arc.

The notation $m\overset{\frown}{AB}$ will be used to denote the measure of arc $\overset{\frown}{AB}$.

An immediate consequence of this definition is the following theorem.

THEOREM 4.6.1.   If two chords of a circle are congruent, then their corresponding minor arcs have the same measure.

Proof. Suppose that in Figure 4.6.3 chords $\overline{AB}$ and $\overline{CD}$ are congruent. If we draw radii $\overline{OA}$, $\overline{OB}$, $\overline{OC}$, and $\overline{OD}$, we will see that $\triangle OAB$ and $\triangle OCD$ are congruent (SSS). Consequently, $m\angle AOB = m\angle COD$, so that by the definition of arc measure we have $m\overset{\frown}{APB} = m\overset{\frown}{CQD}$.

The converse of Theorem 4.6.1 is about as easy to prove.

THEOREM 4.6.2.   If two minor arcs are congruent, then so are the corresponding chords.

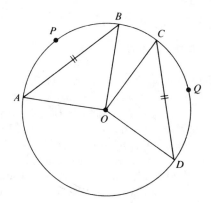

Figure 4.6.3

The proof is left as an exercise.

Since arc measure is defined in terms of angle measures, one might expect that many properties concerning angles carry over to arcs. One property that is particularly useful concerns the addition of arc measures. SMSG Postulate 13 (the angle addition postulate) allows us to measure an angle by adding the measures of adjacent angles that comprise it. The corresponding property concerning arcs provides the motivation for the next theorem.

THEOREM 4.6.3.   *Arc Addition Theorem*: If $\overparen{APB}$ and $\overparen{BQC}$ are arcs of the same circle sharing only the endpoint B, then $m\overparen{APB} + m\overparen{BQC} = m\overparen{ABC}$.

Proof. To begin, let's suppose that $\overparen{APB}$, $\overparen{BQC}$, and $\overparen{ABC}$ are all minor arcs of a circle centered at O (Figure 4.6.4). Since B is contained in minor arc $\overparen{ABC}$ it follows that B, and hence $\overline{OB}$, are in the interior of $\angle AOC$. Applying the angle addition postulate, we have

$$m\angle AOC = m\angle AOB + m\angle BOC \qquad (7)$$

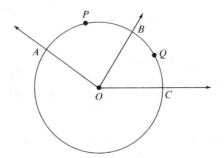

Figure 4.6.4

But since arcs $\overparen{APB}$, $\overparen{BQC}$, and $\overparen{APC}$ are all minor arcs we know that $m\angle AOC = m\overparen{ABC}$, $m\angle AOB = m\overparen{APB}$, and $m\angle BOC = m\overparen{BQC}$. Substituting appropriately in Equation (7) gives

$$m\overparen{ABC} = m\overparen{APB} + m\overparen{BQC}$$

which completes this portion of the proof.

Of course, it need not be the case that all of arcs $\overparen{APB}$, $\overparen{BQC}$, and $\overparen{ABC}$ are minor. For other cases minor alterations in the proof are needed, but in each case the equality holds. The other possible configurations are addressed in the exercise set at the end of this section.

While central angles provide a starting point for the study of angles associated with circles, they are by no means the only angles that need to be considered. In particular, inscribed angles and the arcs they intercept also play important roles. These terms are defined in the following ways:

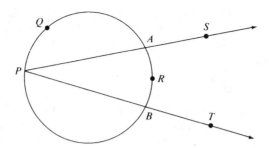

Figure 4.6.5 ∠SPT inscribed in arc $\overset{\frown}{AQB}$ with intercepted arc $\overset{\frown}{ARB}$.

DEFINITION.   *Inscribed Angle:* An angle ∠SPT is said to be inscribed in an arc $\overset{\frown}{AQB}$ if and only if (i) the vertex P of the angle is a point of $\overset{\frown}{AQB}$, (ii) one ray of the angle contains point A, and (iii) the other ray contains point B (Figure 4.6.5).

DEFINITION.   *Intercepted Arc:* An angle ∠SPT, is said to intercept an arc $\overset{\frown}{ARB}$ if and only if (i) both A and B are points of the angle, (ii) each ray of the angle contains at least one endpoint of $\overset{\frown}{ARB}$ and (iii) excepting A and B, each point on $\overset{\frown}{ARB}$ lies in the interior of ∠APB (see Figure 4.6.5).

This definition of the term ''intercepted arc'' allows for a number of different configurations, four of which are shown in Figure 4.6.6.

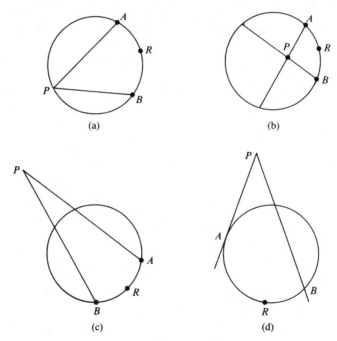

Figure 4.6.6 Arc $\overset{\frown}{ARB}$ intercepted by ∠APB.

As an introduction to the relationships between the measures of angles and their intercepted arcs, consider Figure 4.6.7a, in which $\overline{AB}$ is a diameter and $m\overset{\frown}{CDB} = 60°$. We wish to determine $m\angle CAB$, the inscribed angle whose intercepted arc measures 60°.

We begin by constructing radius $\overline{CO}$ (Figure 4.6.7b). Since $\angle COB$ is a central angle, we know that $m\angle COB = 60°$. Further, since $\angle AOC$ and $\angle COB$ form a linear pair, we may conclude that $m\angle AOC = 120°$. Next, we will consider $\triangle AOC$. Since $\overline{OA}$ and $\overline{OC}$ are radii of the same circle, $\triangle AOC$ is an isosceles triangle, so that $m\angle OAC = m\angle OCA$. Further, since the sum of the measures of the interior angles of Euclidean triangles is 180° (Theorem 4.2.1), we may conclude that $m\angle OAC + m\angle OCA + m\angle AOC = 180°$. Substituting, we see that $2(m\angle CAO) + 120° = 180°$, so that $m\angle CAO = 30°$.

The choice of 60° for $m\overset{\frown}{CDB}$ was of course arbitrary. The same argument can easily be applied using a measure of $x°$ for $\overset{\frown}{CDB}$. Furthermore, a similar approach can be used regardless of whether the center, O, of the circle is an interior point of $\angle CAB$, an exterior point of $\angle CAB$ (see Figure 4.6.8), or a point contained by $\angle CAB$. These consideration are left as exercises. Together, however, they imply the following, very useful, theorem.

THEOREM 4.6.4.   *The Inscribed Angle Theorem:* The measure of an angle inscribed in an arc is one-half the measure of its intercepted arc.

The inscribed angle theorem has two immediate corollaries:

COROLLARY 4.6.5.   An angle inscribed in a semicircle is a right angle.

COROLLARY 4.6.6.   Angles inscribed in the same or congruent arcs are congruent.

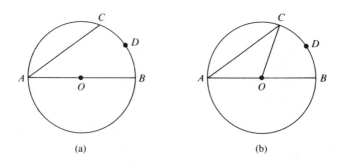

(a)   (b)

Figure 4.6.7

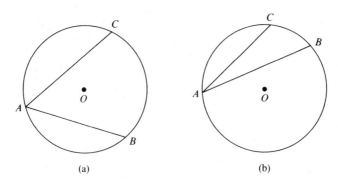

Figure 4.6.8   (a) O in the interior of $\angle CAB$, (b) O in the exterior of $\angle CAB$.

Theorem 4.6.4 provides us with a means of determining the measure of the angle formed by two intersecting chords providing the chords intersect at a point that lies on the circle itself. There are, however, other ways that pairs of chords (and the lines that contain them) can be configured: They can intersect in the interior of the circle, or in the exterior of the circle, or they can fail to intersect. Of course, if they fail to intersect, there is no angle to measure, which leaves us to consider the remaining two possibilities.

First, consider the case where the chords intersect in the interior of the circle as shown in Figure 4.6.9. We wish to determine the measure of $\angle 1$, which is neither a central angle nor an inscribed angle. Although it is clearly true that $m\angle 1 = m\angle 2$ (why?), we do not yet have a means by which we can determine the measure of these angles. The theorem that follows addresses this issue.

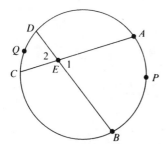

Figure 4.6.9

THEOREM 4.6.7.   *The Two-Chord Angle Theorem:* If two chords intersect in the interior of a circle to determine an angle, the measure of the angle is the average of the measures of the arcs intercepted by the angle and its vertical angle.

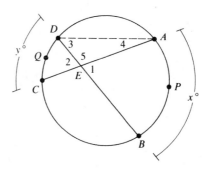

Figure 4.6.10

Proof. In order to determine m∠1 we will draw $\overline{AD}$, and label ∠1 through ∠5 as shown in Figure 4.6.10. Applying Theorem 4.6.4, we see that m∠3 = $\frac{1}{2}x°$ and m∠4 = $\frac{1}{2}y°$. Now consider $\triangle AED$. Since ∠1 is an exterior angle of this triangle, it has a measure that is equal to the sum of the measures of the remote interior angles (Corollary 4.2.2). From this we see that

$$m\angle 1 = \tfrac{1}{2}x° + \tfrac{1}{2}y° = \frac{(x + y)°}{2}$$

which is the average of the measures of the two intercepted arcs.

The next case to consider involves two chords contained in lines that intersect outside the circle, that is, the case of intersecting secants (Figure 4.6.11).

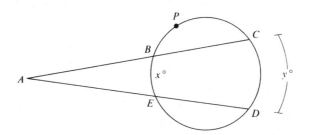

Figure 4.6.11

THEOREM 4.6.8. *The Two-Secant Angle Theorem:* If two secants intersect at a point in the exterior of a circle, the measure of the angle at the point of intersection is one-half the positive difference of the two intercepted arcs.

Proof. In order to determine m∠1 in Figure 4.6.12, we will draw chord $\overline{EC}$. Then, m∠2 = $\frac{1}{2}x°$ and m∠3 = $\frac{1}{2}y°$. Since ∠3 is a remote exterior angle with respect to ∠1 and ∠2 we know (Corollary 4.2.2) that

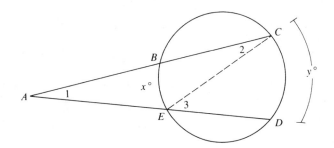

Figure 4.6.12

$$m\angle 1 + m\angle 2 = m\angle 3$$

or

$$m\angle 1 = m\angle 3 - m\angle 2$$

Then, substituting for the measures of $\angle 2$ and $\angle 3$, we see that $m\angle 1 = \frac{1}{2}y° - \frac{1}{2}x° = \frac{1}{2}(y - x)°$, which completes the proof of the theorem.

Thus far we have considered only angles involving chords and secants. Other configurations involve tangent lines and angles between tangents, secants, and chords. These considerations constitute a natural extension of the preceding discussions, since if in Figure 4.6.11 we allow secant $\overleftrightarrow{AC}$ to rotate counterclockwise, points B and C "slide" toward each other along the upper arc of the circle until the secant in effect becomes a tangent at point P (Figure 4.6.13).

Before considering this case, however, we must first prove a related theorem concerning chords and tangents.

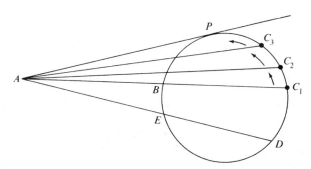

Figure 4.6.13

THEOREM 4.6.9.   *The Tangent-Chord Angle Theorem:* If $\overleftrightarrow{AB}$ is tangent to $C(O,OA)$ at point A, and if $\overline{AC}$ is a chord such that the $m\overparen{APC} = x°$ (Figure 4.6.14) then $m\angle BAC = \frac{1}{2}x°$.

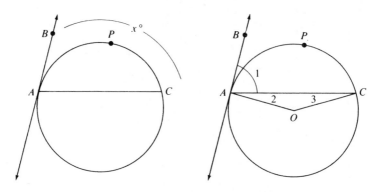

Figure 4.6.14   Figure 4.6.15

Proof. Figure 4.6.15 is a copy of Figure 4.6.14 in which radii $\overline{OA}$ and $\overline{OC}$ have been drawn.

Since $\angle AOC$ is a central angle, its measure is $x°$, the same as the measure of its intercepted arc. By Theorem 4.5.6, we know that radius $\overline{OA}$ is perpendicular to $\overrightarrow{AB}$, so that

$$m\angle OAB = m\angle 1 + m\angle 2 = 90°$$

In addition, since $\triangle OAC$ is isosceles (why?), we know that $m\angle 2 = m\angle 3$, so that

$$m\angle 2 = m\angle 3 = \frac{(180 - x)°}{2}$$

Consequently, we see that

$$m\angle 1 + \frac{(180 - x)°}{2} = 90°$$

or

$$m\angle 1 = m\angle BAC = 90° - \frac{(180 - x)°}{2} = \tfrac{1}{2}x°$$

which completes this part of the proof.

Two additional possible configurations have yet to be considered: one in which chord $\overline{AC}$ contains O, and another in which O is in the interior of $\angle BAC$. The proofs of these two configurations are immediate consequences of the case proved above and are left as exercises.

An extension of Theorem 4.6.9 allows us to determine the measure of angles between secants and tangents in the following way. Suppose that $\overrightarrow{AD}$ is a secant to $C(O,OB)$ and $\overrightarrow{AB}$ is tangent to $C(O,OB)$ at point B. Suppose

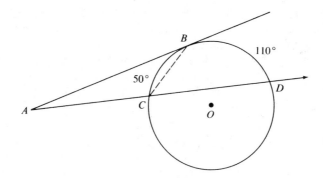

Figure 4.6.16

also that $\text{m}\overset{\frown}{BC} = 50°$ and $\text{m}\overset{\frown}{BD} = 110°$, as shown in Figure 4.6.16. If we draw chord $\overline{BC}$, we know that $\text{m}\angle BCD = 55°$ (Theorem 4.6.4), so that $\text{m}\angle ACB = 180° - 55° = 125°$. Also, $\text{m}\angle ABC = 25°$ (Theorem 4.6.9). Combining, we find that

$$\text{m}\angle BAC + 25° + 125° = 180°$$

so that

$$\text{m}\angle BAC = 30°$$

Note that $(\text{m}\overset{\frown}{BD} - \text{m}\overset{\frown}{BC})°/2 = 30°$, an expression that implies that $\text{m}\angle BAC$ can be computed from the measures of $\overset{\frown}{BD}$ and $\overset{\frown}{BC}$. This is no coincidence and is formalized in the following theorem:

THEOREM 4.6.10.   *The Tangent-Secant Angle Theorem:* If line $\overleftrightarrow{AB}$ is tangent to $C(O,OB)$ at point B, and if $\overleftrightarrow{AD}$ is a secant line to $C(O,OB)$ (see Figure 4.6.16), then $\text{m}\angle BAD$ is one-half the positive difference of the two intercepted arcs.

The proof of this theorem is merely a generalization of the example preceding the statement of the theorem and is left as an exercise.

One other theorem can now be proved rather easily and is stated as Theorem 4.6.11. In order to devise a strategy for constructing the proof of this theorem, it may be worthwhile for you to initially use numerical examples, as was done prior to the tangent-secant angle theorem. The steps in the proof closely follow the steps of the specific numerical examples.

THEOREM 4.6.11.   *The Two-Tangent Angle Theorem:* The measure of an angle formed by two tangents drawn to a circle is one-half the positive difference of the measures of the intercepted arcs.

Theorems 4.6.7 through 4.6.11 can actually be unified into a single, more comprehensive result by noting that in each case where the vertex of the angle is a point in the exterior of a circle, the measure of the angle is one-half the positive difference of the intercepted arcs. The two-chord angle theorem (Theorem 4.6.7), on the other hand, applies to any angle whose vertex is in the interior of a circle. In this situation, the measure of the angle is one-half the *sum* of the intercepted arcs. Finally, the inscribed angle and the tangent-chord angle theorems each involve angles with only one intercepted arc, which may be interpreted as having a second arc of measure 0°, so that one-half the sum of the arcs is equal to one-half the positive difference. In each case, then, we see that the measure of an angle related, in one of these ways, to a circle can be determined by the formula

$$m\angle A = \left| \frac{m(\text{arc } 1) \pm m(\text{arc } 2)}{2} \right|$$

with the choice of the + or − being determined by whether point A is in the interior or the exterior of the circle.

The following corollary is an immediate consequence of the tangent-chord angle theorem:

COROLLARY 4.6.12.   Tangent segments drawn to a circle from the same exterior point are congruent. (*Note:* A tangent segment is a line segment connecting an exterior point to a point of tangency.)

Theorem 4.6.11 completes the set of theorems that concern the measures of angles related to chords, secants, and tangents. Corollary 4.6.12 is the first result to deal with the lengths of line segments. Still to be considered are the lengths of line segments contained within chords and secants. For example, Figure 4.6.17a depicts a circle with a pair of chords ($\overline{AB}$ and $\overline{CD}$) that intersect at an interior point P. Suppose we know the lengths of

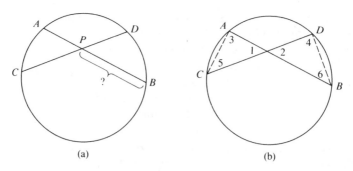

(a)   (b)

Figure 4.6.17

$\overline{AP}, \overline{CP}$, and $\overline{PD}$. Do we have sufficient information to determine the length of $\overline{PB}$? The answer is yes, and the argument uses the strictly Euclidean notion of similarity of triangles.

In order to explore several length relationships of the line segments involved, we will draw $\overline{AC}$ and $\overline{DB}$ (Figure 4.6.17b). The key is to recognize that $\triangle PAC$ and $\triangle PDB$ are similar (Theorem 4.4.5), since $\angle 1$ and $\angle 2$ are vertical angles, and $\angle 3$ and $\angle 4$ are inscribed angles with the same intercepted arc (likewise for $\angle 5$ and $\angle 6$). Therefore pairs of corresponding sides of these triangles are proportional. In particular, we have the following proportion:

$$\frac{PA}{PD} = \frac{PC}{PB}$$

so that

$$PA \times PB = PD \times PC \qquad \qquad (8)$$

In general, the relationship between the lengths of the segments of the chords is shown in Equation (8) and is stated as the next theorem.

THEOREM 4.6.13.   *The Chord-Segment Product Theorem:* If two chords intersect within a circle, the product of the lengths of the segments of one chord is equal to the product of the lengths of the segments of the other chord.

Next we'll consider the lengths of segments from a pair of secants drawn to a circle from a single point in the exterior of the circle (Figure 4.6.18a). The objective is to determine a relationship between the length of each secant and the segments into which it is partitioned by the circle. We will again identify similar triangles that can be used to generate a proportion of the type applied in the chord-segment product theorem. To accomplish this, we will draw $\overline{BD}$ and $\overline{EC}$ (see Figure 4.6.18b), and claim that $\triangle AEC \approx$

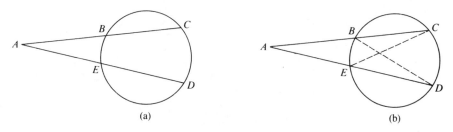

(a)   (b)

Figure 4.6.18

$\triangle ABD$ (why?). From this similarity, the following proportion can be derived:

$$\frac{AE}{AB} = \frac{AC}{AD}$$

or

$$AE \times AD = AB \times AC \qquad (9)$$

Notice that in Equation (9) $\overline{AE}$ and $\overline{AB}$ are the portions of the two secants ($\overleftrightarrow{AC}$ and $\overleftrightarrow{AD}$, respectively) that are in the exterior of $C(O,OA)$ (for this reason $\overline{AE}$ and $\overline{AB}$ are called external segments). We see then that on both sides of Equation (9) we have the product of the length of a secant segment and the length of its external segment. That relationship is summarized in the following theorem.

THEOREM 4.6.14.   *The Secant Segment Product Theorem*: If two secant segments are drawn to a circle from the same exterior point, then the product of the length of the secant segment and the length of its external portion is the same for both secants.

Finally, we consider the relationship between lengths of segments involving a secant and a tangent (Figure 4.6.19a). As we saw earlier in the discussion of angle relationships, this can be thought of as a limiting case of two secants in which one secant rotates toward the exterior of the circle. Consequently, it would make sense to think that the relationship between the lengths in this case should be analogous to the relationship stated in Theorem 4.6.14. Before proceeding further, it might provide a valuable experience for the reader to attempt to conjecture the relationship that exists between $\overline{AB}$, $\overline{AD}$, and $\overline{AC}$ in Figure 4.6.19a.

To derive the relationship deductively, we will focus on a pair of similar triangles produced by drawing $\overline{BC}$ and $\overline{BD}$ (Figure 4.6.19b). We now

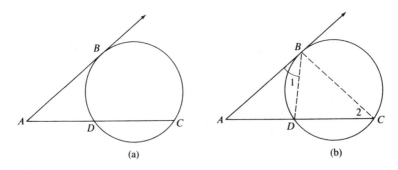

(a)   (b)

Figure 4.6.19

need to show that $\triangle ABC$ and $\triangle ADB$ are in fact similar. Clearly, the angle with its vertex at A is common to both triangles. Second, note that $m\angle 1 = \frac{1}{2}(m\widehat{BD})$ (Theorem 4.6.9) and that $m\angle 2 = \frac{1}{2}(m\widehat{BD})$ (Theorem 4.6.4), so that $m\angle 1 = m\angle 2$. We may then apply the AA similarity condition to conclude that $\triangle ABC \approx \triangle ADB$. A proportion that results from this similarity is

$$\frac{AB}{AD} = \frac{AC}{AB}$$

which is equivalent to

$$AD \times AC = AB^2$$

which is a result summarized in the final theorem of this sequence.

THEOREM 4.6.15.   *The Tangent Secant Segment Theorem:* If a tangent segment and a secant segment are drawn to the same circle from the same exterior point, the product of the length of the secant and the length of its external segment is equal to the square of the length of the tangent segment.

Besides merely allowing us to determine lengths of line segments related to circles, the preceding three theorems also provide us with a means for assigning a number to each point in a plane with respect to a circle in the same plane. This number is called the power of the point and is defined in the following way:

DEFINITION.   *Power of a Point With Respect to a Circle:* The power of a point X with respect to a circle $C(O,OA)$ is the product of the signed distances[10] from point X to any two points of $C(O,A)$ with which it is collinear. The power of point X will be denoted by $P(X)$.

To illustrate the idea of the power of a point, consider the unit circle $C(O,OA)$ and the point X shown in Figure 4.6.20. Here X has been chosen so that it is collinear with points A and B, the endpoints of a diameter of $C(O,OA)$. Consequently, the power of X is the product $XA \times XB$, where XA and XB represent the signed distances from X to A and from X to B.

To begin, suppose that $XA = +1$; then since B is in the same direction from X that A is, its signed distance will be positive also, so that the power of X with respect to O is given by

$$P(X) = (+1)(+3) = +3$$

---

[10] The ruler placement postulate allows us to coordinatize each line in such a way as to discuss "signed distances." If X is assigned the coordinate 0, the signed distance from X to any point is the coordinate assigned to that point.

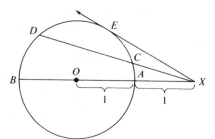

Figure 4.6.20

On the other hand, if $XA = -1$, we see that $XB = -3$, so that the power of X is $(-1)(-3) = +3$. We see then that the power of a point does not depend on which direction is positive and which is negative. Each point in the exterior of the circle will have a power that is positive. Note also that the choice of points A and B is somewhat arbitrary. The Two-Secant Segment Theorem guarantees that $XC \times XD = XA \times XB = 3$ for any choices of C and D that are collinear with X, so that every point exterior to $C(O,OA)$ has one and only one power value. Beyond that, if $\overline{XE}$ is a tangent segment, $XE^2 = XA \times XB = 3$, so that $XE = \sqrt{3}$.

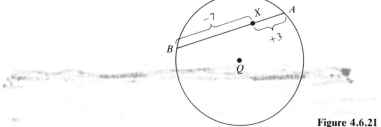

Figure 4.6.21

Now suppose that X is a point in the interior of some other circle Q as in Figure 4.6.21. If $XA = +3$, we must assign to XB a negative value, for example -7, since it is measured in the opposite direction, so that the power for X is given by

$$P(X) = (+3)(-7) = -21$$

It should be clear that the power of any point within a circle will be negative, since the distances to collinear points on the circle will be measured in opposite directions, and hence have opposite signs. Also, note that the two-chord segment theorem guarantees that every chord through X will be partitioned into segments with lengths having the same product, so that each interior point has one and only one power value.

This temporarily leaves the points of the circle itself without power values. Can we define a power value for the points of the circle? Can the power of a point that is on $C(O,OA)$ be positive? Negative? This issue is left as an exercise.

EXERCISE SET 4.6

1. Prove Theorem 4.6.2.
2. Prove Theorem 4.6.3 for the case where $\overset{\frown}{ABC}$ is a semicircle.
3. Prove Theorem 4.6.3 for the case where $\overset{\frown}{ABC}$ is a major arc.
4. Prove Theorem 4.6.3 for the case where $\overset{\frown}{APB}$ is a major arc.
5. Prove that if $\angle CAB$ is inscribed in $\overset{\frown}{CAB}$, where $\overline{AB}$ is a diameter of $C(O,OA)$ (see Figure 4.6.7a), and if the measure of $\overset{\frown}{CDB} = x°$, then the measure of $\angle CAB = \frac{1}{2}x°$ [This is the case of Theorem 4.6.4 in which the center of $C(O,OA)$ is *on* $\angle CAB$].
6. Prove Theorem 4.6.4 for the case where the center O of $C(O,OA)$ is an interior point of $\angle CAB$ (see Figure 4.6.8a).
7. Complete the proof of Theorem 4.6.4 by proving that the theorem applies to the case where the center O of $C(O,OA)$ is an exterior point of $\angle CAB$ (see Figure 4.6.8b).
8. Prove Corollary 4.6.5.
9. Define what is meant by congruent arcs and then prove Corollary 4.6.6.
10. Prove Theorem 4.6.8 for the case in which $\overline{AC}$ contains the center O of the circle.
11. Complete the proof of Theorem 4.6.9 by proving that it is true when point O is in the interior of $\angle BAC$.
12. Provide a proof for the tangent-secant angle theorem (Theorem 4.6.10) by generalizing the numeric argument that precedes the statement of the theorem.
13. Prove the two tangent angle theorem (Theorem 4.6.11).
14. Prove Corollary 4.6.12.
15. Which, if any, of the circle-angle theorems (Theorems 4.6.4 through 4.6.11) do you think depend on the Euclidean parallel postulate or its equivalent. Do you think that any of these theorems could have been included as a neutral theorem in Chapter 3? Explain why or why not.
16. Complete the proof for the chord segment product theorem (Theorem 4.6.13).
17. Show that in Figure 4.6.18b $\triangle AEC$ and $\triangle ABD$ are similar.
18. Complete the proof of Theorem 4.6.13 by deriving the proportion $AE/AB = AC/AD$ using the similarity of $\triangle AEC$ and $\triangle ABD$.
19. Explain why the power of a point P with respect to a circle $C(O,OA)$ [i.e., $P(P)$] is zero if P is a point of the circle.
20. Let P be a point in the same plane as circle $C(O,OA)$, where $OA = r$. If OP represents the signed distance from O to P, show that the power of point P with respect to $C(O,OA)$ is $|OP^2 - r^2|$.
21. Given a circle $C(O,OA)$ in a plane α, define the following function F that maps the points in α to numbers in the following way:

$$F(P) = \frac{\text{Power of } P}{|\text{Power of } P|} = \frac{P(P)}{|P(P)|}, \text{ if } P(P) \neq 0$$

$F(P) = 0$, if $P(P) = 0$. Next we will define the relation, R, on the points in α in the following way:

$$R = \{(P_1, P_2) \mid F(P_1) = F(P_2)\}$$

Show that R is an equivalence relation and describe the equivalence classes that result from this relation.

22. Suppose that P is the function $P(X)$ that assigns to each point X in a plane its power value with respect to some fixed circle $C(O, OA)$. Is P a one-to-one function? Explain why or why not.

4.7 SOME EUCLIDEAN RESULTS CONCERNING TRIANGLES

In Sections 4.5 and 4.6 we investigated several important properties of circles. In this section we continue the investigation of Euclidean results by discussing another important geometric shape, the triangle.

Triangles play an especially important role in the study of geometry, since all polygons can be partitioned into triangular regions. Consequently, many of the theorems concerning triangles can be extrapolated to statements concerning polygons in general. For example, in Theorem 4.2.1 we proved that the sum of the measures of the interior angles of a triangle is 180°. A direct consequence of this is Theorem 4.2.19, which specifies the sum of the measures of the interior angles of any convex polygon.

To begin the study of properties of triangles, we will recall a result from earlier in this chapter—the median concurrence theorem (Theorem 4.2.6):

THEOREM 4.2.6.   The three medians of a triangle are concurrent (at a point called the *centroid* of the triangle).

Theorem 4.2.6 is one of several theorems that specify the concurrence of lines related to triangles. A second such theorem involves the perpendicular bisectors of the sides of a triangle and is stated as follows:

THEOREM 4.7.1.   The three perpendicular bisectors of the sides of a triangle are concurrent (at a point called the *circumcenter*).

Proof. To see this, consider $\triangle ABC$ in Figure 4.7.1.

Theorem 4.5.1 assures us that there is a unique circle containing each of the three points A, B, and C. Let O be the center of this circle, so that $\overline{OA} \cong \overline{OB} \cong \overline{OC}$. Since O is equidistant from A and B, it must lie on the perpendicular bisector of $\overline{AB}$ (Theorem 3.2.5). Likewise, O is on the perpendicular bisectors of $\overline{BC}$ and $\overline{AC}$, so that the three perpendicular bisectors are concurrent at point O. In addition, we know that the point of concurrence is

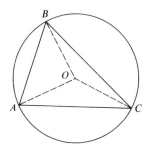

Figure 4.7.1

the center of the circle that circumscribes the triangle, commonly referred to as the circumcenter.

Another concurrence theorem involves the bisectors of the interior angles of a triangle. Before proceeding with the proof of this concurrence, we will need a definition and a preliminary theorem.

DEFINITION.   *Distance from a Point to a Line:* The distance from a point P to a line l is the distance from P to the foot of the perpendicular from P to l.

Informally this means that the distance from a point to a line is the shortest distance from the point to the line, since any other segment from P to l could serve as the hypotenuse of a right triangle having the perpendicular as a leg. Since the hypotenuse is the longest side of a right triangle (a consequence of Theorem 4.4.8, the Pythagorean theorem), the perpendicular distance is the shortest from the point to the line. Thus the definition is consistent with our intuitive notion of distance from a point to a line.

With this term denoted, the following preliminary proof can be stated and proved.

THEOREM 4.7.2.   A point is on the bisector of an angle if and only if it is equidistant from the sides of the angle.

Proof. To begin, let $\overrightarrow{BQ}$ be the bisector of $\angle ABC$ as shown in Figure 4.7.2. Choose any point P on BQ and drop perpendiculars $\overline{PM}$ and $\overline{PN}$ to $\overrightarrow{BA}$ and $\overrightarrow{BC}$, respectively. We wish, first, to show that $\overline{PM} \cong \overline{PN}$. To do this, we will show that $\triangle PMB$ and $\triangle PNB$ are congruent. Note that each of $\triangle PMB$ and $\triangle PNB$ is a right triangle, and that these triangles share hypotenuse $\overline{BP}$. Note also, that $\angle MBP \cong \angle NBP$, since $\overline{BP}$ is contained in the bisector of $\angle MBN$. Applying Theorem 3.3.4 (the AAS congruence condition), we may conclude that $\triangle PMB \cong \triangle PNB$. From this congruence we may conclude that $\overline{PM} \cong \overline{PN}$.

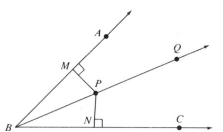

Figure 4.7.2

To complete the proof of the biconditional, we need to show that if P is equidistant from the sides of an angle, then it is on the bisector of the angle. This argument also involves showing the congruence of triangles $\triangle PMB$ and $\triangle PNB$, but requires the use the hypotenuse-leg congruence condition. The remaining details of the proof are left as an exercise.

Theorem 4.7.2. provides the basis for the following theorem.

THEOREM 4.7.3.   The three bisectors of the interior angles of a triangle are concurrent (at a point called the *incenter*).

Proof. Consider a $\triangle ABC$ shown in Figure 4.7.3a. Let Q be the point where the bisector of $\angle A$ intersects $\overline{BC}$, and let R be the point where the bisector of $\angle C$ intersects $\overline{AB}$.

It seems obvious that $\overrightarrow{AQ}$ and $\overrightarrow{CR}$ intersect at some point P in the interior of $\triangle ABC$. To see this, we recall that $m\angle CAB + m\angle CBA < 180°$ (Theorem 3.5.2). Since $m\angle 1 = \frac{1}{2}(m\angle CAB)$ and $m\angle 2 = \frac{1}{2}(m\angle ACB)$, we may conclude that $m\angle 1 + m\angle 2 < 90° < 180°$. Then, applying Euclid's fifth postulate (see Theorem 3.4.4) to $\overrightarrow{AQ}$ and $\overrightarrow{CR}$ crossed by transversal $\overleftrightarrow{AC}$, we may conclude that $\overrightarrow{AQ}$ and $\overrightarrow{CR}$ intersect on the B side of $\overleftrightarrow{AC}$, at some point that we will call P. In a similar fashion (Exercise Set 4.7, Problem 19), we can show that P is on the A side of $\overleftrightarrow{BC}$ and the C side of $\overleftrightarrow{AB}$, making P an interior point of $\triangle ABC$.

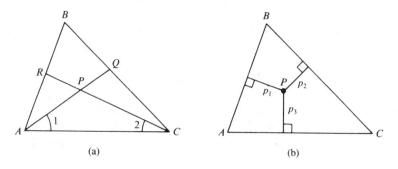

(a)   (b)

Figure 4.7.3

Continuing with the proof, we see then that the bisectors of $\angle A$ and $\angle C$ intersect at a point we shall call P. From P we will draw the perpendiculars (p_1, p_2, and p_3) to sides $\overline{AB}$, $\overline{BC}$, and $\overline{AC}$, respectively (Figure 4.7.3b). Since P is on the bisector of $\angle A$, we know that $p_1 = p_3$. Since P is on the bisector of $\angle C$, we know that $p_3 = p_2$. Combining, we see that $p_1 = p_2$, which means that P is equidistant from $\overline{BA}$ and $\overline{BC}$. By Theorem 4.7.2 we may conclude that P is on the bisector of $\angle ABC$, which shows that the three angle bisectors are concurrent at point P.

Since P, the point of concurrence of the three bisectors of the interior angles of $\triangle ABC$, is equidistant from the three sides of the triangle, a circle can be constructed centered at P and containing the feet of the perpendiculars to the sides. This circle is said to be *inscribed* within the circle, and P is called the *incenter* for the triangle (see Exercise Set 4.7, Problem 3).

At this point, we have considered the concurrence of medians, perpendicular bisectors, and angle bisectors for triangles. A fourth concurrence result involves the altitudes for a triangle. To begin, we will recall what is meant by an altitude of a triangle.

DEFINITION.   *Altitude:* An altitude of a triangle is a perpendicular line segment from a vertex of a triangle to the opposite side (extended if necessary).

Considering the pattern initiated by Theorems 4.2.5, 4.7.1, and 4.7.3, the following theorem should come as no surprise.

THEOREM 4.7.4.   The lines containing the three altitudes of a triangle are concurrent.

Proof. Consider $\triangle ABC$ shown in Figure 4.7.4a. We wish to show that the three altitudes of $\triangle ABC$ are concurrent. Figure 4.7.4(b) shows these three altitudes, $\overline{AF_1}$, $\overline{BF_2}$ and $\overline{CF_3}$. (F_1, F_2, and F_3 are called the feet of the altitudes.) Note that since $\triangle ABC$ is an obtuse triangle, sides $\overline{AB}$ and $\overline{AC}$ needed to be extended in order to locate the feet of the altitudes.[11] Through A we will draw a line parallel to $\overleftrightarrow{BC}$. We will call this line l_1. In similar fashion, we will draw lines through B and C parallel to $\overleftrightarrow{AC}$ and $\overleftrightarrow{AB}$, respectively, and call these lines l_2 and l_3. Lines l_1, l_2 and l_3 will intersect pairwise at points X, Y, and Z as shown in Figure 4.7.4b. (Why can't they be parallel?) Our objective is to show that the lines that contain the altitudes of $\triangle ABC$ also serve as the perpendicular bisectors of $\triangle XYZ$.

For example, consider $\overrightarrow{AF_1}$, the ray that contains the altitude from A to

[11] The altitude concurrence theorem applies to all triangles, so there is no reason to assume that $\triangle ABC$ is obtuse. Exercise 6 at the end of this section considers acute triangles.

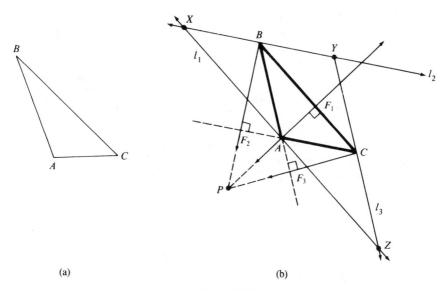

(a)   (b)

Figure 4.7.4

$\overline{BC}$ in $\triangle ABC$. Since l_1 is parallel to $\overleftrightarrow{BC}$, and since $\overleftrightarrow{AF_1} \perp \overleftrightarrow{BC}$, we may conclude that $\overleftrightarrow{AF_1} \perp \overleftrightarrow{XZ}$. Left to show is that A is the midpoint of $\overline{XZ}$. To do this, we will identify a pair of parallelograms. First, consider $\square XACB$. Since $\overline{XB}$ is contained in l_2, and since $l_2 \parallel \overleftrightarrow{AC}$, we see that $\overleftrightarrow{XB} \parallel \overleftrightarrow{AC}$. Similarly, since $\overline{XA}$ is contained in l_1 and $l_1 \parallel \overleftrightarrow{BC}$, we see that $\overleftrightarrow{XA} \parallel \overleftrightarrow{BC}$. As a quadrilateral with opposite sides parallel, we see that $\square XACB$ is a parallelogram, so that $XA = BC$ (Theorem 4.2.3). Using a similar argument, we can show that $\square BAZC$ is a parallelogram, so that $BC = AZ$. Combining these two equalities, we find that $XA = AZ$, so that A is the midpoint of $\overline{XZ}$, making $\overleftrightarrow{AF_1}$ the perpendicular bisector of $\overline{XZ}$.

To complete this proof, we need to show that $\overleftrightarrow{BF_2}$ is the perpendicular bisector of $\overline{XY}$ and that $\overleftrightarrow{CF_3}$ is the perpendicular bisector of $\overline{YZ}$. This process involves identifying parallelograms which show that B and C are the midpoints of $\overline{XY}$ and $\overline{YZ}$, respectively (the details are left as exercises).

We complete the proof by noting that $\triangle XYZ$ is itself a triangle. Therefore the perpendicular bisectors of its sides must be concurrent at some point P (Theorem 4.7.1), which then is also the point of concurrence of the altitudes of $\triangle ABC$.

The point of concurrence of the altitudes of a triangle is called the *orthocenter* of the triangle. The orthocenter *does not* serve as the center of a circle related to the triangle (as the circumcenter and the incenter do). However, the points A, B, C, and P (see Figure 4.7.4b) form what is called an *orthocentric set*, since a triangle whose vertices are any three of these points will have as its orthocenter the remaining point.

For example, in the proof of the theorem we showed that P is the orthocenter of $\triangle ABC$. It is also true, however, that A serves as the orthocenter of $\triangle PBC$, B serves as the orthocenter of $\triangle APC$, and C serves as the orthocenter of $\triangle ABP$. Problem 7 in Exercise Set 4.7 explores this phenomenon in greater detail.

One more theorem is generally included in the sequence of concurrence theorems discussed so far. This theorem is similar to Theorem 4.7.3 in that it deals with angle bisectors. It differs from that theorem, however, in that two of the angle bisectors involve exterior angles.

THEOREM 4.7.5.   Each angle bisector of a triangle, if sufficiently extended, is concurrent with the bisectors of the exterior angles at the remaining two vertices. The point of concurrence is called the *excenter* of the triangle.

Proof. Consider $\triangle ABC$ shown in Figure 4.7.5a. Here b_1 is the bisector of the interior angle at A, and b_2 is the bisector of the exterior angle at B. Note first that b_1 and b_2 cannot be parallel, since if they were, $m\angle 1 = m\angle 2$ (alternate interior angles), which would imply $m\angle BAC = m\angle DBA$, meaning that $\triangle ABC$ would have an exterior angle at B with the same measure as the interior angle at A—a result that contradicts the exterior angle theorem (Theorem 3.2.6). We conclude therefore that b_1 and b_2 intersect at some point P.

Next draw the three perpendiculars from P to sides $\overline{AB}$, $\overline{AC}$, and $\overline{BC}$ (extended if necessary) and label the feet of these perpendiculars F_1, F_2, and

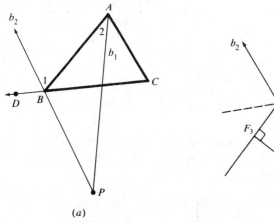

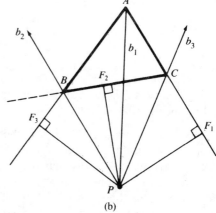

(a)   (b)

Figure 4.7.5

F_3 (Figure 4.7.5b). We see that $\triangle PBF_3 \cong \triangle PBF_2$ (AAS; PB is the bisector of $\angle F_3BF_2$, $\angle PF_3B$ and $\angle PF_2B$ are right angles, and $PB = PB$), so that $PF_3 = PF_2$. Also, since P is on the bisector of $\angle F_3AF_1$, we have $PF_3 = PF_1$ (Theorem 4.7.2). If we now draw $\overline{CP}$, we can show that $\triangle PF_2C \cong \triangle PF_1C$ (hypotenuse-leg congruence condition). From this congruence we conclude that $m\angle F_2CP = m\angle F_1CP$, so that $\overline{CP}$ is the angle bisector of the exterior angle at C and clearly contains P, the point of intersection of b_1 and b_2. Consequently, we can say that b_1, b_2 and b_3 are concurrent at point P, which we call the excenter of $\triangle ABC$.

COROLLARY 4.7.6.   Each excenter of a triangle serves as the center of a circle that is externally tangent to a side of the triangle and the extensions of the other two sides.

The proof of this corollary is a consequence of the fact that $\overline{PF_1} \cong \overline{PF_2} \cong \overline{PF_3}$. The details are left as an exercise.

The next theorem involves an interesting and useful length relationship between the orthocenter and the circumcenter of a triangle. It may be helpful to refer to Figure 4.7.6a as you read this theorem.

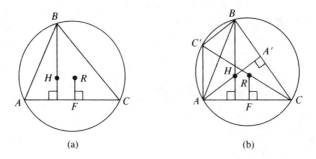

(a)   (b)

Figure 4.7.6

THEOREM 4.7.7.   The distance from a vertex to the orthocenter of a triangle is twice the distance from the circumcenter to the midpoint of the side opposite that vertex.

Proof. In Figure 4.7.6a, R is the circumcenter of $\triangle ABC$ and, as such, centers the circle containing A, B, and C. And H is the orthocenter of $\triangle ABC$. We wish to show that $BH = 2RF$.

Since the circumcenter R is the point of concurrence of the perpendicular bisectors of the sides of $\triangle ABC$, if F is the midpoint of $\overline{AC}$ then $\overline{RF} \perp \overline{AC}$. Also, since H represents the orthocenter of $\triangle ABC$, $\overline{BH} \perp \overline{AC}$, so that $\overline{BH} \parallel \overline{RF}$. Next draw diameter $\overline{CC'}$ and complete $\triangle C'AC$ by drawing $\overline{C'A}$, as shown in Figure 4.7.5b. Since ∠$C'AC$ is inscribed in a semicircle, its measure is 90° (Theorem 4.6.4), so that m∠$C'AC$ = m∠RFC. Additionally, ∠RCF ≅ ∠$C'CA$, so that △$C'CA$ and △RCF are similar. Since we know the ratio AC/FC = 2, we may also conclude that $C'A/RF$ = 2, so that $C'A$ = $2RF$.

Our next task is to show that $\overline{C'A}$ ≅ $\overline{BH}$, so that $BH = 2RF$ as stated in the theorem. To do this we will show that □$AC'BH$ is a parallelogram, so that its opposite sides ($\overline{C'A}$ and $\overline{BH}$) are of equal length. To do this, note first that since H is the orthocenter of △ABC, $\overline{BH}$ (extended) is perpendicular to $\overline{AC}$. Since $\overline{C'A}$ ⊥ $\overline{AC}$, we see that $\overline{C'A}$ ∥ $\overline{BH}$. Along the same lines, we note that ∠$C'BC$ is inscribed in a semicircle so that $\overline{C'B}$ ⊥ $\overline{BC}$, and since $\overline{AH}$ is an altitude of △ABC (why?), $\overline{AA'}$ ⊥ $\overline{BC}$. Consequently, $\overline{AH}$ ∥ $\overline{C'B}$, so that □$C'BHA$ is a parallelogram and $BH = C'A = 2RF$, as was to be demonstrated.

This proof has, at least in its diagram, assumed that △ABC is an acute triangle, so that the circumcenter, R is in the interior of △ABC. The proof proceeds in essentially the same manner if angle B is obtuse. This case is left as an exercise.

Most of the theorems discussed so far in this section have specified concurrence of lines related to triangles. Theorem 4.7.7 will allow us to prove a statement concerning the collinearity of points related to a triangle.

THEOREM 4.7.8.   The orthocenter, the circumcenter, and the centroid of a given circle are collinear (on a line called the *Euler line* for the triangle).

Proof. Suppose that △ABC has orthocenter H and circumcenter R as shown in Figure 4.7.7. The perpendicular from R to $\overline{AC}$ will bisect $\overline{AC}$ (why?), so that M is the midpoint of $\overline{AC}$ and, consequently, $\overline{BM}$ is the median from B to $\overline{AC}$. We assert that there is a point K such that $\overline{BM}$ ∩ $\overline{HR}$ = K (see Exercise Set 4.7, Problem 12). Since $\overline{RM}$ ∥ $\overline{BF}$, we see that △RMK ≈ △HBK (AAA). Also, by the previous theorem, we know that $BH = 2RM$, so that corresponding sides are in a 2:1 ratio. In particular, $\overline{BK}$ and $\overline{KM}$ are in a 2:1 ratio. If we then recall that the centroid

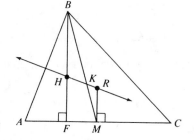

Figure 4.7.7

of a triangle is located two-thirds of the way from a vertex to the midpoint of
the opposite side, we may conclude that K is the centroid of the triangle and,
as such, the centroid, orthocenter, and circumcenter are collinear (on $\overleftrightarrow{HR}$),
and that the centroid is located two-thirds of the way from the orthocenter to
the circumcenter.

We conclude our investigation of properties of triangles with two final
results: the theorems of Menelaus and of Ceva. Although the theorems are
quite similar, their proofs are separated, timewise, by about 1600 years.
First, we shall look at the earlier of the two results, which is due to the Greek
astronomer-geometer Menelaus of Alexandria.

Menelaus (c. 100 A.D.) contributed much to the study of geometry
some 300 years after Euclid. In particular, his three-volume work *Sphaerica*
revised much of Euclid's work on plane geometry so that it applied to figures
on a sphere, e.g. spherical geometry. One of the results derived for spherical triangles that has a planar equivalent, and which was overlooked by
Euclid, involves the collinearity of points chosen on the sides of triangles.

To begin, consider $\triangle ABC$ as shown in Figure 4.7.8. Points X, Y, and Z
in this figure are called *Menelaus points*, by which we mean that they are on
lines containing the sides of the triangle without being vertices of the triangle. The first question we will address is, Under what conditions will X, Y,
and Z be collinear?

In order to investigate this, let's first assume that X, Y, and Z are
collinear and contained in line l, as shown in Figure 4.7.8. Next we will
draw perpendiculars $\overline{AR}$, $\overline{CS}$, and $\overline{BT}$ to l as shown in Figure 4.7.9. For
convenience, we will use p_1, p_2, and p_3 to represent the lengths of $\overline{AR}$, $\overline{CS}$,
and $\overline{BT}$, respectively.

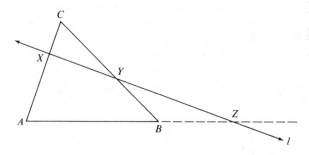

Figure 4.7.8

Notice that we have the following three pairs of similar triangles
(AAA—justify each of these):

$$\triangle ARZ \approx \triangle BTZ$$

$$\triangle XAR \approx \triangle XCS$$

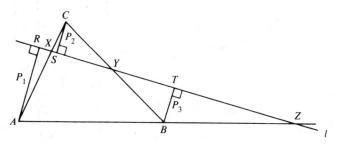

Figure 4.7.9

and

$$\triangle YCS \approx \triangle YBT$$

From these similarities come the following three proportions:

$$\frac{AZ}{ZB} = \frac{p_1}{p_3}$$

$$\frac{BY}{YC} = \frac{p_3}{p_2}$$

and

$$\frac{CX}{XA} = \frac{p_2}{p_1}$$

Now we compute the product of the left-hand members of these ratios and replace each with its equivalent value on the right.

$$\frac{AZ}{ZB} \times \frac{BY}{YC} \times \frac{CX}{XA} = \frac{p_1}{p_3} \times \frac{p_2}{p_1} \times \frac{p_3}{p_2} = 1 \tag{10}$$

indicating that if X, Y, and Z are collinear, then the product of the ratios shown is 1.

Menelaus' theorem is usually stated in a slightly altered form that takes into account directed distances. In particular, we see in Figure 4.7.8 that measurement from A to Z is in a direction opposite that of the measurement from Z to B, so that if we consider $\overline{AZ}$ to have a positive length, it would make sense to consider the length of $\overline{ZB}$ to be negative, or vice versa. In either case, the ratio AZ/ZB would have a negative value. It should be clear that $\overline{CX}$ and $\overline{XA}$ are measured in the same direction, so that the ratio CX/XA is positive, and likewise for BY/YC. Consequently, the product shown in Equation (10) is, in terms of directed distances, not 1 but rather -1. Another possible configuration for Menelaus points is shown in Figure 4.7.10. In this configuration all three ratios are negative, so that the product in

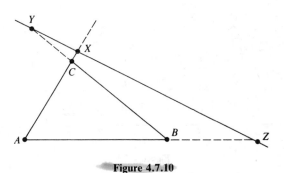

Figure 4.7.10

Equation 10 is still negative. With this in mind, we state the theorem of Menelaus as follows:

THEOREM 4.7.9.   *The Theorem of Menelaus:* If $\triangle ABC$ is a triangle in which X, Y, and Z are collinear Menelaus points on sides $\overline{AC}$, $\overline{BC}$, and $\overline{AB}$ respectively, then

$$\frac{AZ}{ZB} \times \frac{BY}{YC} \times \frac{CX}{XA} = -1$$

Using directed measurements for the lengths of the segments that comprise the ratios.

To conclude this section, we will consider the companion to the theorem of Menelaus: Ceva's theorem. Ceva's theorem involves a product of ratios, much like that of Menelaus. However, in Ceva's theorem the ratios result from the concurrence of certain lines (called Cevian lines or simply cevians) rather than collinearity of (Menelaus) points. A Cevian line is any line that connects a vertex of a triangle to a nonvertex point on the side opposite. The theorem is stated as follows:

THEOREM 4.7.10.   *Ceva's Theorem:* If concurrent Cevian lines $\overrightarrow{AX}$, $\overrightarrow{BY}$, and $\overrightarrow{CZ}$ are drawn from the vertices of $\triangle ABC$ (Figure 4.7.11), then

$$\frac{AY}{YC} \times \frac{CX}{XB} \times \frac{BZ}{ZA} = 1$$

In one sense, Ceva's Theorem is a generalized version of the median concurrence theorem, since if we choose X, Y, and Z to be the midpoints of the sides of $\triangle ABC$, each of the ratios in the product given above is 1, so that the product of the ratios is 1 also. Ceva's theorem goes beyond this, however, and applies to Cevian lines other than medians. The proof proceeds in the following manner:

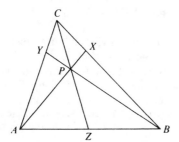

Figure 4.7.11

Proof. We will use Menelaus' theorem to help prove Ceva's theorem in the following way. Consider $\triangle BYC$ (within $\triangle ABC$) as a triangle in its own right. The points X, P, and A are collinear, one on each side of $\triangle BYC$, with $\overline{CY}$ extended as necessary. We may then apply Menelaus' theorem to the segments shown to obtain the following product:

$$\frac{BP}{PY} \times \frac{YA}{AC} \times \frac{CX}{XB} = -1 \tag{11}$$

Similarly, if we consider $\triangle BYA$, we see that points Z, P, and C are collinear, one on each side of the triangle. Again applying Menelaus' theorem, we find

$$\frac{AC}{CY} \times \frac{YP}{PB} \times \frac{BZ}{ZA} = -1 \tag{12}$$

Multiplying the left and right sides of Equations (11) and (12), we have the following product.

$$\frac{BP}{PY} \times \frac{YA}{AC} \times \frac{CX}{XB} \times \frac{AC}{CY} \times \frac{YP}{PB} \times \frac{BZ}{ZA} = (-1)(-1) = 1$$

Simplifying, we see then that

$$\frac{YA}{CY} \times \frac{BZ}{ZA} \times \frac{CX}{XB} = 1$$

which, except for a minor rearrangement of terms, represents the product specified by the theorem.

The converse of Ceva's Theorem is also a theorem, making the result a biconditional. The proof of the converse proceeds indirectly. It assumes that the product of the ratios is 1, that two of the Cevian lines meet at P, but that the third one does not, and it shows that this results in a contradiction. The details of the proof of the converse are left as an exercise.

EXERCISE SET 4.7

1. Complete the proof of Theorem 4.7.2 by proving that if a point is equidistant from the sides of an angle, it lies on the bisector of the angle.

2. In the proof of Theorem 4.7.3 it was implied that point R is contained in $\overline{AB}$ (see Figure 4.7.3a). Explain how we know that R is between A and B so that $R \in \overline{AB}$.

3. Suppose that P is the point of intersection of two angle bisectors of a triangle and that r is the distance from P to a side of the triangle. Will the circle centered at P and having radius r contain any points in the exterior of the triangle? Explain why or why not.

4. In the proof of Theorem 4.7.4 (the altitude concurrence theorem), it was stated that "lines l_1, l_2, and l_3 intersect pairwise at points X, Y, and Z" (see Figure 4.7.4b). Explain why these lines cannot be parallel.

5. Complete the proof of the altitude concurrence theorem by showing that B is the midpoint of $\overline{XY}$ and C is the midpoint of $\overline{YZ}$ in Figure 4.7.4b.

6. Show that the proof of the altitude concurrence theorem applies to acute triangles as well as to obtuse triangles.

7. On a separate sheet of paper, draw an acute triangle $\triangle ABC$ and locate its orthocenter, P, by drawing the three altitudes. Next outline $\triangle ABP$. Locate point Q, the orthocenter of $\triangle APB$. How are points Q and C related? Explain.

8. Suppose M, N, and P are the midpoints of sides $\overline{AB}$, $\overline{AC}$, and $\overline{BC}$ of $\triangle ABC$. Show that the centroid of $\triangle MNP$ is the same point as the centroid of $\triangle ABC$.

9. Suppose that l_1 and l_2 are two altitudes of the same triangle. If H is the orthocenter of this triangle, prove that the product of the lengths into which H divides l_1 is the same as the product of the lengths into which H divides l_2.

10. Prove that each excenter of a triangle serves as the center of a circle that is externally tangent to a side of the triangle and the extensions of the other two sides.

11. Show that Theorem 4.7.7 is valid for obtuse triangles by assuming that, in Figure 4.7.5, $m\angle B > 90°$.

12. Figure 4.7.12 shows a triangle $\triangle ABC$, its orthocenter H, and its circumcenter R. F is the foot of the perpendicular from H to $\overline{AC}$. M is the midpoint of $\overline{AC}$. We wish to show that $\overleftrightarrow{HR}$ and $\overleftrightarrow{MB}$ intersect at some point K (see Figure 4.7.7)

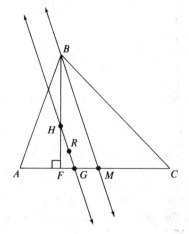

Figure 4.7.12

since if we establish the existence of K the proof of Theorem 4.7.8 (the Euler line) will be complete. To do this, we will proceed indirectly and assume that $\overleftrightarrow{HR} \parallel \overleftrightarrow{MB}$ as shown here.

If we apply Theorem 3.2.5 (Pasch's Theorem) to $\triangle BFM$ and $\overleftrightarrow{HR}$ we may conclude that $\overleftrightarrow{HR}$ intersects $\overline{FM}$ at some point G such that $F = G = M$. Now consider $\triangle HFG$ and $\overleftrightarrow{RM}$. Complete the proof by showing that $\overleftrightarrow{RM}$ cannot intersect $\overline{HF}$ or $\overline{FG}$ thereby contradicting Theorem 3.2.5 and indicating that the assumption that $\overleftrightarrow{HR} \parallel \overleftrightarrow{MB}$ is false.

13. The theorem of Menelaus is often stated as a biconditional, since its converse is also true. State and prove the converse of Theorem 4.7.9. (*Hint:* Assume that the product of the ratios is -1 and then draw a line that contains two of the points X, Y, and Z. Show that a contradiction results if this line fails to contain the third point.)

14. State and prove the converse of Ceva's theorem.

In Problems 15 through 19, determine the missing length x for each of the triangles.

15.

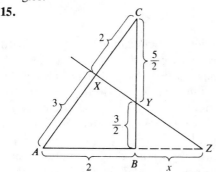

16.

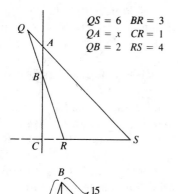

$$QS = 6 \quad BR = 3$$
$$QA = x \quad CR = 1$$
$$QB = 2 \quad RS = 4$$

17.

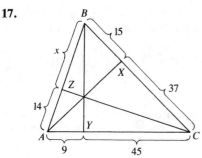

18.

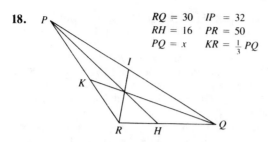

$$RQ = 30 \quad IP = 32$$
$$RH = 16 \quad PR = 50$$
$$PQ = x \quad KR = \tfrac{1}{3} PQ$$

19. In the proof of Theorem 4.7.3 we showed that the bisectors of $\angle A$ and $\angle C$ (see Figure 4.7.3a) intersected at a point, P, that is on the B side of $\overline{AC}$. Explain how we know that P is also on the A side of $\overline{BC}$ *and* the C side of $\overline{AB}$, so that it is an interior point of $\triangle ABC$.

4.8 THE NINE-POINT CIRCLE

Sections 4.5 through 4.7 developed a series of Euclidean results dealing with circles and triangles. There are actually many other theorems that deal with properties of these figures. In fact, there are so many that not all can be included in this survey of Euclidean results. However, in this section one further, notable, Euclidean property concerning triangles and circles is developed—the existence for each triangle of a *nine-point circle,* a circle that contains nine points related to a triangle.[12]

The nine-point circle was investigated by a number of geometers simultaneously in the early nineteenth century. Included in this group were Charles J. Brianchon (1783–1864), Jean Victor Poncelet (1788–1867), and Karl Wilhelm Feurbach (1800–1834). The nine-point circle is sometimes referred to as *Feurbach's circle* or (incorrectly) as *Euler's circle*, the latter likely because of the close relationship between the circle and the Euler line (Theorem 4.7.8), although there is no evidence that Euler himself investigated the nine-point circle.

To begin the discussion of the nine-point circle, consider $\triangle ABC$ with orthocenter H and circumcenter S as shown in Figure 4.8.1a. Also shown in this figure are F_1 (the foot of the altitude from B to $\overline{AC}$), M_1 (the midpoint of $\overline{AC}$), and Q (the midpoint of $\overline{BH}$).

Theorem 4.5.1 assures us that there is a unique circle that contains F_1, M_1, and Q. Our first objective is to describe this circle by identifying its center and specifying the length of its radius.

To do this, we will draw altitude $\overline{BF_1}$ and $\overline{SM_1}$, $\overline{SB}$, $\overline{QM_1}$, and $\overline{HS}$ as shown in Figure 4.8.1b. Note that $\overleftrightarrow{SM_1}$ is the perpendicular bisector of $\overline{AC}$,

---

[12] If a triangle is isosceles or equilateral, two or more of the nine points may coincide, so that the circle may not contain nine *distinct* related points. Investigations of these special cases are included as exercises.

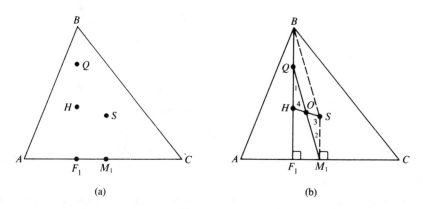

Figure 4.8.1

since it contains the circumcenter S of $\triangle ABC$ and the midpoint M_1 of side $\overline{AC}$ (see Theorem 4.7.1). Note also that $\overline{SB}$ is a radius of the circle that contains the three vertices of $\triangle ABC$.

Since Q is the midpoint of $\overline{HB}$, we know that $\overline{HQ} = \frac{1}{2}HB$. In addition, $M_1S = \frac{1}{2}HB$ (Theorem 4.7.7). Combining these, we see that $\overline{M_1S} = \overline{HQ}$. Since lines $\overleftrightarrow{HQ}$ and $\overleftrightarrow{M_1S}$ are both perpendicular to $\overleftrightarrow{AC}$, we know they are parallel (Corollary. 3.4.2), so that $m\angle 1 = m\angle 2$ (Theorem 3.4.5). Since $\angle 3$ and $\angle 4$ are vertical angles, we know also that $m\angle 3 = m\angle 4$. Together these facts imply that $\triangle OHQ \cong \triangle OSM_1$ (Theorem 3.3.4). From this we may conclude that $\overline{HO} = \overline{OS}$ (so that O is the midpoint of $\overline{HS}$) and that $\overline{OQ} = \overline{OM_1}$ (so that O is also the midpoint of $\overline{QM_1}$).

Next we will consider the right triangle $\triangle QF_1M_1$ and the circle $\mathscr{C}$ that circumscribes it. Since $\angle QF_1M_1$ is a right angle, the arc it intercepts within $\mathscr{C}$ is $180°$ (Theorem 4.6.4), which means that $\overset{\frown}{QF_1M_1}$ is a semicircle and that $\overline{QM_1}$ is a diameter of circle $\mathscr{C}$. Since O is the midpoint of diameter $\overline{QM_1}$, it serves as the center of the circle containing F_1, M_1, and Q.

In order to complete the description of the circle containing M_1, F_1, and Q, we need to specify the length of the radius $\overline{OQ}$ as it relates to $\triangle ABC$. We will do this using a pair of similar triangles, namely, $\triangle OHQ$ and $\triangle SHB$. In order to show that these triangles are in fact similar, we need only note that they share $\angle BHS$ and that the lengths of the sides on both rays of $\angle BHS$ are in a $2:1$ ratio so that the triangles are similar by Theorem 4.4.6 (SAS similarity theorem). As a result, $\overline{OQ}$ (a radius of the circle that contains Q, M_1, and F_1) is one-half the length of $\overline{SB}$, a radius of the circle that circumscribes $\triangle ABC$, the triangle with which we began the discussion. Beyond this, we have seen that the center O of the circle containing M_1, F_1, and Q is the midpoint of $\overline{HS}$, the segment that joins the orthocenter H and the circumcenter S of $\triangle ABC$.

A reasonable question one might ask at this point is, How does the

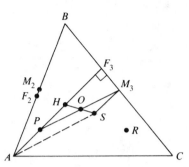

Figure 4.8.2

circle discussed in the preceding paragraphs (namely, the circle containing M_1, F_1, and Q) relate to the nine-point circle we are supposed to be investigating? The answer is that the circle described above *is* the nine-point circle. We now need to specify the *other* six points.

To do this, we simply reorient the view of $\triangle ABC$ given in Figure 4.8.1. This time we will draw altitude $\overline{AF_3}$ (from A to $\overline{BC}$), locate M_3 (the midpoint of $\overline{BC}$), and locate P (the midpoint of $\overline{AH}$) as shown in Figure 4.8.2.

In much the same manner as was used earlier, we may show that $\triangle OHP \cong \triangle OSM_3$, so that the point where $\overline{PM_3}$ intersects $\overline{HS}$ is the midpoint O of $\overline{HS}$, the same point found earlier. This point will serve as the center of the circle containing P, F_3 and M_3, so that for the circle containing these three points to be the same circle as the one that contains Q, F_1, and M_1, we need only to show that the two circles also have radii of the same length. This can be done by showing $OP = \frac{1}{2}SA$ (making use of the similarity of $\triangle OHP$ and $\triangle SHA$) and by using the fact that $SA = SB$, since each is a radius of the circle that circumscribes $\triangle ABC$. The details of this argument are left as an exercise, as is the task of showing that R, M_2, and F_2 lie on the circle also. Summarizing all this, we have the following theorem.

THEOREM 4.8.1.   *The Nine-Point Circle:* The nine points consisting of (i) the midpoints of the sides of a triangle (in this illustration $\triangle ABC$), (ii) the feet of the three altitudes of this triangle, and (iii) the midpoints of the segments formed by joining the orthocenter to the vertices of the triangle are concyclic. The center of this nine-point circle is the midpoint of the segment joining the orthocenter and the circumcenter of $\triangle ABC$, and the length of the radius of this circle is one-half the length of the radius of the circle that circumscribes $\triangle ABC$.

EXERCISE SET 4.8

In each of the triangles in Problems 1 through 4, H is the orthocenter and S is the circumcenter. Locate the center and identify the length of the radius of the nine-point circle associated with each triangle. Use a compass to draw the nine-point circle for each triangle.

1.

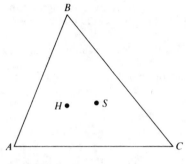

2.

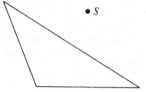

3.

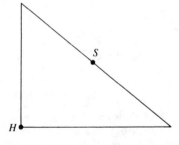

4.

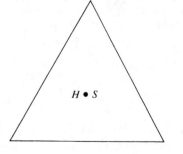

5. Provide the details to show that the circle containing M_3, F_3, and P in Figure 4.8.2 has the same center and radius as the circle that contains M_1, F_1, and Q.

6. Complete the proof of the nine-point circle theorem by showing that the circle containing M_2, F_2, and R in Figure 4.8.2 is the same circle that contains M_1, F_1, and Q.

7. In Problem 1, H is the orthocenter and S is the circumcenter of $\triangle ABC$. Copy this diagram twice.
   (a) Use a ruler and compass to draw, on copy 1, the nine-point circle for $\triangle ABC$.
   (b) Use a ruler and compass to draw, on copy 2, the nine-point circle for $\triangle HBC$.
   How do the circles from parts (a) and (b) compare? Can you explain this result?

8. Explain why the Euler line for a triangle *always* contains a diameter for the nine-point circle.

9. Is the following statement true or false: All triangles that have the same orthocenter and the same circumcircle have the same nine-point circle. If the statement is true, explain how you can be sure. If it is false, give a counterexample.

10. Shown here is $\triangle ABC$, its circumcircle, and its nine-point circle. Choose a point X on the circumcircle and construct $\overline{HX}$, which will intersect the nine-point circle at a point we shall call Y. Prove that $\overline{HY} \cong \overline{YX}$.

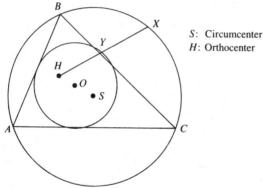

S: Circumcenter
H: Orthocenter

11. Do you think that Theorem 4.8.1, the nine-point circle theorem, could have been included in Chapter 3 (neutral geometry)? If not, where is the Euclidean parallel postulate (or an equivalent form of it) used in the proof of the theorem? Is it possible that an alternate proof could be formulated that makes no use of a form of the Euclidean parallel postulate? Explain how you might investigate this possibility.

4.9 EUCLIDEAN CONSTRUCTIONS

Having surveyed many of the significant theorems from Euclidean geometry, we now turn our attention to a separate, but related, area that was of interest to Euclid and other Greek geometers—constructions. In order to appreciate the importance of the role of constructions in the foundations of

Euclidean geometry, we need only review the first three postulates set forth by Euclid in the *Elements*.

1. To draw a straight line from any point to any point.
2. To produce a finite straight line (or line segment) continuously in a straight line.
3. To describe (or draw) a circle with any center and distance (or radius).

These three postulates allow us to perform the three simplest constructions and imply that we have available for use two implements: a straightedge and a compass.[13] In this section we will extend this list of simple postulated constructions to a longer list, including some that are fairly complex. Throughout we will restrict ourselves to the use of the two Euclidean tools.

The straightedge available to us must be thought of as a ruler without length markings, since none of the postulates provides us with the ability to measure lengths. Thus we may use our straightedge to connect a pair of given points (Postulate 1) or to extend a given line segment arbitrarily in a straight line (Postulate 2). We cannot, however, use the straightedge to extend a line segment 3 inches (or any other specified length), since a Euclidean straightedge has no unit of length marked on it.

A Euclidean compass is actually slightly different from the modern compass with which we are familiar. In Euclid's time, a compass could be used to draw a circle at a given point, but when lifted would collapse so that lengths could not be transferred via the compass. For this reason, the Euclidean compass is often called a *collapsing compass*, and this is why the proof of Proposition 2 from the *Elements*[14] seems so tedious compared to the way in which it is done in most current textbooks. Fortunately, any construction that can be done with a modern compass can also be done with a collapsing compass, albeit with considerably more effort. To illustrate this, we will show that a circle with a specified radius can be copied to a location with a given center using only a straightedge and a collapsing compass.

THEOREM 4.9.1.   *The Compass Equivalence Theorem:* A circle $C(B,r)$ can be congruently copied (using only a straightedge and a collapsing compass) so that a given point A serves as the center of the copy.

Proof. Consider $C(B,r)$ and point A as shown in Figure 4.9.1a. Our objective is two construct $C(A,r)$ using only a straightedge and a

---

[13] These two items, which date back to the time of Plato, are usually referred to as the Euclidean tools.

[14] Proposition 2 states the following: "To place at a given point (as an extremity) a straight line [segment] equal [in length] to a given straight line [segment]." See Figure 2.2.2 and the discussion that follows it.

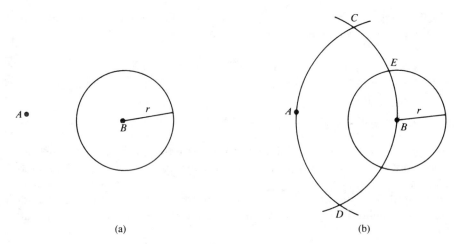

(a)   (b)

Figure 4.9.1

collapsing compass. We begin by constructing circles $C(A,AB)$ and $C(B,BA)$ which intersect at points we will call C and D. One intersection of $C(B,r)$ and $C(A,AB)$ will be called E (Figure 4.9.1b). Next we construct circle $C(C,CE)$ and use P to represent the intersection of this circle with $C(B,BA)$, as shown in Figure 4.9.2. We claim that $AP = r$, so that $C(A,AP)$ is the circle required by the statement of Theorem 4.9.1.

To show that $AP = r$, it will suffice to show that $\overline{AP} \cong \overline{BE}$, since $\overline{BE}$ and r are radii of the same circle. Our approach will be to show that $\triangle APC \cong \triangle BEC$ by SAS (SMSG Postulate 15). Clearly, $\overline{CP} \cong \overline{CE}$ and $\overline{AC} \cong \overline{BC}$ (Why?), so that it remains for us to show that $m\angle PCA = m\angle ECB$. In order to do this, we must consider, for a moment, another pair of congruent triangles.

Consider $\triangle PCB$ and $\triangle ECA$. These triangles are congruent by SSS

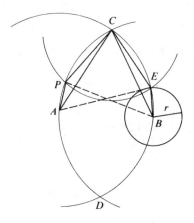

Figure 4.9.2

(Exercise Set 4.9, Problem 1), so that m∠ PCB = m∠ ECA. From each of these angles we subtract m∠ ACB, so that

$$m\angle PCB - m\angle ACB = m\angle ECA - m\angle ACB$$

But using a suitable substitution, we may replace these differences by their equivalent angle measures to find that

$$m\angle PCA = m\angle ECB$$

so that △APC ≅ △BEC, from which we may conclude that $AP = BE = r$, so that $C(A,AP)$ is a copy of $C(B,r)$ as stated in the theorem.

Theorem 4.9.1 provides us with the capability to copy circles from place to place using a collapsing compass in essentially the same way that a modern compass is used. Henceforth, we will assume the use of a modern compass, as the constructions proceed more efficiently in that manner.

Many of the most useful Euclidean constructions are standard in secondary school curricula and will not be elaborated on here. A few, however, are central to Euclidean constructions and are described below. The proofs of these procedures are elementary, depending mostly on the congruence conditions for triangles, and are left as exercises.

CONSTRUCTION 1.   To construct the perpendicular bisector of a line segment.

To construct the perpendicular bisector of line segment $\overline{AB}$ (Figure 4.9.3) we will

1. Construct $C(A,AB)$ and $C(B,BA)$. Here C and D will be used to label the points of intersection of these two circles.
2. Draw line $\overleftrightarrow{CD}$. $\overleftrightarrow{CD}$ is the perpendicular bisector of $\overline{AB}$.

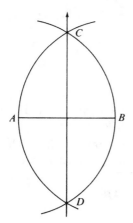

Figure 4.9.3

CONSTRUCTION 2.   To construct a perpendicular l_p from a given point P to a line l not containing the P.

To construct a perpendicular from a point P to a line l we will

1. Construct circle $C(P,r)$ using any positive length r such that $C(P,r)$ intersects l in two points which will be called A and B.
2. Construct the perpendicular bisector of $\overline{AB}$ using Construction 1. The perpendicular bisector will be l_p.

CONSTRUCTION 3.   To construct a line l_p perpendicular to a given line l at a given point P on l.

To construct a perpendicular to l at P we will

1. Construct the circle $C(P,r)$, where r is any positive length. A and B will denote the intersections of $C(P,r)$ and l.
2. Construct the perpendicular bisector of $\overline{AB}$ using Construction 1. The perpendicular bisector of $\overline{AB}$ is l_p.

CONSTRUCTION 4.   To construct the angle bisector of an angle $\angle ABC$.

To construct the bisector of $\angle ABC$ (Figure 4.9.4) we will

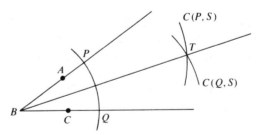

Figure 4.9.4

1. Construct $C(B,r,)$ where r is any positive length. P and Q will denote the intersection of $C(B,r)$ with $\overrightarrow{BA}$ and $\overrightarrow{BC}$, respectively.
2. Construct circles $C(P,s)$ and $C(Q,s)$ using any length s such that $C(P,s)$ $\cap$ $C(Q,s) \neq \varnothing$. Here T will denote any intersection of $C(P,s)$ and $C(Q,s)$ that is in the interior of $\angle ABC$.
3. Construct $\overrightarrow{BT}$. $\overrightarrow{BT}$ is the bisector of $\angle ABC$.

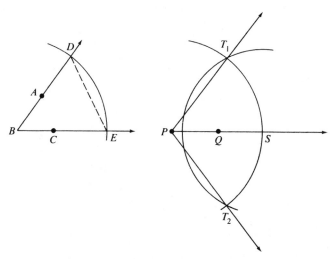

Figure 4.9.5

CONSTRUCTION 5.   To a copy of $\angle ABC$ on ray $\overrightarrow{PQ}$, with P as the vertex of the copy.

To copy $\angle ABC$ onto $\overrightarrow{PQ}$ (Figure 4.9.5) we will

1. Construct $C(B,r)$, where r is any positive length. D and E will denote the intersections of $C(B,r)$ with $\overrightarrow{BA}$ and $\overrightarrow{BC}$, respectively.
2. Construct $C(P,r)$. S will denote the intersection of $C(P,r)$ and $\overrightarrow{PQ}$.
3. Construct $C(S,DE)$, where DE represents (as usual) the distance from D to E. Here T_1 and T_2 will be used to denote the two intersections of $C(S,DE)$ with $C(P,r)$.
4. Each of $\angle T_1PS$ and $\angle T_2PS$ is a copy of $\angle ABC$.

We now look at some extensions of these basic constructions. For example, while Construction 1 can be used to bisect a given line segment into a pair of congruent pieces, it is often useful to be able to produce more than two congruent pieces from a given segment. The following construction addresses this need.

CONSTRUCTION 6.   To partition a line segment $\overline{AB}$ into three (or more) congruent pieces.

The procedure used to partition a segment into three congruent pieces generalizes easily to partitions of four or more pieces and involves the following steps (Figure 4.9.6):

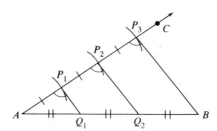

Figure 4.9.6

1. Choose any point C not on $\overleftrightarrow{AB}$ and construct $\overrightarrow{AC}$.
2. Construct $C(A,r)$ for any positive length r. Use P_1 to denote the intersection of $C(A,r)$ and $\overrightarrow{AC}$.
3. Construct $C(P_1,r)$. Use P_2 to denote the intersection of $C(P_1,r)$ and $\overrightarrow{AC}$ such that $P_2 \neq A$.
4. Construct $C(P_2,r)$. Use P_3 to denote the intersection of $C(P_2,r)$ and $\overrightarrow{AC}$ and construct $\overline{P_3B}$ such that $P_3 \neq P_1$.
5. Copy $\angle AP_3B$ onto $\overrightarrow{P_2A}$ with P_2 as the vertex. Use Q_2 to denote the intersection of the copy of $\angle AP_3B$ and $\overline{AB}$.
6. Repeat step 5 using P_1 as the vertex of the copy of $\angle AP_3B$. Use Q_1 to denote the intersection of this copy and $\overline{AB}$. Then Q_1 and Q_2 trisect segment $\overline{AB}$ into congruent segments.

Construction 6 can also be used to partition a given segment into a given ratio of lengths. For example, suppose that a $\overline{AB}$ is to be divided into two pieces with a ratio of $4:3$. In order to do this we would merely have to divide the segment into seven pieces using a generalized version of the procedure given in Construction 6, use the first four of these pieces for the first segment, and use the last three for the second segment, giving the $4:3$ ratio needed. Beyond this, if we denote a given segment to represent a unit length, we can use the same procedure to find a line segment representing any fractional part of the unit segment.

To illustrate the second of these procedures, suppose that $\overline{AB}$ (Figure 4.9.7) is chosen to be a unit length.

We can extend $\overline{AB}$ to $\overleftrightarrow{AB}$ and use the ruler placement postulate to make the correspondence $A \to 0$ and $B \to 1$, creating a number line containing $\overline{AB}$ as the unit segment from 0 to 1. Suppose we wish to locate the fraction $\frac{2}{7}$ on

Figure 4.9.7

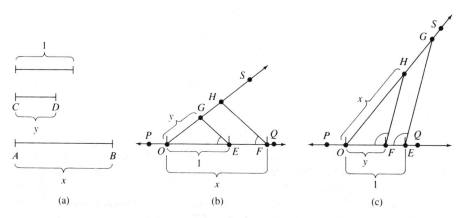

Figure 4.9.8

the number line we have created. To do this, we partition $\overline{AB}$ into seven parts and count two, after which we will have located the point corresponding to the rational number $\frac{2}{7}$. It should be clear that there is nothing special about $\frac{2}{7}$. In general, we can now locate any rational number[15] on a number line, given a unit length with which to start.

Euclidean constructions can also be used to determine lengths corresponding to sums, differences, products, and quotients of lengths of segments on a number line. Given $\overline{AB}$ and $\overline{CD}$ with lengths x and y, respectively (Figure 4.9.8a) we can easily construct segments corresponding to the sum, $x + y$, and the difference, $x - y$, of the lengths of these segments (see Exercise Set 4.9, Problem 19).

Somewhat less obvious is the technique by which we may construct segments corresponding to the product and quotient of lengths x and y. To determine a length corresponding to xy, we will first copy $\overline{AB}$ onto a number line $\overrightarrow{PQ}$ scaled with a unit length $\overline{OE}$, starting at the origin O. Then $\overline{OF}$ will represent the copy of $\overline{AB}$ as shown in Figure 4.9.8b. Next we will choose a point S, not contained in $\overrightarrow{PQ}$, and construct $\overrightarrow{OS}$. We then make a copy of $\overline{CD}$ on $\overrightarrow{OS}$ with O as one endpoint, using G to denote the other endpoint. Next we construct $\overline{GE}$ and copy $\angle GEP$ onto the G side of $\overrightarrow{FP}$ with F as the vertex. If we use H to denote the intersection of $\overrightarrow{OS}$ and the copy of $\angle GEP$, we can show that $HO = xy$ using the following argument:

To begin, note that $\triangle GOE \approx \triangle HOF$ (AA). From this, we find the proportion

$$\frac{GO}{HO} = \frac{EO}{FO}$$

[15] Recall that a rational number is a number that can be expressed as a ratio of the form a/b, where a and b are integers and b is not 0.

Comparing the products of means and extremes, we have

$$HO \times EO = GO \times FO$$

But substituting for these quantities, we find

$$HO \times 1 = (y)(x)$$

or

$$HO = xy$$

A related procedure, which also uses similar triangles, can be applied to identify a segment whose length is the quotient x/y. To do this, we can make a copy of $\overline{CD}$ on $\overrightarrow{PQ}$, and a copy of $\overline{AB}$ on $\overrightarrow{OS}$. These copies are shown as $\overline{OF}$ and $\overline{OH}$ in Figure 4.9.8c. If we then construct $\overline{FH}$ and copy $\angle HFP$ at E as shown, we can use the similarity of $\triangle GEO$ and $\triangle HFO$ to show that $\overline{GO}$ has length x/y. The details of the proof are left as an exercise.

Thus far we have seen how it is possible to construct line segments that represent rational numbers and their sums, differences, products, and quotients. Since the set of rational numbers is closed for addition and multiplication, we cannot use the preceding procedures to construct line segments with irrational lengths. However, it is not hard to imagine that at least some irrational lengths can be constructed using Euclidean tools. For example, since we have techniques that allow us to construct perpendiculars and copy lengths, it is an easy task to construct a square with sides of unit length (see Exercise Set 4.9, Problem 22). Using the Pythagorean theorem (Theorem 4.4.8), we can deduce that the length of each diagonal of this square is $\sqrt{2}$, an irrational value. Since the techniques described above for sums, differences, products, and quotients don't require that the lengths be rational, we can then construct lengths that represent values such as $3 + \sqrt{2}$, $5\sqrt{2}$ and $4 - 7\sqrt{2}$. In general, we now have the ability to represent lengths of the form $a + b\sqrt{2}$, where a and b represent lengths that are constructible (i.e., rational numbers or previously produced lengths of the form $a + b\sqrt{2}$).

It can be shown that not all irrational numbers can be represented as $a + b\sqrt{2}$. A logical next question might be, Which other irrational lengths can be constructed using Euclidean tools? The answer to the question is, Many but not all, and is worth investigating.

To begin, we will need a technique allowing us to construct line segments that represent the square roots of numbers other than 2. A general technique for constructing a segment whose length[16] is $\sqrt{x}$ proceeds as follows:

---

[16] Some lengths representing the square roots of integers can be constructed as diagonals rectangles in much the same way that $\sqrt{2}$ was constructed. For example, $\sqrt{13}$ can be constructed as the diagonal of a rectangle having sides of length 2 and 3. Most irrational lengths, however, cannot be constructed this way. For example, try using this method to construct a line segment with length $\sqrt{3}$.

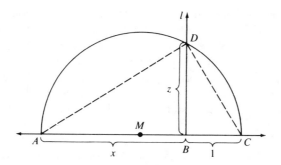

Figure 4.9.9

Begin with a segment $\overline{AB}$ whose length is x. Append a segment of unit length onto $\overline{AB}$ to produce a segment $\overline{AC}$ whose length is $x + 1$. Now locate the midpoint M of $\overline{AC}$. (The perpendicular bisector can be used to locate M.) Next, construct the semicircle $C[M,\frac{1}{2}(x + 1)]$. This circle will contain points A and C (verify this) as shown in Figure 4.9.9. At point B draw a line l perpendicular to $\overline{AB}$ and use D to name the point of intersection of the semicircle and l. We claim that the length, z, of $\overline{BD}$ is $\sqrt{x}$.

To see this, note that $\triangle DBA$ and $\triangle CBD$ are similar (why?) and give the following proportion for corresponding sides:

$$\frac{CB}{DB} = \frac{DB}{BA}$$

Substituting for these quantities, we have

$$\frac{1}{z} = \frac{z}{x}$$

from which we get $z^2 = x$, or $z = \sqrt{x}$.

From the discussions above, we see that numbers, x, of the form $x = a + b\sqrt{c}$ can be constructed for any rational values of a, b, and c. In addition, if x_1, x_2, and x_3 are all numbers of the form $a + b\sqrt{c}$, numbers such as $x_1 + x_2\sqrt{x_3}$ can also be constructed even though the coefficients are not rational. The ability to construct numbers such as these allows us, in a sense, to do some algebra in a geometric fashion. For example, suppose that we wish to construct a line segment whose length is a solution to the quadratic equation $2x^2 - 4x + 1 = 0$. From algebra we know that the solutions to quadratic equations of the form $ax^2 + bx + c = 0$ can be determined using the formula

$$x = \frac{-b \pm \sqrt{b^2 - 4ac}}{2a}$$

For $a = 2$, $b = -4$, and $c = 1$, the formula suggests solutions

$$x_1 = 1 + \tfrac{1}{2}\sqrt{2} \qquad \text{and} \qquad x_2 = 1 - \tfrac{1}{2}\sqrt{2}$$

each of which is, we now know, constructible using Euclidean tools. Similarly, the solutions to any quadratic equation with constructible coefficients can be constructed providing that they are real numbers.

A related question that naturally arises concerns the constructibility of solutions to equations of higher degree. For example, can solutions to third, fourth, and higher-degree equations be constructed by Euclidean means? The answer is yes, in at least some cases. For example, since the solutions to many cubic equations are rational (e.g., $6x^3 - 17x^2 - 5x + 6 = 0$ has solutions $x = 3$, $\frac{1}{2}$ and $-\frac{2}{3}$), in these cases the lengths can be constructed. However, the solutions to many cubic equations are *not* rational, and the question remains as to whether segments with these lengths can be constructed. Perhaps the simplest of the cubic equations with irrational solutions is $x^3 = 2$. The solution to this equation, $\sqrt[3]{2}$, is certainly irrational, but can it be constructed? It turns out that the solution to even this simplest of cubic equations is not constructible, although the proof of this fact is difficult and more suited to a book concerning abstract algebra than to a text on the foundations of geometry.[17] But the fact that $\sqrt[3]{2}$ (and other numbers like it) is not constructible has some significant ramifications that ultimately led to the solution of several problems that had interested geometers since Euclid's time—the impossibility of certain constructions.

From Euclid's time until about 200 years ago, three construction problems occupied the time and effort of untold numbers of geometers. Their simplicity and seeming close relationship to other constructions made them appear to be reasonable candidates for construction procedures similar to those discussed earlier in this chapter which had been successfully solved thousands of years ago. These three constructions are:

1. To construct a cube with twice the volume of any given cube.
2. To trisect any given angle.
3. To construct a square having the same area as any given circle.

Construction 1 requires that we start with a cube of any given size and construct a companion cube whose area is twice that of the original cube. In order to simplify the discussion, let's begin with a cube of side 1 unit (so that its volume is 1 cubic unit) (Figure 4.9.10 a) and attempt to construct a cube with a volume twice a great, that is, 2 cubic units (Figure 4.9.10b).

Suppose for a moment that we have completed this construction successfully. Then we have constructed a cube with sides of length y whose volume is 2 cubic units, so that $y^3 = 2$, or $y = \sqrt[3]{2}$. Therefore we have constructed a line segment whose length is $\sqrt[3]{2}$. But, as mentioned earlier, $\sqrt[3]{2}$ is not constructible, so that is not possible to construct a line segment of

---

[17] For a detailed discussion of the nonconstructibility of $\sqrt[3]{2}$, see Moise, *Elementary Geometry from an Advanced Standpoint*, 2nd Ed., Addison Wesley Pubs., 1974, *pp.* 233–247.

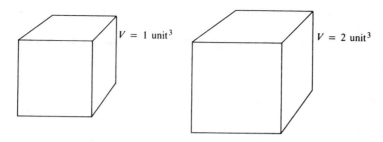

Figure 4.9.10

that length and consequently it is not possible to "double the cube." What this means is that those geometers who strove for over 2000 years to successfully complete this construction, were doomed to failure since the construction is impossible. It should be noted, of course, that even though all the attempts to complete this construction were unsuccessful, many side results were achieved during the attempts that turned out to be mathematically significant. Therefore these efforts were not entirely in vain.

The other two classic constructions listed above are also impossible, in the general case, since each requires that a line segment be constructed whose length is not constructible. These problems are investigated in the exercise set that follows.

EXERCISE SET 4.9

1. Show that, in Figure 4.9.2, $\triangle PCB \cong \triangle ECA$.
2. Prove that, in Construction 1, $\overleftrightarrow{CD}$ is the perpendicular bisector of $\overline{AB}$.
3. Construct the perpendicular bisector of each of the following segments:

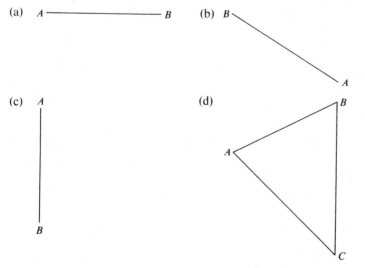

4. In Construction 2 show that the perpendicular bisector of $\overline{AB}$ contains point P.

5. In each of the following figures, construct a perpendicular from point P to line l.

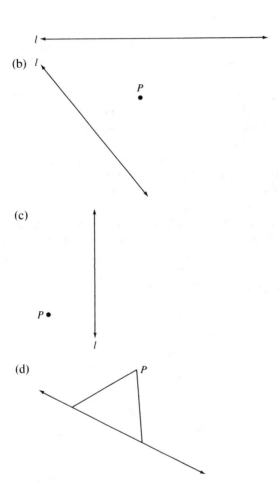

(a)
(b)
(c)
(d)

6. Show that the procedure given as Construction 3 produces a perpendicular to l at P.

7. In each of the following figures construct a perpendicular to line l at point P:

(a)

(b)

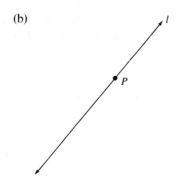

8. In Construction 4, prove that $\overleftrightarrow{BT}$ is the bisector of $\angle ABC$.

9. Construct the bisector of each of the following angles.

(a)   (b)

(c)   (d)

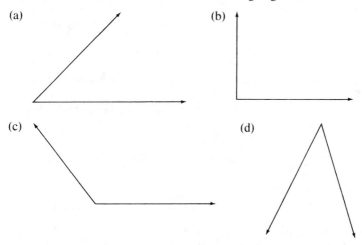

10. In Construction 4 $C(P,s)$ and $C(Q,s)$ actually intersect in two points, at least one of which will be in the interior of $\angle ABC$. (See the accompanying diagram where these two points are labeled R and R'.) Prove that B, R, and R' are collinear.

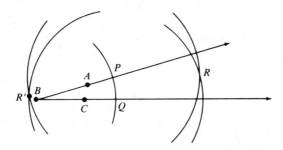

11. Prove that, in Construction 5, $\angle T_1PS$ and $\angle T_2PS$ are each congruent to $\angle ABC$.

12. In each of the following figures make a copy of $\angle ABC$ onto $\overrightarrow{PQ}$ with P as the vertex of the copy.

(a)

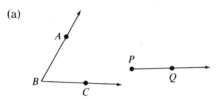

(b)

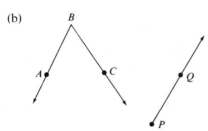

13. Prove that, in Construction 6, Q_1 and Q_2 partition $\overline{AB}$ into three congruent segments.

14. Explain how the procedure given in Construction 6 could be generalized in order to partition a segment $\overline{AB}$ into more than three congruent pieces.

15. Partition $\overline{AB}$ into three congruent pieces.

16. Partition $\overline{AB}$ into five congruent pieces.

17. Divide each of the following segments into the indicated ratios.
   (a) $2:3$
   (b) $4:2$

18. Locate the following fractions on the number lines shown:
   (a) $\frac{3}{5}$
   (b) $\frac{7}{3}$

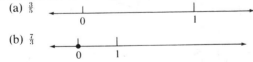

19. Explain how we can use Euclidean tools to identify lengths corresponding to $x + y$ and $x - y$ in Figure 4.9.8.

20. Prove that, in Figure 4.9.8c, $GO = x/y$.

21. Using the lengths x and y shown, construct a line segment with each of the following lengths.

————————————————————   ———————————
x y

(a) $2x + y$ (b) $x^2 + y^2$ (c) $xy/3$

22. Use the Euclidean tools to construct a unit square using the segment below as the unit length.

——————————————

23. Accept as fact that a line segment of length $\sqrt{\pi}$ cannot be constructed. Explain how this implies that it is impossible to construct a square having the same area as a circle whose radius is of unit length.

24. It can be shown that in order to trisect an angle that measures 60°, a line segment must be constructed whose length is a solution to the equation

$$8x^3 - 6x - 1 = 0$$

Use a numerical method (such as Newton's method from calculus) to approximate a positive solution to this equation. Do you think a 60° angle can be bisected using only a compass and a straightedge? Explain why or why not.

25. In order to trisect ∠ ABC shown, a high school sophomore once offered the following construction:

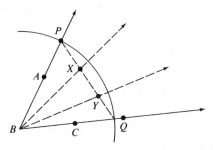

(i) Construct $C(B,r)$ for any positive length r. Use P and Q to denote the intersections of $C(B,r)$ and $\overrightarrow{BA}$ and $\overrightarrow{BC}$ respectively.
(ii) Construct line segment $\overline{PQ}$.
(iii) Use Construction 3 to locate the trisection points of $\overline{PQ}$, and use X and Y to denote these trisection points.
(iv) Rays $\overrightarrow{BX}$ and $\overrightarrow{BY}$ trisect ∠ ABC.

Does this procedure work? Explain why or why not. Try it on several angles and use your protractor to check the results.

CHAPTER 4 SUMMARY

4.2 The Parallel Postulate (and Some Implications)

Median of a Triangle: A median of a triangle is a line segment that has as its endpoints a vertex of the triangle and the midpoint of the opposite side of the triangle.

Parallelogram: A quadrilateral is a parallelogram if and only if both pairs of opposite sides are parallel.

Theorem 4.2.1. The sum of the measures of the interior angles of a triangle is 180°.

Corollary 4.2.2. The measure of an exterior angle of a triangle is *equal to* the sum of the measures of the two remote interior angles.

Theorem 4.2.3. The opposite sides of a parallelogram are congruent.

Theorem 4.2.4. If a transversal intersects three parallel lines in such a way as to make congruent segments between the parallels, then every transversal intersecting these parallel lines will do likewise.

Corollary 4.2.5. If a transversal crosses three or more parallel lines in such a way as to result in congruent segments between the parallels, then every transversal will do likewise.

Theorem 4.2.6. The three medians of a triangle are concurrent.

Corollary 4.2.7. Any pair of medians of a triangle intersect at a point that is two-thirds of the distance from any vertex to the midpoint of the opposite side.

Theorem 4.2.8. Two lines parallel to the same line are parallel to each other.

Theorem 4.2.9. If a line intersects one of two parallel lines, then it intersects the other.

Theorem 4.2.10. Each diagonal of a parallelogram partitions the parallelogram into a pair of congruent triangles.

Theorem 4.2.11. The diagonals of a parallelogram bisect each other.

Theorem 4.2.12. If the diagonals of a quadrilateral bisect each other, then the quadrilateral is a parallelogram.

Theorem 4.2.13. If a line segment has as its endpoint the midpoints of two sides of a triangle, then the segment is contained in a line that is parallel to the third side and the segment is one-half the length of the third side.

Theorem 4.2.14. The diagonals of a rhombus are perpendicular. (A *rhombus* is a quadrilateral in which all four sides are congruent.)

Theorem 4.2.15. If the diagonals of a quadrilateral bisect each other and are perpendicular, then the quadrilateral is a rhombus.

Theorem 4.2.16. The median to the hypotenuse of a right triangle is one-half the length of the hypotenuse.

Theorem 4.2.17.   If, in a right triangle, one of the angles measures 30°, then the side opposite this angle is one-half the length of the hypotenuse.

Theorem 4.2.18.   If one leg of a right triangle is one-half the length of the hypotenuse, then the angle opposite that leg has a measure of 30°.

Theorem 4.2.19.   The sum of the measures of the interior angles of a convex *n*-gon is $(n - 2) (180°)$.

Theorem 4.2.20.   The sum of the exterior angles (one at each vertex) of a convex *n*-gon is 360°.

4.3 Congruence and Area

Theorem 4.3.1.   Proposition 35, Book I.   Parallelograms that share a common base and have sides opposite this base contained in the same (parallel) line are equal in area.

Theorem 4.3.2.   The area of a parallelogram is the product of the measure (length) of its base and the measure of its height.

Theorem 4.3.3.   The area of a right triangle is one-half the product of its legs.

Theorem 4.3.4.   The area of a triangle is one-half the product of any base and the corresponding height.

Theorem 4.3.5.   The area of a trapezoid is the product of its height and the arithmetic mean of its bases.

Theorem 4.3.6.   The area of a rhombus is one-half the product of the lengths of the diagonals.

4.4 Similarity

Similar Polygons: Two polygons $P_1 P_2 P_3 \ldots P_n$ and $P_1' P_2' P_3' \ldots P_n'$ are similar if and only if

(i) $m \angle P_i = m \angle P_i'$, for all $i = 1$ to n, and

(ii)
$$\frac{P_1 P_2}{P_1' P_2'} = \frac{P_2 P_3}{P_2' P_3'} = \ldots \ldots = \frac{P_n P_1}{P_n' P_1'}$$

Theorem 4.4.1.   Similarity of polygons is an equivalence relation.

Theorem 4.4.2.   The Basic Proportionality Theorem.   If a line parallel to one side of a triangle intersects the other two sides in two different points, then it divides these sides into segments that are proportional.

Corollary 4.4.3. If a line parallel to one side of a triangle intersects the other two sides in different points, then it cuts off segments that are proportional to the sides of the triangle.

Theorem 4.4.4. If a line *l* intersects two sides of a triangle in different points so that it cuts off segments that are proportional to the sides, then the line is parallel to the third side.

Theorem 4.4.5. The AAA Similarity Theorem. If the three interior angles of one triangle are congruent to the three interior angles of a second triangle, then the triangles are similar.

Theorem 4.4.6. SAS Similarity Theorem. If an angle of one triangle is congruent to an angle from another triangle, and if the sides that surround this angle are proportional, then the triangles are similar.

Theorem 4.4.7. SSS Similarity Theorem. If the three sides of one triangle are proportional to the three sides of a second triangle, then the triangles are similar.

Theorem 4.4.8. The Pythagorean Theorem. If a and b are the measures of the legs of a right triangle, and if c is the measure of the hypotenuse, then $a^2 + b^2 = c^2$.

4.5 Some Euclidean Results Concerning Circles

Circle: A circle is a set of points each of which is equidistant from a given point. The given point is called the *center* of the circle, and the common distance is called the *radius* of the circle.

Chord: A chord of a circle is a line segment joining two of the points of the circle.

Diameter: A diameter of a circle is a chord that contains the center of the circle.

Secant: A secant to a circle is a line that contains exactly two points of the circle.

Tangent: A tangent to a circle is a line that contains exactly one point of the circle.

Theorem 4.5.1. In Euclidean geometry three distinct, noncollinear points determine a unique circle.

Theorem 4.5.2. If $\overline{AB}$ is a diameter of a circle, and if $\overline{CD}$ is another chord of the same circle that is not a diameter, then $AB > CD$.

Theorem 4.5.3. If a diameter of a circle is perpendicular to a chord of the circle, then the diameter bisects the chord.

Theorem 4.5.4.   If a diameter of a circle bisects a chord of the circle (which is not a diameter), then the diameter is perpendicular to the chord.

Theorem 4.5.5.   The perpendicular bisector of a chord of a circle contains a diameter of the circle.

Theorem 4.5.6.   If a line is tangent to a circle, then it is perpendicular to the radius drawn to the point of tangency.

4.6 More Euclidean Results Concerning Circles

Central Angle: Any angle whose vertex is the center of a circle is called a central angle for the circle.

Semicircle: If line $\overleftrightarrow{AB}$ contains a diameter of $C(O,OA)$, and if H represents either half-plane defined by $\overleftrightarrow{AB}$ (see SMSG Postulate 9), then the union of points A, B, and all points of $C(O,OA)$ that lie in H is called a semicircle of $C(O,OA)$ (see Figure 4.6.2a).

Minor Arc: If A and B are points of $C(O,OA)$ that are not endpoints of the same diameter, then the union of A, B, and all points of $C(O,OA)$ that are in the interior of central angle $\angle AOB$ is called a minor arc of C (O,OA) (see Figure 4.6.2b).

Major Arc: If A and B are points of $C(O,OA)$ that are not endpoints of the same diameter, then the union of A, B, and all points of $C(O,OA)$ that are in the exterior of angle $\angle AOB$ is called a major arc of $C(O,OA)$ (see Figure 4.6.2c).

Measure of Arc AB: The degree measure of an arc $\overset{\frown}{AB}$ of a $C(O,OA)$ is (i) The degree measure of the angle $\angle AOB$ if $\overset{\frown}{AB}$ is a minor arc, (ii) 180° if $\overset{\frown}{AB}$ is a semicircle, and (iii) 360° minus the measure of $\angle AOB$ if $\overset{\frown}{AB}$ is a major arc.

Inscribed Angle: An angle $\angle APB$ is said to be inscribed in an arc $\overset{\frown}{AQB}$ if and only if (i) the vertex P of the angle is a point of $\overset{\frown}{AQB}$, (ii) one ray of the angle contains point A, and (iii) the other ray contains point B (see Figure 4.6.5).

Intercepted Arc: An angle $\angle APB$ is said to intercept an arc $\overset{\frown}{ARB}$ if and only if (i) both A and B are points of the angle, (ii) each ray of the angle contains at least one endpoint of arc $\overset{\frown}{ARB}$, and (iii) excepting A and B, each point on arc $\overset{\frown}{ARB}$ lies in the interior of $\angle APB$ (see Figure 4.6.5).

Power of a Point With Respect to a Circle: The power of a point X with respect to a circle $C(O,OA)$ is the product of the signed distances from point X to any two points of $C(O,OA)$ with which it is collinear. The power of point X will be denoted by $P(X)$.

Theorem 4.6.1.   If two chords of a circle are congruent, then their corresponding minor arcs have the same measure.

Theorem 4.6.2.   If two minor arcs are congruent, then so are the corresponding chords.

Theorem 4.6.3.   *Arc Addition Theorem.*   If $\overset{\frown}{APB}$ and $\overset{\frown}{BQC}$ are arcs of the same circle sharing only the endpoint B, then $m\overset{\frown}{APB} + m\overset{\frown}{BQC} = m\overset{\frown}{ABC}$.

Theorem 4.6.4.   *The Inscribed Angle Theorem.*   The measure of an angle inscribed in an arc is one-half the measure of its intercepted arc.

Corollary 4.6.5.   An angle inscribed in a semicircle is a right angle.

Corollary 4.6.6.   Angles inscribed in the same or congruent arcs are congruent.

Theorem 4.6.7.   *The Two-Chord Angle Theorem.*   If two chords intersect to form an angle within a circle, the measure of the angle is the average of the measures of the arcs intercepted by the angle and its vertical angle.

Theorem 4.6.8.   *The Two-Secant Angle Theorem.*   If two secants intersect at a point in the exterior of a circle, the measure of the angle at the point of intersection is one-half the positive difference of the two intercepted arcs.

Theorem 4.6.9.   *The Tangent-Chord Angle Theorem.*   If line $\overleftrightarrow{AB}$ is tangent to $C(O,OA)$ at point A, and if $\overline{AC}$ is a chord such that the measure of $APC = x°$ (see Figure 4.6.14a), then $m\angle BAC = \frac{1}{2}x°$.

Theorem 4.6.10.   *The Tangent-Secant Angle Theorem.*   If line $\overleftrightarrow{AB}$ is tangent to $C(O,OB)$ at point B, and if $\overrightarrow{AD}$ is a secant line to $C(O,OB)$ (see Figure 4.6.16), then $m\angle BAD$ is one-half the positive difference of the measure of the two intercepted arcs.

Theorem 4.6.11.   *The Two-Tangent Angle Theorem.*   The measure of an angle formed by two tangents drawn to a circle is one-half the positive difference of the measures of the intercepted arcs.

Corollary 4.6.12.   Tangent segments drawn to a circle from the same exterior point are congruent. (*Note:* A tangent segment is a line segment connecting an exterior point to a point of tangency.)

Theorem 4.6.13.   *The Chord Segment Product Theorem.*   If two chords intersect within a circle, the product of the lengths of the segments of one chord is equal to the product of the lengths of the segments of the other chord.

Theorem 4.6.14.   *The Secant Segment Product Theorem.*   If two secant segments are drawn to a circle from the same exterior point, then the product of the length of the secant segment and the length of its external portion is the same for both secants.

Theorem 4.6.15.   The Tangent-Secant Segment Theorem. If a tangent segment and a secant segment are drawn to the same circle from the same exterior point, the product of the length of the secant and the length of its external segment is equal to the square of the length of the tangent.

4.7 Some Euclidean Results Concerning Triangles

Distance from a Point to a Line: The distance from a point P to a line l is the distance from P to the foot of the perpendicular from P to l.

Altitude: An altitude of a triangle is a perpendicular line segment from a vertex of the triangle to the opposite side (extended if necessary).

Theorem 4.7.1.   The three perpendicular bisectors of the sides of a triangle are concurrent (at a point called the *circumcenter*).

Theorem 4.7.2.   A point is on the bisector of an angle if and only if it is equidistant from the sides of the angle.

Theorem 4.7.3.   The three bisectors of the interior angles of a triangle are concurrent (at a point called the *incenter*).

Theorem 4.7.4.   The lines containing the three altitudes of a triangle are concurrent.

Theorem 4.7.5.   Each angle bisector of a triangle, if sufficiently extended, is concurrent with the bisectors of the exterior angles at the remaining two vertices. The point of concurrence is called the *excenter* of the triangle.

Corollary 4.7.6.   Each excenter of a triangle serves as the center of a circle that is externally tangent to a side of the triangle and the extensions of the other two sides.

Theorem 4.7.7.   The distance from a vertex to the orthocenter of a triangle is twice the distance from the circumcenter to the midpoint of the side opposite that vertex.

Theorem 4.7.8.   The orthocenter, the circumcenter, and the centroid of a given circle are collinear (on a line called the *Euler line* for the triangle).

Theorem 4.7.9.   The Theorem of Menelaus.   If $\triangle ABC$ is a triangle in which X, Y, and Z are collinear Menelaus points on sides $\overline{AC}$, $\overline{BC}$, and $\overline{AB}$, respectively, then

$$\frac{AZ}{ZB} \times \frac{BY}{YC} \times \frac{CX}{XA} = -1$$

using directed measurements for the lengths of the segments that comprise the ratios.

Theorem 4.7.10. *Ceva's Theorem.* If concurrent Cevian lines $\overline{AX}$, $\overline{BY}$, and $\overline{CZ}$ are drawn from the vertices of $\triangle ABC$ (see Figure 4.7.11), then

$$\frac{AY}{YC} \times \frac{CX}{XB} \times \frac{BZ}{ZA} = 1$$

4.8 The Nine-Point Circle

Theorem 4.8.1. *The Nine-Point Circle.* The nine points consisting of (i) the midpoints of the sides of a triangle (in this illustration $\triangle ABC$), (ii) the feet of the three altitudes of this triangle, and (iii) the midpoints of the segments formed by joining the orthocenter to the vertices of the triangle, are concyclic. The center of this nine-point circle is the midpoint of the segment joining the orthocenter and the circumcenter of $\triangle ABC$, and the length of the radius of this circle is one-half the length of the radius of the circle that circumscribes $\triangle ABC$.

4.9 Euclidean Constructions

Theorem 4.9.1. *The Compass Equivalence Theorem.* A circle $C(B,r)$ can be congruently copied (using only a straightedge and a collapsing compass) so that a given point A serves as the center of the copy.

Construction 1. To construct the perpendicular bisector of a line segment.

Construction 2. To construct a perpendicular l_p from a given point P to a line l not containing P.

Construction 3. To construct a line l_p perpendicular to a given line l at a given point P on l.

Construction 4. To construct the angle bisector of an angle $\angle ABC$.

Construction 5. To make a copy of $\angle ABC$ on ray PQ with P as the vertex of the copy.

Construction 6. To partition a line segment $\overline{AB}$ into three (or more) congruent pieces.

SIDE TRIPS

Analytic
and Transformational
Geometry

5.1 INTRODUCTION

In previous chapters we briefly studied several different branches of geometry, among them the affine geometry of Euclid, the elliptic geometry of Riemann, and the hyperbolic geometry of Bolyai and Lobachevsky. In this chapter we will take a short side trip. While remaining within the context of Euclidean geometry, we will explore several new methods of solving geometric problems.

In Section 5.2 we will investigate analytic geometry and make use of the real numbers and the power of algebra to assist us in proving geometric theorems. Section 5.3 provides us with an overview of transformational geometry which, while formalizing Euclid's concept of superposition, makes use of the function concept and provides us with another powerful method of geometric theorem proving. In Section 5.4, we combine the methods of the two previous sections and investigate analytical tranformations and their properties. Finally, in Section 5.5, we will briefly investigate the transformation of inversion, an unusual transformation of the plane which will have important applications in Chapter 6.

These methods are, as we will see, not new branches of geometry but simply alternative techniques for solving problems. It is our hope that after reading and studying this chapter, you will have at your disposal a variety of different methods for attacking the proofs of geometric theorems. The reader should be aware that we do not propose that these methods are easier or even better than the synthetic methods of previous chapters. At times an analytic proof may appear to be easier than a corresponding synthetic or transformational proof, while at other times an analytic proof may be virtually impossible and a synthetic or transformational proof quite easy. It is left to you, as a geometric problem solver, to decide which method suits your needs best.

5.2 ANALYTIC GEOMETRY

Historical Perspectives

In this section we will investigate the properties of *analytic geometry*[1] which, as we will see, is not a new branch of geometry but a method of solving geometric problems. The essence of analytic geometry is the use of the one-to-one correspondence between points on a line and the real numbers (SMSG Postulate 3) to establish a coordinate system for points in the plane, thereby allowing the use of algebra and analysis to prove theorems in geometry.

Mathematical historians have differing opinions as to who should be credited with the "invention" of analytic geometry. The ancient Greeks, as we can see in Euclid's *Elements*, made significant use of geometric algebra. The Romans and the Egyptians, as well as the Greeks, used a form of coordinates in map making. Apollonius (262–190 B.C.) employed the geometric equivalent of analytic equations to develop his geometry of conic sections, while Nicole Oresme (1323–1382) anticipated the laws of graphing a dependent variable against an independent variable.

However, the majority of historians agree that the most significant contributions to the development of analytic geometry were made during the seventeenth century by French mathematicians René Descartes (1596–1650) and Pierre Fermat (1601–1665). In "La Geometrie," the third appendix of a philosophical treatise on universal science, Descartes outlined the arithmetization of geometry. In spite of the fact that "La Geometrie" does not contain an explicit set of coordinate axes, it did provide the fundamental basis for the method that eventually evolved into modern analytic geometry. For that reason, we commonly refer to the coordinatization of a plane as a *Cartesian coordinate system*.

Simultaneously, Fermat investigated other aspects of analytic geometry, developed the equations of a general straight line and of a circle, and studied the properties of hyperbolas, ellipses, and parabolas. As opposed to Descartes, who began with a geometric concept and developed an algebraic

---

[1] Analytic geometry is also known as *coordinate geometry*.

equation that represented it, Fermat began with an algebraic representation and described its geometric counterpart.[2]

In the remainder of this section we will discuss coordinatization of the Euclidean plane and develop procedures by which familiar algebraic principles can be used to prove geometric theorems.

Coordinatization of the Plane

Before we begin our coordinatization of the Euclidean plane we need to consider the following theorem:

THEOREM 5.2.1.   Two intersecting lines determine a unique plane.

Proof. Let l and m be two distinct lines that intersect. By a previous result, the intersection of the two distinct lines is a point P. Now let Q be any other point on l and let R be any other point on m. SMSG Postulate 1 guarantees that P, Q, and R are noncollinear points, and thus by SMSG Postulates 6 and 7, l and m lie in exactly one plane.

As a result of Theorem 5.2.1, we may choose any two intersecting lines that lie in the Euclidean plane to uniquely determine the plane. Therefore we will choose line x, and for convenience line y, which is perpendicular[3] to x at some point O. These lines will be called the *x-axis* and *y-axis*, respectively. The ruler postulate (SMSG Postulate 3) allows us to coordinatize each of these lines with real numbers, and the ruler placement postulate (SMSG Postulate 4) allows us to place the coordinate system such that both lines have the real number zero at point O. Now it should be clear that every point on the lines x and y can be represented by an ordered pair of real numbers: $(r_x,0)$ for each point on x and $(0,r_y)$ for each point on y. In particular, point O, has real coordinates $(0,0)$ and is called the *origin*. Since each coordinate is a real number, we will, for the sake of brevity, denote points on the x axis and y axis by simply $(x,0)$ and $(0,y)$, respectively (See Figure 5.2.1). Now given any *other* point P in the plane, we can construct a unique perpendicular from P that intersects the x axis at $(r_x,0)$ and a unique perpendicular intersecting the y axis at $(0, r_y)$, thus creating a unique rectangle and allowing us to represent, in a unique way, point P with the ordered pair of real numbers (r_x,r_y) or simply (x,y). (See Figure 5.2.1.) With the use of this procedure, every point in the plane has a unique ordered pair representation cleverly referred to as its *rectangular coordinates*.

---

[2] A more complete history of the development of analytic geometry can be found in Howard Eves, *An Introduction to the History of Mathematics*, 3rd ed. (New York: Holt, Rinehart and Winston, 1969), p. 258.

[3] It is not essential that lines x and y be perpendicular, but as we will see, it will simplify some future computations.

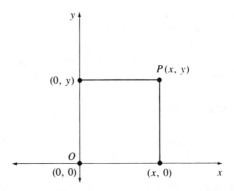

Figure 5.2.1

Distance in the Plane

Using the coordinatization scheme depicted in Figure 5.2.1, the ruler postulates and the Pythagorean theorem, we can now determine the distance between any two points in the plane.

Example 5.2.1

As a result of the ruler postulates, the distance between any two points that lie on lines parallel to the x or y axis is determined by calculating the absolute value of the difference in their x or y coordinates, respectively. For example, if $P = (x_1, y_1)$ and $Q = (x_2, y_1)$, then $PQ = |x_1 - x_2|$. Similarly, if $P = (x_1, y_1)$ and $R = (x_1, y_2)$, then $PR = |y_1 - y_2|$ (Figure 5.2.2).

The distance between any two points that lie on a line that is oblique to the x or y axis can now be determined with the help of the following theorem:

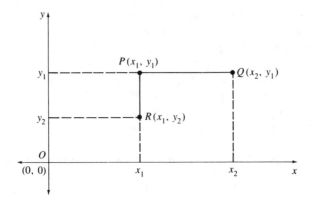

Figure 5.2.2

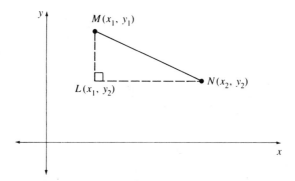

Figure 5.2.3

THEOREM 5.2.2.   If $M = (x_1, y_1)$ and $N = (x_2, y_2)$ are any two points in the plane, then $MN = \sqrt{(x_1 - x_2)^2 + (y_1 - y_2)^2}$ (Figure 5.2.3).

Proof. It is easy to see that $\triangle LMN$ is a right triangle and that by the Pythagorean theorem, $(MN)^2 = (ML)^2 + (LN)^2$. Now by the previous example, $LN = |x_1 - x_2|$ and $ML = |y_1 - y_2|$ hence by substitution, $(MN)^2 = (|x_1 - x_2|)^2 + (|y_1 - y_2|)^2$. Finally, solving for MN, we find that $MN = \sqrt{(x_1 - x_2)^2 + (y_1 - y_2)^2}$.

Example 5.2.2

If $M = (3, -1)$ and $N = (5, 2)$, then $MN = \sqrt{(3 - 5)^2 + (-1 - 2)^2} = \sqrt{13}$.

The next theorem allows us to find the coordinates of a point of division, P, on a line segment $\overline{AB}$ such that the ratio AP/PB is equal to some arbitrarily chosen positive real number.

THEOREM 5.2.3.   Given $A = (x_1, y_1)$, $B = (x_2, y_2)$, and r which is any positive real number, if P is on $\overline{AB}$ such that A-P-B and $AP/PB = r$, then P has coordinates

$$\left(\frac{x_1 + rx_2}{1 + r}, \frac{y_1 + ry_2}{1 + r} \right)$$

(See Figure 5.2.4.)

Proof. Let $P = (p, q)$. This proof requires that we consider several cases. For the sake of brevity, we shall assume that $x_1 < p < x_2$ and $y_1 < q < y_2$. Other cases are left as exercises. Let M and N be the projections of P onto $\overline{AC}$ and $\overline{BC}$ as indicated. Using our assumption, $AM = p - x_1$, $PM = q - y_1$, $PN = x_2 - p$, and $BN = y_2 - q$. Now since $\triangle AMP \sim \triangle PNB$ and $AP/PB = r$, we have $AM/PN = (p - x_1)/(x_2 - p) = r$. Solving for p, we find

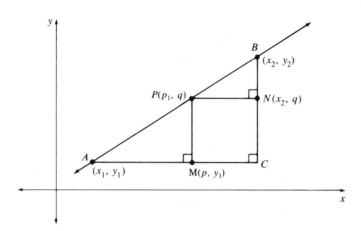

Figure 5.2.4

that $p = (x_1 + rx_2)/(1 + r)$. Similarly, $PM/BN = (q - y_1)/(y_2 - q) = r$, and $q = (y_1 + ry_2)/(1 + r)$.

Example 5.2.3

In the previous theorem, if P is the *midpoint* of $\overline{AB}$, then $AP/PB = 1$, and thus the coordinates of P are:

$$\left(\frac{x_1 + x_2}{2}, \frac{y_1 + y_2}{2}\right).$$

Analytic Equations of Straight Lines and Circles

Having identified each point in the plane with an ordered pair of real numbers, it seems reasonable to ask, Can all the points on any straight line be identified in some way? Before we can answer this question, we need two additional definitions. The *inclination* of a line l, which is not parallel to the x axis, is the smallest positive angle θ measured counterclockwise from the x axis to l (Figure 5.2.5). If l is parallel to the x axis, its inclination is defined to be zero.

The *slope* m of any line will be defined to be the tangent of the angle of inclination. If $A = (x_1, y_1)$ and $B = (x_2, y_2)$ (see Figure 5.2.5), and if m is the slope of line l, then $m = \tan \theta = (y_2 - y_1)/(x_2 - x_1)$.

THEOREM 5.2.4.   If l is any nonvertical[4] line having slope m, then the coordinates of every point $P = (x, y)$ on l must satisfy an equation of the form $y = mx + k$, where k is a real number.

---

[4] We use here the common notion that any line that is perpendicular to the x axis is called a vertical line.

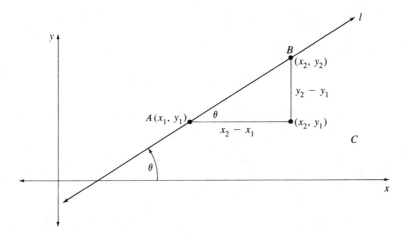

Figure 5.2.5

Proof. Let $l = AB$ as in Figure 5.2.5. Recalling similar triangles, it is easy to see that the slope of any line can be found by calculating the ratio of "the change in the y coordinates" to "the change in the x coordinates, in the same order" for any two points on l. Now if we let $P = (x, y)$ be any other point on l, we find that $(y - y_1)/(x - x_1) = (y_2 - y_1)/(x_2 - x_1)$. Therefore $y - y_1 = [(y_2 - y_1)/(x_2 - x_1)](x - x_1)$ or, since $m = (y_2 - y_1)/(x_2 - x_1)$, $y - y_1 = m(x - x_1)$. Solving for y, we find $y = mx + (y_1 - mx_1)$, and letting $k = y_1 - mx_1$, we get $y = mx + k$ as required.

Example 5.2.4

(a) If $A = (2,4)$ and $B = (1, -1)$ then $m = [4 - (-1)]/(2 - 1) = 5$ and the equation[5] of $\overrightarrow{AB}$ is $y - 4 = 5(x - 2)$ or $y = 5x - 6$.
(b) If $R = (3,1)$ and $S = (6,1)$, the angle of inclination is zero and $\overrightarrow{RS}$ is called a horizontal line. $m = \tan 0° = 0$, and the equation of $\overrightarrow{RS}$ is $y - 1 = 0(x - 3)$ or $y = 1$.

Vertical lines present a special problem since the angle of inclination is 90° and $\tan 90°$ is undefined. But if we observe that all points on a vertical line have the same x coordinate, then all vertical lines have equations of the form $x = k$, where k is the x coordinate of any point on the line.

Example 5.2.5

If $M = (3, -1)$ and $N = (3, 7)$, then $\overrightarrow{MN}$ is a vertical line and the equation of $\overrightarrow{MN}$ is $x = 3$.

---

[5] We could have used $y - 1 = 5[x - (-1)]$, which also gives us $y = 5x - 6$.

Two classifications of lines, parallel lines, and perpendicular lines, will prove to be of significant importance. The next two theorems provide the analytic character of parallel and perpendicular lines.

THEOREM 5.2.5.   Parallel lines are lines having the same slope. The proof is left as an exercise.

THEOREM 5.2.6.   Let l_1 and l_2 be two nonvertical lines with slopes m_1 and m_2, respectively. If l_1 and l_2 are perpendicular, then[6] $m_2 = -1/m_1$.

Proof. Let l_1 have inclination α and l_2 have inclination β. Since l_2 is perpendicular to l_1, $\beta = \alpha + 90°$. Now since $m_1 = \tan \alpha$ and $m_2 = \tan \beta$, we have $m_2 = \tan \beta = \tan(\alpha + 90°) = -\cot \alpha = -1/m_1$.

The previous theorem is a special case of the following theorem which allows us to calculate the angle between two lines.

THEOREM 5.2.7.   The angle α, measured counterclockwise from line l_2, whose slope is m_2, to line l_1, whose slope is m_1, is

$$\alpha = \arctan\left[\frac{m_1 - m_2}{1 + m_1 m_2}\right].$$

Proof. Let the inclination of l_1 be θ_1 and the inclination of l_2 be θ_2 (Figure 5.2.6). Therefore $m_1 = \tan \theta_1$ and $m_2 = \tan \theta_2$. Since θ_1 is an exterior angle, $\theta_1 = \theta_2 + \alpha$ and we find that $\alpha = \theta_1 - \theta_2$. Now $\tan \alpha = \tan(\theta_1 - \theta_2) = (\tan \theta_1 - \tan \theta_2)/(1 + \tan \theta_1 \tan \theta_2)$. Substituting appropriately, we find $\tan \alpha = (m_1 - m_2)/(1 + m_1 m_2)$ and $\alpha = \arctan[(m_1 - m_2)/(1 + m_1 m_2)]$.

[6] At times it will prove beneficial to multiply each side of this equation by m_1 and state that two lines are perpendicular if and only if $m_1 m_2 = -1$.

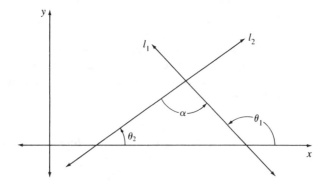

Figure 5.2.6

Example 5.2.6

Let l_1 be represented by $y = 5x + 1$ and l_2 by $y = \frac{2}{3}x + 1$; then $m_1 = 5$ and $m_2 = \frac{2}{3}$. Therefore $\tan \alpha = (5 - \frac{2}{3})/[1 + \frac{2}{3}(5)] = 1$. Hence $\alpha = \arctan 1 = 45°$.

Finally, the following theorem allows us to analytically describe any circle in the Euclidean plane:

THEOREM 5.2.8.   Any point $P = (x,y)$ on a circle whose center is $O = (h,k)$ and whose radius has length r must satisfy an equation of the form $(x - h)^2 + (y - k)^2 = r^2$.

Proof. Let $O = (h,k)$ be any point in the plane and r be any positive real number. First, we recall that any circle is defined as the set of all points $P = (x,y)$ whose distance from O is r. Now by Theorem 5.2.2, $OP = r = \sqrt{(x - h)^2 + (y - k)^2}$, and squaring both sides of the equation, we have $r^2 = (x - h)^2 + (y - k)^2$.

The following section will provide the opportunity to use the results of the previous theorems as an alternative method of proving geometric theorems.

Applications of Analytic Geometry

As we mentioned earlier, analytic geometry is not a different branch of geometry but is a method that can be used to prove theorems. The remainder of this section is devoted to demonstrating the proofs of several common theorems using representative analytical techniques.

THEOREM 5.2.9.   The opposite sides of a parallelogram are congruent.

Proof. We must first coordinatize a general parallelogram $\square ABCD$ in the plane. In order to simplify computation,[7] let $A = (a,0)$, $B = (0,0)$, and $C = (b,c)$, with the restrictions that $a > 0$, $c > 0$, and $b \geq 0$. Now if we let $D = (x,y)$, with $x > b$ and $y \geq 0$, we must find x and y such that $\square ABCD$ is a parallelogram (Figure 5.2.7). By definition of a parallelogram,[8] $\overline{AB} \parallel \overline{CD}$ and $\overline{CB} \parallel \overline{AD}$. Since $\overleftrightarrow{AB}$ is a horizontal line and $\overline{AB} \parallel \overline{CD}$, we can easily see that $y = c$. Now $m_{\overleftrightarrow{BC}} = (c - 0)/(b - 0)$ and $m_{AD} = (y - 0)/(x - a)$, and since $\overleftrightarrow{BC} \parallel \overleftrightarrow{AD}$, we know that $m_{\overline{BC}} = m_{\overline{AD}}$, hence $y/(x - a) = c/b$. Substituting for y and

---

[7] Since, as it will be seen in subsequent sections of this chapter, any arbitrary figure could be repositioned, the location of the figure in this way does not affect the generality of the argument.

[8] Experience will show us that other methods exist for coordinatizing figures, but for the purposes of general applicability in theorem proving, it will always prove best to use the definition.

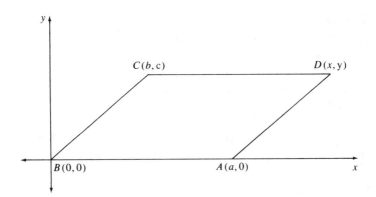

Figure 5.2.7

solving for x, we see that $x = b + a$, and therefore $D = (b + a,c)$. With the parallelogram coordinatized, we can now use the distance formula to show that $\overline{AB} \cong \overline{CD}$ and $\overline{BC} \cong \overline{AD}$. $AB = a$ and $CD = (b + a) - b = a$; therefore $\overline{AB} \cong \overline{CD}$. Also, $BC \cong \sqrt{(b - 0)^2 + (c - 0)^2} = \sqrt{b^2 + c^2}$ and $AD = \sqrt{(b + a - a)^2 + (c - 0)^2} = \sqrt{b^2 + c^2}$, and $\overline{BC} \cong \overline{AD}$.

THEOREM 5.2.10.   The diagonals of a parallelogram bisect each other.

Proof. As in the proof of Theorem 5.2.9, we first coordinatize a general parallelogram $\square PQRS$. However, in this case we will double each of the coordinates, which, as we will see, simplifies the algebra but does not affect its generality. Therefore let $P = (2a,0)$, $Q = (0,0)$, $R = (2b,2c)$, and $S = (2(b + a), 2c)$ (Figure 5.2.8). Now we will show that both diagonals share the same midpoint, hence bisect each other. Using the result of Example 5.2.3, the midpoint of $\overline{PR}$ is $[(2b + 2a)/2, (2c + 0)/2] = (b + a, c)$ and the midpoint of $\overline{QS}$ is $[2(b + a)/2, (2c + 0)/2] = (b + a, c)$.

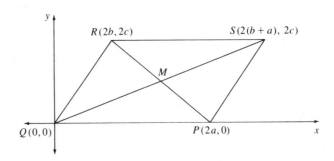

Figure 5.2.8

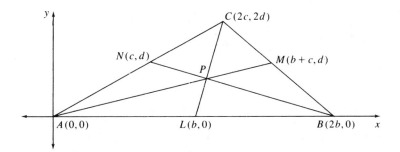

Figure 5.2.9

THEOREM 5.2.11.   The medians of a triangle are concurrent.

Proof. Coordinatize $\triangle ABC$ as follows. $A = (0,0)$, $B = (2b,0)$, and $C = (2c, 2d)$, with the restrictions that $b > 0$, $c \geq 0$, and $d > 0$ (Figure 5.2.9). If we let L, M, and N be the midpoints of sides $\overline{AB}$, $\overline{BC}$, and $\overline{CA}$, respectively, then $\overline{CL}$, $\overline{AM}$, and $\overline{BN}$ are the medians of $\triangle ABC$. Since $M = (b + c, d)$ and $A = (0,0)$, the equation of $\overleftrightarrow{AM}$ is $y = d/(b + c)x$. Similarly, the equation of $\overleftrightarrow{CL}$ is $y = 2d(x - b)/(2c - b)$ and the equation of $\overleftrightarrow{BN}$ is $y = d(x - 2b)/(c - 2b)$. Solving the equations of $\overline{AM}$ and $\overline{CL}$ simultaneously, we find that they share a common point $P = [2(c + b)/3, 2d/3]$. To complete our proof, it suffices to show that P is on $\overline{BN}$, that is, that the coordinates of P satisfy the equation of $\overleftrightarrow{BN}$. Substituting $2(c + b)/3$ for x in the equation $y = d(x - 2b)/(c - 2b)$, we get

$$y = \left(\frac{d}{c - 2b}\right)\left(\frac{2(c + b) - 2b}{3}\right) = \left(\frac{d}{c - 2b}\right)\left(\frac{2c - 4b}{3}\right) = \frac{2d}{3}$$

which is the y coordinate for P; hence P is on $\overline{BN}$, and all three medians are concurrent.

Note that this method provides an additional bonus, since it also shows that P is two-thirds of the distance from each vertex to the midpoint of the opposite side. The reader should verify this fact using Theorem 5.2.3.

THEOREM 5.2.12.   The base angles of an isosceles trapezoid are congruent.

Proof. Let $A = (a,0)$, $B = (0,0)$, and $C = (b,c)$, where $a > 0$, $b > 0$, and $c > 0$ (Figure 5.2.10). Now we let $D = (x,y)$ and find x and y such that $\square ABCD$ is an isosceles trapezoid. Since $\overline{AB}$ must be parallel to $\overline{CD}$, we see that $y = c$. Now since $\square ABCD$ is an isosceles trapezoid, $\overline{AD}$ is not parallel to $\overline{BC}$ and $\overline{AD} \cong \overline{BC}$. Using the distance formula, $BC = \sqrt{b^2 + c^2}$ and $AD = \sqrt{(x - a)^2 + c^2}$. Since $BC = AD$, we find that $\sqrt{(x - a)^2 + c^2} = \sqrt{b^2 + c^2}$,

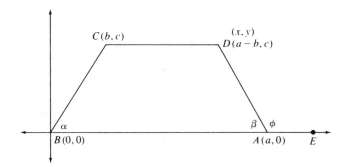

Figure 5.2.10

which implies that $(x - a)^2 = b^2$. Solving for x, we find that $x = a + b$ or $x = a - b$. We must let $x = a - b$, otherwise, $\overline{AD}$ is parallel to $\overline{BC}$ and $\square ABCD$ is a parallelogram.[9] We must now show that $\angle ABC \cong \angle BAD$. Let $\angle CBA = \alpha$, $\angle BAD = \beta$, and $\angle DAE = \phi$. Since ϕ is the angle of inclination of $\overrightarrow{AD}$, $m_{\overline{AD}} = c/-b = \tan \phi$, and since α is the angle of inclination of $\overrightarrow{BC}$, $m_{\overline{BC}} = c/b = \tan \alpha$. Now, $\beta = 180° - \phi$ and $\tan \beta = \tan(180° - \phi) = [0 - (-c/b)]/(1 + 0) = c/b$. Therefore $\tan \alpha = \tan \beta$ and either $\alpha = \beta$ or $\alpha = \beta + 180°$. And $\alpha \neq \beta + 180°$; otherwise, the angle sum of trapezoid $\square ABCD$ is greater than 360°, therefore $\alpha = \beta$, and the proof is complete.

The following theorem represents an analytical view of Theorem 4.5.1 which was proven using synthetic methods in Chapter 4.

THEOREM 5.2.13.   Any three noncollinear points lie on a circle.

Proof. Let $A = (0,0)$, $B = (2a,2b)$, and $C = (2c,2d)$, where a and b are not both zero, c and d are not both zero, and $ad \neq bc$ (for noncollinearity) (Figure 5.2.11). The general analytic formula for a circle is $(x - h)^2 + (y - k)^2 = r^2$, where $O = (h,k)$ is the center of the circle and r is the length of its radius. Since the center of a circle is equidistant from all points on the circle, we can find the coordinates of the center by finding the intersection of the perpendicular bisectors of $\overline{AC}$ and $\overline{AB}$ (see Theorem 3.2.6). First, we will find the equation of each of the perpendicular bisectors. Since $m_{\overline{AC}} = d/c$ and the midpoint of $\overline{AC} = (c,d)$, the equation of the perpendicular bisector of $\overline{AC}$ is $y = (-c/d)(x - c) + d$. Similarly, since $m_{\overline{AB}} = b/a$ and the midpoint of $\overline{AB} = (a,b)$ we find that the equation of the perpendicular bisector of $\overline{AB}$ is $y = (-a/b)(x - a) + b$. Solving the two equations simultaneously, we find that

$$x = \frac{b(c^2 + d^2) - d(a^2 + b^2)}{bc - ad}$$

[9] See the coordinatization of the parallelogram in Theorem 5.2.9.

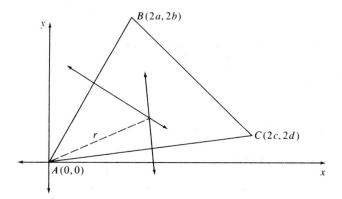

Figure 5.2.11

and solving for y, we get

$$y = \frac{c(a^2 + b^2) - a(c^2 + d^2)}{bc - ad}$$

Therefore the coordinates of the center (h,k) of our required circle are

$$\left(\frac{b(c^2 + d^2) - d(a^2 + b^2)}{bc - ad}, \frac{c(a^2 + b^2) - a(c^2 + d^2)}{bc - ad} \right)$$

Now we find r^2 by using the distance formula to calculate the distance between (h,k) and $(0,0)$. Substituting, we find

$$r^2 = \frac{[b(c^2 + d^2) - d(a^2 + b^2)]^2}{(bc - ad)^2} + \frac{[c(a^2 + b^2) - a(c^2 + d^2)]^2}{(bc - ad)^2}$$

and simplifying,

$$r^2 = \frac{(a^2 + b^2)(c^2 + d^2)[(a - c)^2 + (b - d)^2]}{(bc - ad)^2}$$

and our equation becomes

$$\{x - [b(c^2 + d^2) - d(a^2 + b^2)]\}^2 + \{y - [c(a^2 + b^2) - a(c^2 + d^2)]\}^2$$
$$= \frac{(a^2 + b^2)(c^2 + d^2)[(a - c)^2 + (b - d)^2]}{(bc - ad)^2}$$

Clearly, $(0,0)$ lies on the circle. It only remains to show that B and C lie on the circle, which is left as an exercise (see Exercise Set 5.2, Problem 13).[10]

[10] An alternate proof for this theorem, which makes use of the distance formula, can be found in Exercise Set 5.2, Problem 14.

EXERCISE SET 5.2

1. Find the distance between each of the following pairs of points:
   (a) $(5,1)$ and $(5, -3)$ **(b)** $(7,-2)$ and $(4, -2)$
   (c) $(4,0)$ and $(0,3)$ **(d)** $(-1,3)$ and $(2,4)$.

2. If $A = (2a,0)$, $B = (0,0)$, and $C = (a, a\sqrt{3})$, where $a \neq 0$, prove that $\triangle ABC$ is equilateral.

3. The proof of Theorem 5.2.3 refers to "other cases." What are the other cases? Justify the theorem in each of these cases.

4. If $P = (2,3)$ and $Q = (4,5)$,
   (a) Find the midpoint of $\overline{PQ}$.
   (b) Find point R such that Q is the midpoint of $\overline{PR}$.
   (c) Find the point S such that P is the midpoint of $\overline{SQ}$.

5. If $P = (1,1)$ and $Q = (4,5)$, find R such that P-R-Q and
   (a) $PR/RQ = \frac{2}{3}$ **(b)** $PR/RQ = \frac{3}{2}$ **(c)** $PR/RQ = 1$ **(d)** $PR/RQ = \sqrt{2}$

6. **(a)** If $A = (4,0)$, $B = (0,0)$, $C = (2,3)$, and $D = (6,3)$, show that $\square ABCD$ is a parallelogram.
   (b) If $M = (-4,0)$, $N = (-3,-2)$, $O = (3,1)$, and $P = (2,3)$, show that $\square MNOP$ is a rectangle.
   (c) If $P = (1,1)$, $Q = (3,5)$, $R = (-1,7)$, and $S = (-3,3)$, prove that $\square PQRS$ is a square.

7. Write the equations of each of the following straight lines:
   (a) A line passing through the points $P = (2,3)$ and $Q = (5, -1)$.
   (b) A line whose slope is $\frac{2}{3}$ and which passes through the point $(-1,1)$
   (c) A line that is the perpendicular bisector of $\overline{PQ}$ where
       (i) $P = (2,3)$ and $Q = (8,3)$
       (ii) $P = (2,3)$ and $Q = (2,9)$
       (iii) $P = (2,3)$ and $Q = (5, -1)$

8. Find the angles between each of the following lines:
   (a) $y = (\sqrt{3}/3)x + 1$ and $y = (\sqrt{3})x + 1$
   (b) $y = 2x - 3$ and $y = -\frac{1}{2}x + 2$
   (c) $y = 2x + 3$ and $y = 3x + 4$

9. Find the interior angles of $\triangle ABC$ where $A = (2,-3)$, $B = (4,2)$, and $C = (-5,-2)$.

10. Write the equation of the bisector of $\angle ABC$ where
    (a) $A = (3, 3\sqrt{3})$, $B = (0,0)$, and $C = (2,0)$
    (b) $A = (-1,1)$, $B = (0,0)$, and $C = (1,0)$
    (c) $A = (3, 3)$, $B = (1,1)$, and $C = (3,1)$

11. Complete the proof of Theorem 5.2.3 by listing and proving the remaining cases.

12. Write the equation of the circle containing each of the following sets of three points.
    (a) $A = (0,0)$, $B = (2,0)$, and $C = (0,2)$
    (b) $A = (1,1)$, $B = (3,3)$, and $C = (2,4)$
    (c) $A = (0,1)$, $B = (1,3)$, and $C = (2,5)$

13. Complete the proof of Theorem 5.2.13 by showing that points B and C are on the required circle.

14. Find an alternate proof of Theorem 5.2.13 by equating the distances OA, OB, and OC.

15. Using analytic techniques, prove that the opposite sides of a rectangle are congruent.

16. Using analytic techniques, prove that the opposite angles of a parallelogram are congruent.

17. Using analytic techniques, prove that the diagonals of a rectangle are congruent.

18. Using analytic techniques, prove that the diagonals of a rhombus are perpendicular bisectors of each other.

19. Using analytic techniques, prove that the diagonals of an isosceles trapezoid are congruent.

5.3 TRANSFORMATIONAL GEOMETRY

Introduction

Throughout our study of neutral geometry in Chapter 3 and Euclidean geometry in Chapter 4, we encountered several different definitions of congruence. Two line segments are said to be congruent if they have the same length. Two angles are said to be congruent if they have the same measure. And, two polygons are congruent if there is a one-to-one correspondence between their vertices such that their corresponding sides are congruent and their corresponding angles are congruent.

As important as these definitions are, none seem to have the same intuitive meaning as the Euclidean idea of congruence which is founded on the concept of *superposition*. By implicitly assuming that "geometric figures could be moved without changing their size or shape" and thus congruent figures could be made to coincide, Euclid justified congruence with the "common notion" that "things which coincide with one another are equal to one another."

Whereas Euclid tacitly assumed the ability to move figures in his development of geometry, in the following sections we will provide a mathematically precise method for doing so.

Mappings and Transformations

To begin our investigation of a mathematically precise method for moving geometric figures, we need to recall the following definitions from previous mathematics courses.

DEFINITION.   Let *A* and *B* be sets. Then a *mapping* or *function f from A to B* is a rule that assigns to each element *x* in *A* exactly one element *y* in *B* [denoted $y = f(x)$]. The set *A* is called the *domain of f*, *y* is called the *image of x under f*, and *x* is called the *preimage of y under f*. The set of all images of elements of *A* under *f* (which is some subset of *B*) is called the *range of f*.

DEFINITION.   A mapping f from A to B is *onto* B if for any y in B there is at least one x in A for which $f(x) = y$.

DEFINITION.   A mapping f from A to B is *one-to-one* if each element of the range has exactly one preimage. For example, if a and b are elements of A such that $f(a) = f(b)$, then $a = b$.

Example 5.3.1

(a) $f: \Re \to \Re$ defined as $f(x) = 2x + 3$ is one-to-one and onto.

(b) $g: \Re \times \Re \to \Re \times \Re$ defined as $g(x,y) = (x + 3, y - 2)$ is one-to-one and onto.

(c) $h: \Re \to \Re$ defined as $h(x) = x^2$ is not one-to-one, since -2 and 2 both have the same image under h.

(d) $t: Z \times Z \to Z \times Z$ defined as $t(p,q) = (2p,4q)$ is not onto $Z \times Z$, since $(1,2)$ is in $Z \times Z$ but has no preimage in $Z \times Z$.

DEFINITION.   If $f: A \to B$ and $g: B \to C$ are mappings, then the *composition* (or product) of f and g, $g \circ f$, is a mapping $h: A \to C$ such that $h(x) = g(f(x))$.

Example 5.3.2

(a) If $f: \Re \to \Re$ is defined as $f(x) = 2x + 3$ and $g: \Re \to \Re$ is defined as $g(x) = 4x - 1$, then $(g \circ f)(x) = g(f(x)) = g(2x + 3) = 4(2x + 3) - 1 = 8x + 11$.

(b) $(f \circ g)(x) = f(g(x)) = 2(4x - 1) + 3 = 8x + 1.$[11]

DEFINITION.   If $f: A \to B$ is a one-to-one onto mapping, then $f^{-1}: B \to A$ is called the *inverse* of f if and only if $(f^{-1} \circ f)(x) = x$ for all $x \in A$ and $(f \circ f^{-1})(y) = y$ for all $y \in B$.

Example 5.3.3

(a) If $f: \Re \to \Re$ is defined as $f(x) = 3x + 1$, then $f^{-1}(x) = (x - 1)/3$. (b) If $g: \Re \to \Re$ is defined as $g(x) = x^2$, then $g(x)$ is not one-to-one and thus has no inverse.

With the concept of mapping and with a plane being interpreted as a set of points, we are now prepared to define a transformation of the plane.

DEFINITION.   A *transformation of the plane t* is a one-to-one mapping of points of the plane onto points in the plane.

Example 5.3.4

(a) Let O be any point in the plane, and let the image of O be itself; that is, $t(O) = O$. Now if P is any other point in the plane, let $t(P) = P'$ such that P'

---

[11] The composition of two mappings is not necessarily commutative.

is the midpoint of $\overline{OP}$. Clearly, each point in the plane has a unique image, and each point in the plane is the unique image of some point; therefore t is a transformation of the plane. (b) Consider the coordinate plane. Let P be any point in the plane, and let $t(P) = P'$, where P' is a point 2 units to the right of P and on a line passing through P and parallel to the x axis. Again we find that t is a transformation of the plane.

THEOREM 5.3.1.   The composition of two transformations of the plane is a transformation of the plane. The proof is left as an exercise.

One essential concept in the discussion of transformations of the plane is the idea that while certain geometric properties of points in a plane are preserved under a given transformation, others are not.[12] If a property is preserved under a transformation, then that property is said to be *invariant* under that transformation.[13] For example, if any set of points are collinear and the set of the images of these points under some transformation are also collinear, then collinearity is said to be invariant under that transformation.

In the next section we will investigate a collection of particular transformations under which distance between points is invariant. These transformations will prove to be exactly what Euclid had in mind when he allowed us to "pick up and move geometric figures without changing their size or shape."

Isometries

DEFINITION.   A transformation of the plane for which distance between points is invariant is called an *isometry*. For example, if t is a transformation of the plane and P and Q are any points in the plane such that $t(P) = P'$ and $t(Q) = Q'$, and if $PQ = P'Q'$, then t is called an isometry.

Example 5.3.5

The transformation described in Example 5.3.4b is an isometry. The transformation described in Example 5.3.4a is not an isometry.

THEOREM 5.3.2.   The composition of two isometries is an isometry. The proof is left as an exercise.

It may be of value at this time to discover what other geometric properties, if any, remain invariant under isometries. The following theorems will provide a basis for our further investigation of the properties of isometries.

---

[12] These geometric properties include, among others, distance, betweeness, collinearity, angle measure, and so on.

[13] Felix Klein (1849–1925) conceived of a geometry as the study of those properties of a set of points that are invariant under the transformations in a certain transformation group.

THEOREM 5.3.3.   Collinearity is invariant under an isometry.

Proof. Let A, B, and C be any three distinct collinear points such that A-B-C, and let α be an isometry such that $\alpha(A) = A'$, $\alpha(B) = B'$, and $\alpha(C) = C'$. Since A-B-C, we know that $AB + BC = AC$. Since α is an isometry, we also know that $AB = A'B'$, $BC = B'C'$, and $AC = A'C'$, and substituting, we find that $A'B' + B'C' = A'C'$. Now if A', B', and C' were noncollinear, the triangle inequality would tell us that $A'B' + B'C' > A'C'$; therefore A', B', and C' must be collinear.

THEOREM 5.3.4.   Betweeness is invariant under an isometry. The proof is left as an exercise.

COROLLARY 5.3.5.   The image of a line segment (or ray, angle, or triangle) under an isometry is a line segment (or ray, angle, or triangle). The proof is left as an exercise.

COROLLARY 5.3.6.   The image of a line segment under an isometry is a congruent line segment. The proof is left as an exercise.

THEOREM 5.3.7.   The image of a triangle under an isometry is a congruent triangle.

Proof. Let $\triangle ABC$ be any triangle, and let α be any isometry such that $\alpha(A) = A'$, $\alpha(B) = B'$, and $\alpha(C) = C'$. We must show that $\triangle ABC$ is congruent to $\triangle A'B'C'$. Since α is an isometry, $\alpha(\overline{AB}) = \overline{A'B'}$, $\alpha(\overline{BC}) = \overline{B'C'}$, and $\alpha(\overline{AC}) = \overline{A'C'}$, hence $\overline{AB} \cong \overline{A'B'}$, $\overline{BC} \cong \overline{B'C'}$, and $\overline{AC} \cong \overline{A'C'}$. Therefore by the SSS theorem, $\triangle ABC \cong \triangle A'B'C'$.

COROLLARY 5.3.8.   Angle measure is invariant under an isometry.[14] The proof is left as an exercise.

THEOREM 5.3.9.   The image of a circle under an isometry is a congruent circle. The proof is left as an exercise.

As a result of the previous theorems, we may now replace our collection of congruence definitions by the following definition which in addition replaces Euclid's concept of superposition with a mathematically rigorous concept.

DEFINITION.   Two figures are said to be *congruent* if and only if one is the image of the other under some isometry.

---

[14] A transformation under which angle measure is invariant is said to be *conformal*.

Before we investigate further properties of isometries, we need to consider under what circumstances an isometry is unique. To answer this question, we must first consider the following results:

LEMMA 5.3.10.   An isometry that leaves each of three noncollinear points fixed leaves every point of the plane fixed.[15]

Proof: Let A, B, and C be three distinct noncollinear points, and let α be an isometry such that $\alpha(A) = A$, $\alpha(B) = B$, and $\alpha(C) = C$. By Theorem 5.3.5, $\alpha(\overleftrightarrow{AB}) = \overleftrightarrow{AB}$, $\alpha(\overleftrightarrow{BC}) = \overleftrightarrow{BC}$, and $\alpha(\overleftrightarrow{AC}) = \overleftrightarrow{AC}$ (Figure 5.3.1). Now if P is any point not on any of the three lines, and if Q is on $\overleftrightarrow{AB}$ ($Q \neq A$ and $Q \neq B$), then $\overleftrightarrow{PQ}$ cannot be parallel to both $\overleftrightarrow{AC}$ and $\overleftrightarrow{BC}$; therefore suppose that $\overleftrightarrow{PQ}$ intersects $\overleftrightarrow{AC}$ at point R. Now α leaves Q and R fixed, and thus $\overleftrightarrow{QR}$ is fixed, hence P is fixed, and thus, all points in the plane are fixed.

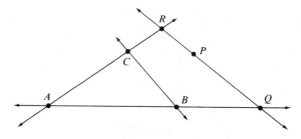

Figure 5.3.1

LEMMA 5.3.11.   Each point in the plane is uniquely determined by its given distances from three noncollinear points. The proof is left as an exercise.

THEOREM 5.3.12.   An isometry is uniquely determined by three noncollinear points and their images under that isometry.

Proof. Lemma 5.3.11 guarantees that at least one isometry is determined by three noncollinear points and their images. Now suppose α and β are isometries that map noncollinear points A, B, and C onto noncollinear points A', B', and C', respectively. Now $\alpha \circ \beta^{-1}$ leaves each of A, B, and C fixed,[16] hence $\alpha \circ \beta^{-1} = I$. Therefore $(\alpha \circ \beta^{-1}) \circ \beta = I \circ \beta$, and thus $\alpha = \beta$.

---

[15] An isometry that leaves every point of the plane fixed is called the *identity* and will be denoted by I.

[16] Since β is an isometry, β^{-1} exists and is defined as a transformation where $(\beta \circ \beta^{-1})(P) = (\beta^{-1} \circ \beta)(P) = P$ for all points P in the plane. The proof that β^{-1} is also an isometry is left as an exercise.

At this point in our discussion, we have discovered a variety of properties that isometries must possess, but we have not provided a mathematically precise characterization of a specific isometry nor have we even
proven that such a transformation exists. We shall remedy this situation
with the following sequence of definitions and theorems.

DEFINITION.   If **PQ** is a vector,[17] then a *translation through vector*
PQ, denoted by T_{PQ}, is a transformation of the plane such that if A is any
point in the plane and $T_{PQ}(A) = A'$, then **AA'** and **PQ** are equal vectors; that
is, $\overline{AA'}$ and $\overline{PQ}$ are parallel and equal in length (Figure 5.3.2). Similarly, in
Figure 5.3.2, $\overline{BB'}$ and $\overline{CC'}$ are congruent and parallel to $\overline{PQ}$, and since
AA' = **PQ**, we can say that $T_{AA'}(\triangle ABC) = \triangle A'B'C'$.

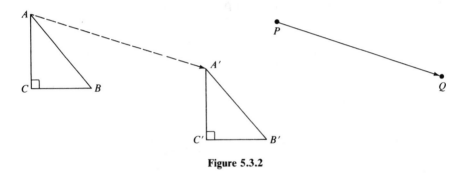

Figure 5.3.2

THEOREM 5.3.13.   A translation is an isometry.

Proof. Let R and S be any two distinct points in the plane, and let T_{PQ}
be a translation such that $T_{PQ}(R) = R'$ and $T_{PQ}(S) = S'$ (Figure 5.3.3). We
must now show that $RS = R'S'$. By the definition of T_{PQ}, $\overline{PQ}$, $\overline{RR'}$, and $\overline{SS'}$
are parallel and congruent, hence $\square RSS'R'$ is a parallelogram and $\overline{RS} \cong$
$\overline{R'S'}$. Therefore $RS = R'S'$.

COROLLARY 5.3.14.   A translation maps a line segment onto a parallel
line segment.[18] The proof is left as an exercise.

We now turn to a second type of transformation which we define as
follows:

DEFINITION.   A *rotation* $R_{P,\theta}$ about a point P through an angle θ[19] is a

[17] For the purposes of this discussion we will refer to a directed line segment as a
vector. For example, vector **PQ** has the same length as $\overline{PQ}$ and the direction indicated by $\overrightarrow{PQ}$,
while vector **QP** has the same length as $\overline{QP}$ and the direction indicated by $\overrightarrow{QP}$.

[18] Parallel line segments are defined to subsets of parallel lines.

[19] By convention, θ is measured in a counterclockwise direction about P.

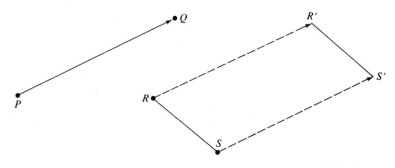

Figure 5.3.3

transformation of the plane where $R_{P,\theta}(P) = P$, and if $A \neq P$, $R_{P,\theta}(A) = A'$ such that $\overline{PA} \cong \overline{PA'}$ and $m\angle APA' = \theta$. Point P is called the *center* of the rotation (Figure 5.3.4). Similarly, in Figure 5.3.4, $\overline{PB} \cong \overline{PB'}$, $\overline{PC} \cong \overline{PC'}$, and $m\angle BPB' = \theta$, $m\angle CPC' = \theta$. A rotation about P through an angle of 180° is called a *half-turn*.

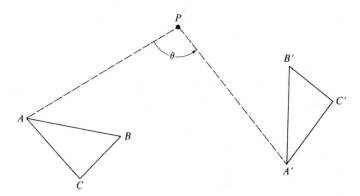

Figure 5.3.4

THEOREM 5.3.15.   A rotation about a point P through an angle θ is an isometry.

Proof. Let A and B be any two points in the plane, and let $R_{P,\theta}$ be a rotation. If $P = A$ or $P = B$, then the proof is trivial; therefore we will assume that $P \neq A$ and $P \neq B$ and let $R_{P,\theta}(A) = A'$ and $R_{P,\theta}(B) = B'$. We must now show that $AB = A'B'$ (Figure 5.3.5). By definition of $R_{P,\theta}$, $\overline{PA} \cong \overline{PA'}$, $\overline{PB} \cong \overline{PB'}$, and $\angle APA' \cong \angle BPB'$. Now by angle subtraction, $\angle APB \cong \angle A'PB'$, hence $\triangle APB \cong \triangle A'PB'$. Therefore $\overline{AB} \cong \overline{A'B'}$ and $AB = A'B'$.

Our discussion of transformational geometry continues with a third isometry called a *reflection in a line* or simply a *reflection*.

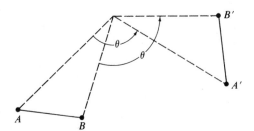

Figure 5.3.5

DEFINITION.   A *reflection* R_l in a line l, is a transformation of the plane where if A is not on l, then $R_l(A) = A'$ such that l is the perpendicular bisector of $\overline{AA'}$, and if P is on l, then $R_l(P) = P$ (Figure 5.3.6).[20] Similarly, in Figure 5.3.6, l is the perpendicular bisector of $\overline{BB'}$ and $\overline{CC'}$.

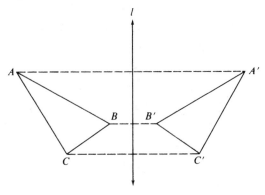

Figure 5.3.6

THEOREM 5.3.16.   A reflection is an isometry. The proof is left as an exercise.

The following series of definitions will assist us in identifying each of the previous isometries and, as we will find in later sections, will be useful in proving theorems using transformational techniques.

DEFINITION.   Any point that is its own image under a transformation is called an *invariant* or a *fixed point*.

Example 5.3.6

(a) The center of a rotation is a fixed point. (b) Any point on the line of reflection of R_l is a fixed point.

---

[20] In Figure 5.3.6 we have assumed that l does not intersect $\triangle ABC$. Other possibilities do exist, and the reader should investigate them. (See Exercise Set 5.3, Problem 8.)

DEFINITION. $\triangle ABC$ has a *clockwise* (or counterclockwise) *orientation* if when we traverse $\triangle ABC$ so as to encounter A, B, and C, (in that order), the direction traversed is clockwise (or counterclockwise).

Example 5.3.7

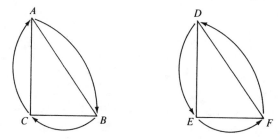

Figure 5.3.7 $\triangle ABC$ has a clockwise orientation, while $\triangle DEF$ has a counterclockwise orientation.

DEFINITION.   If, under a transformation, figure orientation is invariant, then the transformation is called a *direct transformation*. Otherwise, the transformation is an *opposite transformation*.

Example 5.3.8

Translations and rotations are direct isometries, while reflections are opposite isometries.

At this point, we have considered two types of direct isometries, translations and rotations, and one type of opposite isometry, a reflection. It seems reasonable to ask, Do any other isometries exist? The following example will partially answer this question:

Example 5.3.9

Consider $\triangle ABC$ and $\triangle A'B'C'$ as shown in Figure 5.3.8. If $\triangle ABC \cong \triangle A'B'C'$, the previous results indicate that there exists an isometry α that maps $\triangle ABC$ onto $\triangle A'B'C'$, and Theorem 5.3.12 guarantees that this isometry is unique. Since $\triangle ABC$ has a clockwise orientation and $A'B'C'$ has a counterclockwise orientation, we know that α is an opposite isometry and therefore is not a translation or a rotation. Now if α is a reflection, the line of reflection l must be the perpendicular bisector of $\overline{AA'}$, $\overline{BB'}$, and $\overline{CC'}$, which clearly cannot happen for any set of line segments. Therefore α is an opposite isometry that is not a reflection.

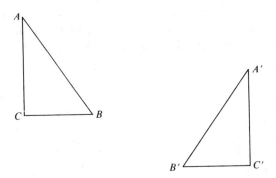

Figure 5.3.8

Earlier in this section we discussed the composition of two isometries. We will now use this definition to create a fourth isometry which, as we will see, is the missing isometry of Example 5.3.9.

DEFINITION.   A *glide reflection* $G_{EH,l}$ is the product of R_l and T_{EH}, where $EH \parallel l$; that is, $G_{EH,l} = R_l \circ T_{EH} = T_{EH} \circ R_l$ (Figure 5.3.9). In the figure $G_{EH,l}(A) = G$, $G_{EH,l}(B) = H$, $G_{EH,l}(C) = I$, and $G_{EH,l}(\triangle ABC) = \triangle GHI$. The reader should note that a glide reflection is an opposite isometry (as a result of R_l) and has no invariant points (as a result of T_{EH}).

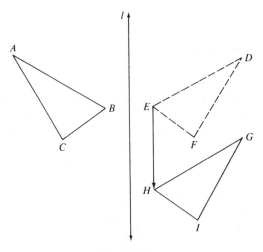

Figure 5.3.9

THEOREM 5.3.17.   A glide reflection is an isometry.

Proof. By definition, a glide reflection is the composition of two isometries and thus, by Theorem 5.3.2, is an isometry itself.

Applications of Isometries to Theorem Proving

As we indicated in the previous section, an isometry is exactly what Euclid had in mind when he allowed us to "pick up and move geometric figures without changing their size or shape." With our definitions of specific isometries, together with our definition stating that two geometric figures are congruent if one figure is the image of the other under some isometry, we are prepared to use these transformations as an alternate method for proving theorems in Euclidean geometry.

Table 5.3.1 briefly summarizes the isometries discussed earlier and may prove helpful by assisting us in identifying the isometries necessary for completing our proofs.

TABLE 5.3.1   Isometries and Their Characteristics

	Translations	Rotations	Reflections	Glide Reflections
Direct or opposite	Direct	Direct	Opposite	Opposite
Number of invariant points	None	One	Points on the line of reflection	None
Line segment maps to a parallel segment	Yes	When θ = 180°	If line is parallel to line of reflection	If line is parallel or perpendicular to line of reflection

We will begin by proving three previously proven theorems for comparative purposes.

THEOREM.   *The Isosceles Triangle Theorem.*

Transformational Proof. Let $\triangle ABC$ be an isosceles triangle with $\overline{AB} \cong \overline{AC}$, and, as in our previous proof of this theorem, let $\overrightarrow{AP}$ be the bisector of $\angle BAC$ (Figure 5.3.10). We will now prove that $\angle ABC \cong \angle ACB$ by showing that $\angle ACB$ is the image of $\angle ABC$ under the reflection $R_{\overrightarrow{AP}}$. By definition, $R_{\overrightarrow{AP}}(A) = A$ and $R_{\overrightarrow{AP}}(P) = P$, and thus $\overrightarrow{AP}$ is fixed. Now let $R_{\overrightarrow{AP}}(B) = X$. Since $\angle BAP \cong \angle CAP$, and since $\overrightarrow{AP}$ is fixed, Corollary 5.3.8 and the angle construction postulate tell us that X must be on $\overrightarrow{AC}$. Since $R_{\overrightarrow{AP}}$ is an isometry, $\overline{AX} \cong \overline{AB}$, hence $\overline{AX} \cong \overline{AC}$, and thus by the segment construction postulate, $X = C$. Therefore $R_{\overrightarrow{AP}}(B) = C$. Similarly $R_{\overrightarrow{AP}}(C) = B$, and thus $R_{\overrightarrow{AP}}(\angle ABC) = \angle ACB$.

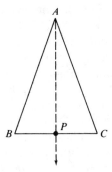

Figure 5.3.10

THEOREM.   *The ASA Triangle Congruence Theorem.*

Transformational Proof. Suppose we are given $\triangle ABC$ and $\triangle DEF$ with $\angle CAB \cong \angle FDE$, $\angle CBA \cong \angle FED$, and $\overline{AB} \cong \overline{DE}$ (Figure 5.3.11). Since $\overline{AB} \cong \overline{DE}$, there are two isometries, one direct and one opposite, that map A onto D and B onto E. Let α be one of these isometries such that $\alpha(C) = C'$, where C' is in the opposite half-plane from F. Since α is an isometry, $\triangle ABC \cong \triangle DEC'$. Therefore $\angle CAB \cong \angle C'DE$, $\angle CBA \cong \angle C'ED$, and consequently, $\angle C'DE \cong \angle FDE$ and $\angle C'ED \cong \angle FED$. Now consider $R_{\overline{DE}}$ and let $R_{\overline{DE}}(C') = X$. Since $R_{\overline{DE}}(D) = D$ and $R_{\overline{DE}}(E) = E$, $R_{\overline{DE}}$ fixes $\overline{DE}$, and since $\angle C'DE \cong \angle XDE$, X must be on $\overrightarrow{DF}$. Similarly, X must be on $\overrightarrow{EF}$. Hence, since F is the only point that lies on both $\overrightarrow{DF}$ and $\overrightarrow{EF}$, $X = F$ and $R_{\overline{DE}}(\triangle DEC') = \triangle DEF$. Now $(R_{\overline{DE}} \circ \alpha)(\triangle ABC) = \triangle DEF$, and since the composition of two isometries is an isometry, $\triangle ABC \cong \triangle DEF$.

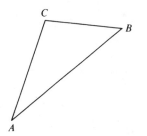
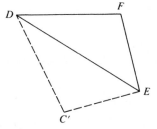

Figure 5.3.11

THEOREM.   The angle sum of any triangle is 180°.

Transformational Proof. Let $\triangle ABC$ be any triangle and consider the translation T_{AB}, where $T_{AB}(A) = B$, $T_{AB}(B) = B'$, and $T_{AB}(C) = C'$ (Figure 5.3.12). By Corollary 5.3.14, $\overline{AC} \parallel \overline{BC'}$, and thus $\angle 3 \cong \angle 4$. Since T_{AB} is an isometry and $T_{AB}(\triangle ABC) = \triangle BB'C'$, $\angle 1 \cong \angle 5$. Now since A, B, and B' are collinear,[21] $m\angle 2 + m\angle 4 + m\angle 5 = 180°$ and, by substitution, $m\angle 2 + m\angle 3 + m\angle 1 = 180°$.

---

[21] By definition of T_{AB}, A, B, and B' must be collinear and A-B-B'.

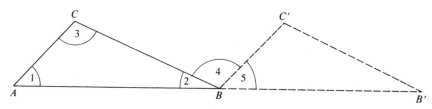

Figure 5.3.12

The following sequence of theorems should provide additional insight into the use of transformational techniques in theorem proving.

LEMMA 5.3.18.   Let $R_{P,180°}$ be denoted by σ_P. Then $\sigma_D \circ \sigma_C \circ \sigma_B \circ \sigma_A = I$ if and only if $\square ABCD$ is a parallelogram.

Proof. First, we will assume that $\sigma_D \circ \sigma_C \circ \sigma_B \circ \sigma_A = I$ and show that $\square ABCD$ is a parallelogram. Let P be any point in the plane such that $\sigma_A(P) = P$ (Figure 5.3.13). By definition of a half-turn, D is the midpoint of $\overline{PP'''}$ and C is the midpoint of $\overline{P''P'''}$; therefore by Theorem 4.1.11, $\overline{DC} \parallel \overline{PP''}$. Using the same theorem again, since A is the midpoint of $\overline{PP'}$ and B is the midpoint of $\overline{P''P'}$, we find that $\overline{AB} \parallel \overline{PP''}$. Thus $\overline{AB} \parallel \overline{DC}$. Similarly, $\overline{BC} \parallel \overline{AD}$, and $\square ABCD$ is a parallelogram. To prove the converse, we will let $\square ABCD$ be a parallelogram and show that $\sigma_D \circ \sigma_C \circ \sigma_B \circ \sigma_A = I$. Let P be any point in the plane such that $\sigma_A(P) = P'$, $\sigma_B(P') = P''$, $\sigma_C(P'') = P'''$, and $\sigma_D(P''') = X$ (Figure 5.3.14). Since $\sigma_D \circ \sigma_C \circ \sigma_B \circ \sigma_A(P) = X$, if we show that $X = P$, we have shown that all the points in the plane are fixed, hence $\sigma_D \circ \sigma_C \circ \sigma_B \circ \sigma_A = I$. Now since D is the midpoint of $\overline{P'''X}$ and C is the midpoint of $\overline{P''P'''}$, $\overleftrightarrow{XP''} \parallel \overleftrightarrow{CD}$. Also, $\overleftrightarrow{PP''} \parallel \overleftrightarrow{AB} \parallel \overleftrightarrow{CD}$. Therefore since there can exist only one line through P'' that is parallel to $\overleftrightarrow{CD}$, $X \in \overleftrightarrow{PP''}$. Now $DC = AB = \frac{1}{2}(PP'')$ and

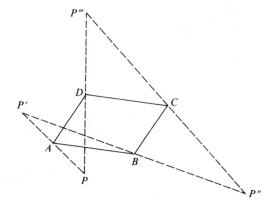

Figure 5.3.13

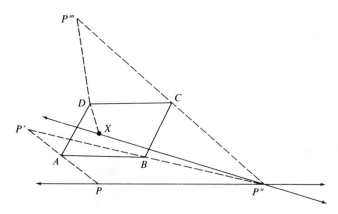

Figure 5.3.14

$DC = \frac{1}{2}(XP'')$; therefore $XP'' = PP''$, and since X must be on the same side of P'' as P (see Exercise Set 5.3, Problem 28), $X = P$, and the proof is complete.

Lemma 5.3.18 proves to be a useful tool when we wish to use transformational techniques to prove that a quadrilateral is a parallelogram. Consider the following theorem.

THEOREM. The quadrilateral created by joining the consecutive midpoints of the sides of any quadrilateral is a parallelogram.

Proof. Let $\square MNOP$ be any quadrilateral, and let Q, R, S, and T be the midpoints of each of its sides (Figure 5.3.15). It should be clear that $\sigma_T \circ \sigma_S \circ \sigma_R \circ \sigma_Q (M) = M$, and that M is a fixed point. Now $\sigma_T \circ \sigma_S \circ \sigma_R \circ \sigma_Q$ is a translation (see Exercise Set 5.3, Problem 29). Since the only translation that has fixed points is the identity, $\sigma_T \circ \sigma_S \circ \sigma_R \circ \sigma_Q = I$, and by the previous lemma, $\square QRST$ is a parallelogram.

The following lemma will prove to be of assistance in connection with the question of concurrency.

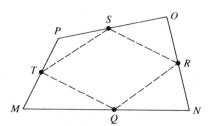

Figure 5.3.15

LEMMA 5.3.19.   The product of three reflections in three concurrent lines is a reflection in a line that passes through the point of concurrency.

Proof. Let l, m, and n be any three concurrent lines, and let P be the point of concurrency. Now $R_l \circ R_m \circ R_n$[22] is an isometry α, and in particular α is an opposite isometry. (Why?) Clearly, $R_l \circ R_m \circ R_n (P) = \alpha(P) = P$; therefore α has at least one fixed point. By previous discussions, α is either a reflection or a glide reflection, but since glide reflections have no fixed points, α must be a reflection in some line k. Furthermore, since P is fixed, P must lie on line k, hence k, l, m, and n are concurrent.

An application of the previous result is demonstrated in the following theorem:

THEOREM.   The perpendicular bisectors of the sides of any triangle are concurrent.

Proof. Let $\triangle ABC$ be any triangle, let l be the perpendicular bisector of $\overline{AC}$, and let m be the perpendicular bisector of $\overline{AB}$ (Figure 5.3.16). First, we observe that l must intersect m at some point P. (Why must this be true?) Now if we consider $\alpha = R_m \circ R_{\overline{AP}} \circ R_l$ it is clear that $\alpha(C) = B$, and by the previous lemma, α must be a reflection in some line n that contains P. Since $R_n (C) = B$, n must be the perpendicular bisector of $\overline{BC}$, and the proof is completed.

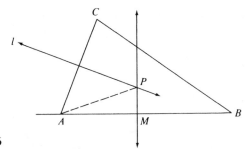

Figure 5.3.16

At this point, the reader is reminded that transformational techniques simply represent an alternative method for proving theorems in the same way that analytic geometry is an alternative method. It should be noted that sometimes a transformational proof may appear to be "easier," while at other times an analytic proof or a synthetic proof may indeed be "easier". It is the problem solver's own creativity that determines which method is the best.

The exercises at the end of this section will provide an opportunity to apply transformational techniques to theorem proving.

---

[22] The order of the reflections has no effect on the result of this proof.

Similarities and Their Applications to Theorem Proving

In the previous sections we formalized Euclid's concept of congruence by superposition with the study of isometries. The reader may perhaps now ask, Can transformational techniques be used to clarify the concept of polygonal similarity? Is it possible to define similar polygons as those that are images of each other under some type of transformation? Certainly, since congruent polygons satisfy our previous definition of similarity, an isometry is a transformation that answers the question in part. But since isometries are distance-preserving transformations, they prove to be insufficient when dealing with similar but noncongruent polygons.

To address this difficulty, we must consider a type of transformation that does not leave distance invariant but which changes distances proportionally. To this end, we introduce the following definition:

DEFINITION.   A *homothety*[23] $H_{O,k}$, where O is a fixed point and k is a nonzero real number, is a transformation of the plane[24] where $H_{O,k}(O) = O$, and if $P \neq O$, then $H_{O,k}(P) = P'$, such that O, P, and P' are collinear and $OP' = k(OP)$. If $k > 1$, then $O\text{-}P\text{-}P'$; if $0 < k < 1$, then $O\text{-}P'\text{-}P$ and if $k < 0$, then $P'\text{-}O\text{-}P$. Also, O is called the *center* of the homothety, and k is called the *constant of proportionality*.

Example 5.3.10

(a) In Figure 5.3.17, $H_{O,2}(\triangle ABC) = \triangle A'B'C'$.

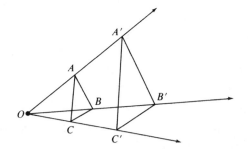

Figure 5.3.17

(b) In Figure 5.3.18, $H_{O,-2}(\triangle DEF) = \triangle D'E'F'$.
(c) In Figure 5.3.19, $H_{O,\frac{1}{2}}(\triangle GHI) = \triangle G'H'I'$.

THEOREM 5.3.20.   Betweenness is invariant under a homothety.

---

[23] This transformation is also called a *dilation* or a *size transformation*.
[24] The reader may wish to verify that $H_{O,k}$ is indeed a transformation of the plane.

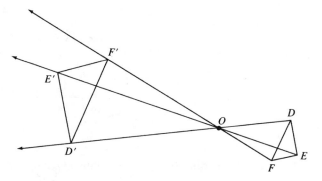

Figure 5.3.18

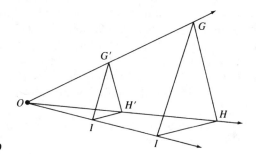

Figure 5.3.19

Proof. Let A, B, and C be three distinct points such that A-B-C, and let $H_{O,k}(A) = A'$, $H_{O,k}(B) = B'$, and $H_{O,k}(C) = C'$. We must now show that A'-B'-C'. Several cases need to be considered. For the purposes of this discussion, we will assume that $k > 0$ and O-A-B-C; other cases are left as exercises. By definition, $OA' = k(OA)$ and $OB' = k(OB)$, and by using the given betweeness relationships, it is easy to show that $A'B' = k(AB)$. In similar fashion, $B'C' = k(BC)$ and $A'C' = k(AC)$. It now suffices to show that $A'B' + B'C' = A'C'$. Suppose that $A'B' + B'C' \neq A'C'$, and in particular that $A'B' + B'C' > A'C'$. Substituting, we have $k(AB) + k(BC) > k(AC)$, and dividing through by k, we find that $AB + BC > AC$, which is a contradiction to the collinearity and betweeness of A, B, and C. If we had assumed that $A'B' + B'C' < A'C'$, we would have arrived at a similar contradiction, and our proof is complete.

COROLLARY 5.3.21.   The image of a line segment (or ray or line) under a homothety is a line segment (or ray or line). The proof is left as an exercise.

THEOREM. 5.3.22.   If O, A, and B are not collinear, and if $H_{O,k}(\overline{AB}) = A'B'$, then $\overline{AB} \parallel \overline{A'B'}$. The proof is left as an exercise.

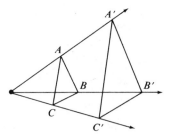

Figure 5.3.20

THEOREM 5.3.23.   A homothety maps triangles onto similar triangles.[25]

Proof. Let $H_{O,k}(\triangle ABC) = \triangle A'B'C'$ (Figure 5.3.20).[26] As shown before, $A'C' = k(AC)$, $B'C' = k(BC)$, and $A'B' = k(AB)$. Therefore we need only show that the corresponding angles of $\triangle ABC$ and $\triangle A'B'C'$ are congruent to complete our proof. By the previous theorem, $\overline{A'C'} \parallel \overline{AC}$ and $\overline{A'B'} \parallel \overline{AB}$. Therefore $\angle CAO \cong \angle C'A'O$ and $\angle BAO \cong \angle B'A'O$. Now by angle subtraction, we have that $\angle CAB \cong \angle C'A'B$. In like fashion, $\angle ABC \cong \angle A'B'C'$ and $\angle BCA \cong \angle B'C'A'$, and the triangles are similar.

COROLLARY 5.3.24.   A homothety is a direct and conformal transformation. The proof is left as an exercise.

Our investigation of homotheties has moved us one step closer to the development of a definition of similarity that is analogous to the transformational definition of congruence discussed earlier in this section. However, in order for one polygon to be the image of another under a homothety, it is a necessary condition that each pair of corresponding sides of the two polygons be either collinear or parallel. If two similar, noncongruent polygons are situated such that at least one set of corresponding sides is neither collinear nor parallel, then a homothety will not map one onto the other. To resolve this seeming difficulty, we consider the following theorem.

THEOREM 5.3.25.   If two triangles are similar and noncongruent, then there exists a transformation S_k, which is the product of an isometry and a homothety, that maps one onto the other.

Proof. Let $\triangle ABC$ be similar to $\triangle A'B'C'$ and let $A'B' > AB$ (Figure 5.3.21). Since $\angle CAB \cong \angle C'A'B'$, there exists some isometry α such that $\alpha(A) = A'$ and $\alpha(B) = P$, where $P \in \overline{A'B'}$. Also, since α is conformal, $\alpha(C) = Q$, where $Q \in \overline{A'C'}$, and as a result we have $\triangle ABC \cong \triangle A'PQ$. Next

---

[25] This theorem can be easily generalized to include polygons.

[26] For the purposes of this discussion we have assumed that $k > 0$ and the relative positions of O, A, B, and C are as shown. Verification of other possibilities is similar.

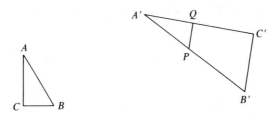

Figure 5.3.21

consider a homothety centered at A' and having a ratio $k = A'B'/A'P = A'B/AB$. Clearly, $H_{A',k}(A') = A'$, and since $(A'B'/A'P)(A'P) = A'B'$, $H_{A',k}(P) = B'$. Now if we let $H_{A',k}(Q) = X$, by definition, $X \in \overline{A'C'}$, and since $\angle ACB \cong \angle A'PQ \cong \angle A'C'B'$, the conformal nature of $H_{A',k}$ ensures that $X \in \overline{B'C'}$. Therefore $X = C'$ and $H_{A',k}(\triangle A'PQ) = \triangle A'B'C'$. Hence $S_k(\triangle ABC) = H_{A',k} \circ \alpha(\triangle ABC) = \triangle A'B'C'$ as required.

The transformation S_k indicated in the previous theorem is described in the following definition:

DEFINITION.   A *similarity* S_k is a transformation of the plane that is the composition of an isometry and a homothety with ratio k.

Example 5.3.11

(a) In Figure 5.3.22, $S_2(\triangle ABC) = H_{P,2} \circ R_{P,90°}(\triangle ABC) = \triangle A'B'C'\cdot$
(b) In Figure 5.3.23, $S_3(\triangle PQR) = H_{P,3} \circ R_l(\triangle PQR) = \triangle P'Q'R'$.

COROLLARY 5.3.26.   A similarity maps a polygon onto a similar polygon. The proof is left as an exercise.

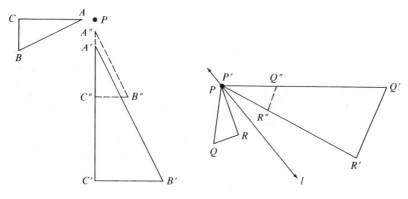

Figure 5.3.22   **Figure 5.3.23**

COROLLARY 5.3.27.   Angle measure and collinearity are invariant under a similarity. The proof is left as an exercise.

We are now prepared to provide a transformational definition of polygonal similarity.

DEFINITION.   Two polygons are similar if and only if one is the image of another under some similarity.[27]

To illustrate the use of similarities in proving, geometric theorems we offer the following previously proven theorems, reproven using transformational techniques.

THEOREM.   The line connecting the midpoints of two sides of a triangle is parallel to the third side, and its measure is equal to one-half the measure of the third side.

Proof. Let S and T be the midpoints of two sides of $\triangle PQR$ (Figure 5.3.24). Now consider a homothety centered at P whose ratio is $PS/PR = 2$. If we let $H_{P,2}(S) = X$, by definition, $X \in \overrightarrow{PR}$, and since $PX = 2(PS) = PR$, $X = R$. In a similar fashion, $H_{P,2}(T) = Q$. Now since $H_{P,2}(\overline{ST}) = \overline{RQ}$, $\overline{ST}$ is parallel to $\overline{RQ}$ by Theorem 5.3.22, and since $RQ = 2(ST)$, we have that $ST = \frac{1}{2}(RQ)$.

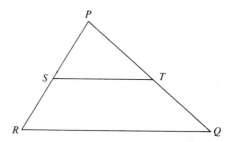

Figure 5.3.24

THEOREM.   The altitude to the hypotenuse of a right triangle determines two triangles, each of which is similar to the original triangle.

Proof. Suppose that $\triangle ABC$ is a right triangle and that $\overline{CD}$ is the altitude to the hypotenuse $\overline{AB}$ (Figure 5.3.25). We must now show that $\triangle CBD$ is similar to $\triangle ABC$. Consider the similarity $S_k = H_{B,k} \circ R_l$, where $l = \overrightarrow{BP}$, the bisector of $\angle CBA$, and k is to be determined. If we can show that $S_k(\triangle CBD) = \triangle ABC$, then the proof will be complete. First, we consider

---

[27] A homothety is a similarity whose isometry is the identity, and an isometry is a similarity whose homothety has a ratio equal to 1 or -1.

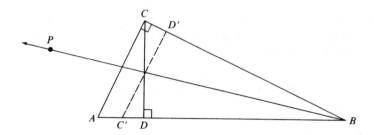

Figure 5.3.25

the image of $\triangle CBD$ under R_l. Now $R_l(B) = B$, $R_l(D) = D'$, where $D' \in \overline{BC}$ (Why?), and $R_l(C) = C'$, where $C' \in \overrightarrow{BA}$. Therefore $R_l(\triangle CBD) = \triangle C'BD'$ and $\triangle CBD \cong \triangle C'BD'$. Now consider the image of $\triangle C'BD'$ under $H_{B,k}$, where $k = BC/BD' = BC/BD$. By definition, $H_{B,k}(B) = B$ and $H_{B,k}(D') = C$. Now let us suppose that $H_{B,k}(C') = X$. Again by definition, $X \in \overrightarrow{BA}$, and since $\angle CDB \cong \angle C'D'B \cong \angle ACB$, $X \in \overrightarrow{CA}$. Therefore $X = A$ and $H_{B,k}(\triangle C'BD') = \triangle ABC$. Now since $S_k = H_{B,k} \circ R_l$, $S_k(\triangle CBD) = \triangle ABC$, and the required triangles are similar. An analogous proof can be used to show that $\triangle ACD$ is similar to $\triangle ABC$.

EXERCISE SET 5.3

Mappings and Transformations

1. Determine which of the following are mappings (functions) with the domains and ranges as indicated:
   (a) $f: \mathfrak{R} \to \mathfrak{R}$ such that $f(x) = 2x - 1$
   (b) $f: \mathfrak{R} \to \mathfrak{R}$ such that $f(x) = 4 - x$
   (c) $f: \mathfrak{R} \to \mathfrak{R}$ such that $f(x) = \pm \sqrt{x}$
   (d) $f: \mathfrak{R} \to \mathfrak{R}$ such that $f(x) = \sin x$
   (e) $f: \mathfrak{R} \to \mathfrak{R}$ such that $y = f(x)$ and $x^2 + y^2 = 16$
   (f) $f: \mathfrak{R} \to \mathfrak{R}$ such that $f(x) = 1/x$
   (g) $f: \mathfrak{R} \to \mathfrak{R}$ such that $f(x) = (x^2 - 4)/(x + 2)$
   (h) $f: \mathfrak{R} \times \mathfrak{R} \to \mathfrak{R} \times \mathfrak{R}$ such that $f(x,y) = (x - 2, y + 3)$
   (i) $f: \mathfrak{R} \times \mathfrak{R} \to \mathfrak{R} \times \mathfrak{R}$ such that $f(x,y) = (\pm\sqrt{x}, \pm\sqrt{y})$
2. Determine which of the following mappings are onto the indicated sets:
   (a) $f: \mathfrak{R} \to \mathfrak{R}$ such that $f(x) = (x - 1)/3$
   (b) $f: \mathfrak{R} \to \mathfrak{R}$ such that $f(x) = x^2$
   (c) $f: \mathfrak{R} \to \mathfrak{R}$ such that $f(x) = \sin x$
   (d) $f: \mathfrak{R} \times \mathfrak{R} \to \mathfrak{R} \times \mathfrak{R}$ such that $f(x,y) = (2x,3y)$
   (e) $f: N \times N \to N \times N$ such that $f(x,y) = (2x,3y)$
3. Which of the mappings in Problem 2 are one-to-one?
4. Which of the mappings in Problem 2 could be transformations of the plane?
5. Prove Theorem 5.3.1.

Isometries

In Problems 6 through 8 use a right triangle $\triangle ABC$ (with right angle at C) as the preimage and accurately draw its image under each of the given isometries.

6. (a) T_{AC}
   (b) T_{AD} where **AD** is not parallel to any side of the triangle
7. $R_{P,90}$ where
   (a) P is on side $\overline{AC}$
   (b) $P = B$
   (c) P is not on $\triangle ABC$
8. R_l where
   (a) Line l contains point C
   (b) Line l intersects sides $\overline{AC}$ and $\overline{BC}$ in distinct points
   (c) Line l is parallel to side $\overline{AC}$
   (d) Line l does not intersect nor is parallel to any side of $\triangle ABC$
9. Prove Theorem 5.3.2.
10. Prove Theorem 5.3.4.
11. Prove Corollary 5.3.5.
12. Prove Corollary 5.3.6.
13. Prove Corollary 5.3.8.
14. Prove Corollary 5.3.9.
15. Prove Lemma 5.3.11.
16. Prove: If β is an isometry, then β^{-1} is an isometry.
17. Prove Corollary 5.3.14.
18. Prove Theorem 5.3.16.
19. Show that a translation T_{PQ} is equivalent to the composition of two reflections in parallel lines l_1 and l_2, where l_1 and l_2 are perpendicular to **PQ**. If d is the distance between l_1 and l_2, show that $PQ = 2d$.
20. In the following figure $\triangle A'B'C'$ is the image of $\triangle ABC$ under a translation. Find the lines of reflection discussed in Problem 19.

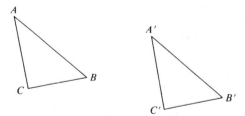

21. Show that a rotation $R_{P,\theta}$ is equivalent to the composition of two reflections in intersecting lines l_1 and l_2. If α is the angle between l_1 and l_2, show that $\theta = 2\alpha$.
22. In the following figure $\triangle A'B'C'$ is the image of $\triangle ABC$ under a rotation. Find the lines of reflection discussed in Problem 21.

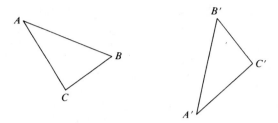

23. Show that a glide reflection is the composition of reflections in three lines. Describe the three lines.

24. In the following figure $\triangle A'B'C'$ is the image of $\triangle ABC$ under a glide reflection. Find the lines of reflection discussed in Problem 23.

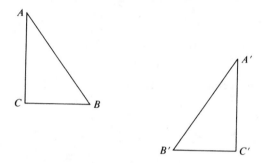

For Problems 25 through 27, a set of elements G and a binary operation $*$ form a *group* if and only if each of the following properties is true.

(i) If $a, b \in G$, then $a * b \in G$.

(ii) If $a, b, c \in G$, then $a * (b * c) = (a * b) * c$.

(iii) G has an identity element $i \in G$ such that if $a \in G$, then $i * a = a * i = a$.

(iv) Each element of G has an inverse under $*$.

25. Prove that the set of all translations with the binary operation, composition, form a group.

26. Prove that the set of all rotations about a given point P with the operation, composition, form a group.

27. Prove that the set of all isometries under the operation, composition, form a group.

Applications of Isometries

28. In the proof of Lemma 5.3.18, we stated that X must be on the same side of P'' as P. Justify this statement. (*Hint:* Suppose that P-P''-X and use the fact that $\overline{AD} \parallel \overline{BC}$.

29. In the proof of the theorem associated with Figure 5.3.15, we stated that $\sigma_T \circ \sigma_S \circ \sigma_R \circ \sigma_Q$ is a translation. Justify this statement.

30. In the theorem associated with Figure 5.3.16, we claimed that lines l and m intersect at some point P. Justify this statement.

31. Given: $\overline{PA} \cong \overline{PC}$, and $\overrightarrow{PB}$ bisects $\angle APC$. Prove: $\overline{AB} \cong \overline{BC}$.

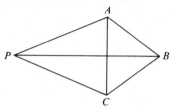

32. Given: $\overline{AD} \cong \overline{AC}$, and $\angle ADE \cong \angle ACB$. Prove: $\triangle ADE \cong \triangle ACB$.

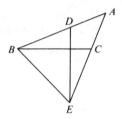

33. Given: $\overleftrightarrow{AP}$ is the perpendicular bisector of $\overline{BC}$, and $\angle CPE \cong \angle BPD$. Prove: $\overline{AE} \cong \overline{AD}$.

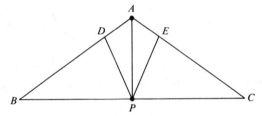

34. Given: $\angle 1 \cong \angle 2$, $\overline{AB} \cong \overline{AC}$, and $\overline{AD} \cong \overline{AE}$. Prove: $\triangle ABD \cong \triangle ACE$.

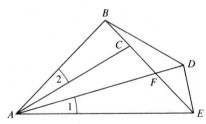

35. Given: $\square JMLK$ is a parallelogram, and $\overline{JN} \parallel \overline{QL}$. Prove: $\overline{QL} \cong \overline{JN}$.

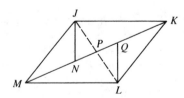

36. Given: $\overline{BC} \cong \overline{BA} \cong \overline{AF}$, $\overline{DC} \cong \overline{DA} \cong \overline{AE}$, and $\overline{BD} \cong \overline{EF}$. Prove: $\triangle BDC \cong \triangle FEA$.

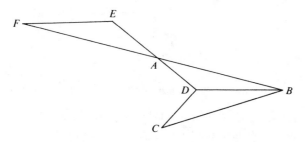

37. Given $\overline{AE} \perp \overline{BD}$, $\overline{BC} \cong \overline{BE}$, and $\overline{AB} \cong \overline{DB}$. Prove: $\triangle ABC \cong \triangle DBE$.

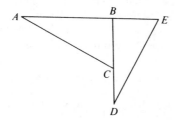

38. Given: C, B, and E are collinear, $\angle ACB \cong \angle FBE$, $\overline{AC} \cong \overline{BF}$, and $\overline{CB} \cong \overline{BE}$. Prove: $\triangle ABC \cong \triangle FEB$.

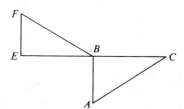

39. Given: $\angle CDB \cong \angle CDA \cong \angle ADB$, and $\overline{CD} \cong \overline{AD} \cong \overline{DB}$. Prove: $\triangle ABC$ is equilateral.

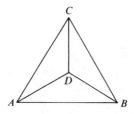

40. Given: P is the midpoint of $\overline{AB}$, Q is the center of circle Q, $\overrightarrow{AC}$ and $\overrightarrow{BD}$ are tangents to circle Q, and $AQFD$ and $BQEC$ are secants to circle Q. Prove: $\overline{BC} \cong \overline{AD}$. See figure at top of following page.

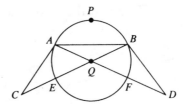

41. Prove that if distinct lines *l* and *m* intersect, then the vertical angles formed are congruent.
42. Prove the converse of the isosceles triangle theorem.
43. Prove the hypotenuse and leg triangle congruence theorem.
44. Prove the hypotenuse and acute angle triangle congruence theorem.
45. If the diagonals of a quadrilateral are perpendicular bisectors of each other, then the quadrilateral is a rhombus.
46. Prove that the opposite angles of a parallelogram are congruent.
47. Prove that the bisectors of the three interior angles of any triangle are concurrent.

Similarities and Their Applications to Theorem Proving

48. Using a right triangle $\triangle ABC$ (right angle at C) as the preimage, draw its image under each of the following similarities:
    (a) $H_{P,2}$ where (i) P is on side $\overline{AC}$, (ii) $P = B$, and (iii) $P \notin \triangle ABC$.
    (b) $S_2 = H_{P,2} \circ T_{AD}$ where vector **AD** is

    (c) $S_2 = H_{P,2} \circ R_{P,90°}$ where (i) P is on side $\overline{AC}$, (ii) $P = B$, and (iii) $P \notin \triangle ABC$.
    (d) $S_2 = H_{P,2} \circ R_l$ where $P \notin \triangle ABC$ and (i) line l intersects sides $\overline{AC}$ and $\overline{BC}$ and (ii) line l does not intersect and is not parallel to any side of $\triangle ABC$.
49. In the proof of Theorem 5.3.20, only one of several cases was addressed. List the other cases necessary for the proof. Justify each of these cases.
50. Prove Corollary 5.3.21.
51. Prove Theorem 5.3.22.
52. Prove Corollary 5.3.24.
53. Prove Corollary 5.3.26.
54. Prove Corollary 5.3.27.
55. In the proof of the theorem associated with Figure 5.3.25, we let $R_l(D) = D'$ and stated that D' must be on $\overline{BC}$. Justify this statement.
56. Given: Right triangle $\triangle ABC$ (right angle at C) and $\overline{DE}$ perpendicular to $\overline{AB}$ at E.
    Prove: $\triangle ABC$ is similar to $\triangle DBE$.

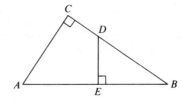

57. Given: $\angle DCA \cong \angle ABC$. Prove: AC is the mean proportional between AB and AD.

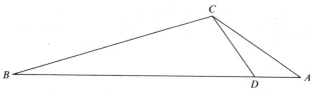

58. Given: $\triangle ABC$ and $\triangle BDM$ are equilateral. Prove: $\triangle ACE$ is similar to $\triangle BDE$.

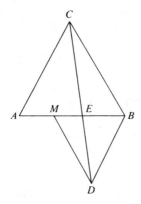

59. Prove that if a line intersects two sides of a triangle in distinct points and is parallel to the third side, then the triangle created is similar to the original triangle.

60. Prove that if a tangent and a secant are drawn to a circle from an external point, the length of the tangent segment is the geometric mean between the length of the entire secant segment and the length of its external segment.

61. Two circles are internally tangent, and the ratio of their circumference is $2:1$. Prove that any chord of the larger circle drawn from the point of tangency is bisected by the smaller circle.

62. Prove the nine-point circle theorem. (*Hint:* Consider a homothety of ratio equal to 2 centered at the orthocenter.)

5.4 ANALYTICAL TRANSFORMATIONS

Introduction

In Section 5.2 we coordinatized the plane with ordered pairs of real numbers and applied algebraic techniques to assist us in solving geometric problems. In the previous section we formalized a definition of congruence through the study of a group of distance-preserving transformations of the plane known as isometries, and then extended that notion to include similar figures by investigating the properties of a group of transformations called similarities.

In this section we will combine aspects of both these techniques to derive analytic equations for the transformations of the plane studied in the last section and to apply these newly found analytical formulas to geometric theorem proving.

Analytical Equations for Isometries

As you recall from the previous section, the image of a point under a translation $T_{\mathbf{PQ}}$ is determined by the magnitude and direction of the vector of translation $\mathbf{PQ}$. Before we begin to derive analytical equations for $T_{\mathbf{PQ}}$, we need to discuss some of the properties of $\mathbf{PQ}$. We begin with the following definitions and theorems.

DEFINITION. If $\mathbf{PQ}$ is a vector in the plane, P is called the *initial point* and Q is called the *terminal point*.[28] The *magnitude* of $\mathbf{PQ}$ is equivalent to the length of line segment $\overline{PQ}$, and the *direction* of $\mathbf{PQ}$ is determined by the counterclockwise angle of inclination of $\overrightarrow{PQ}$.

Example 5.4.1

Let $P = (2,1)$ and $Q = (5,5)$. The magnitude of $\mathbf{PQ}$ is equal to $PQ = \sqrt{(5-2)^2 + (5-1)^2} = 5$, and the direction of $\mathbf{PQ}$ is determined by α, where $\alpha = \arctan \frac{4}{3}$.

DEFINITION. A *position vector* is any vector in the plane whose initial point is the origin.

THEOREM 5.4.1. Any vector $\mathbf{RS}$ in the plane is equal to a position vector $\mathbf{OP}$, and if $R = (a,b)$ and $S = (c,d)$, then $P = (c - a, d - b)$.

Proof. Let $R = (a,b)$, $S = (c,d)$, and $P = (p,q)$ (Figure 5.4.1). We must find the coordinates of P such that $\mathbf{OP}$ is equal in both magnitude and direction to $\mathbf{RS}$. This can be accomplished by finding values for p and q such that $\overline{OP}$ is both congruent and parallel to $\overline{RS}$. We can easily see that this is equivalent to finding the coordinates of P such that $\square ORSP$ is a parallelogram. Employing methods used in Section 5.2, we find that $p = c - a$ and $q = d - b$.

THEOREM 5.4.2. If $A = (x,y)$ is any point in the plane and if $T_{\mathbf{OP}}$ is a translation such that $T_{\mathbf{OP}}(A) = A'$, where $O = (0,0)$ and $P = (h,k)$, then $A' = (x',y')$, where $x' = x + h$ and $y' = y + k$.[29] The proof is left as an exercise.

---

[28] In vector $\mathbf{QP}$, Q is the initial point and P is the terminal point.

[29] If $\mathbf{OP}$ is not a position vector, then Theorem 5.4.1 provides us with an equivalent position vector that can be used.

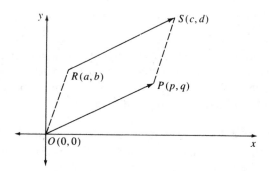

Figure 5.4.1

Example 5.4.2

(a) If $P = (3,4)$, then $T_{OP}(2,-1) = (2 + 3,-1 + 4) = (5,3)$. (b) If $P = (3,4)$, then $T_{PO}(2,-1) = (2 - 3,-1 - 4) = (-1,-5)$.

The following theorem and corollary will provide us with analytical representations for rotations about any point. First, we will consider the special case of a rotation about the origin.

THEOREM 5.4.3.   If $A = (x,y)$ is any point in the plane, and if $R_{O,\theta}$ is a rotation such that $R_{O,\theta}(A) = A'$, where $O = (0,0)$, then $A' = (x',y')$, where $x' = x \cos \theta - y \sin \theta$ and $y' = x \sin \theta + y \cos \theta$.

Proof. Let $A = (x,y)$, $O = (0,0)$, $AO = r$, and $\angle AOQ = \alpha$ (Figure 5.4.2). By analytic methods, we know that $x = r \cos \alpha$ and $y = r \sin \alpha$. Also, by definition of $R_{O,\theta}$, $OA' = OA = r$, and therefore $x' = r \cos (\alpha + \theta)$ and $y' = \sin(\alpha + \theta)$. Now using the familiar formulas for $\cos(\alpha + \theta)$ and $\sin(\alpha + \theta)$, we find that $x' = (r \cos \alpha)(\cos \theta) - (r \sin \alpha)(\sin \theta)$ and $y' = (r \cos \alpha)(\sin \theta) +$

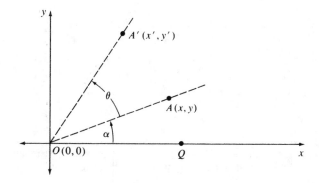

Figure 5.4.2

$(r \sin \alpha)(\cos \theta)$, and by substitution, $x' = x \cos \theta - y \sin \theta$ and $y' = x \sin \theta + y \cos \theta$.

Suppose we now wish to find, analytically, the image of a point that has been rotated about a point other than the origin. The following corollary addresses this question, but more importantly, its proof provides us with a technique we will find useful in future derivations.[30]

COROLLARY 5.4.4.   If $A = (x,y)$ is any point in the plane and $R_{P,\theta}$ is a rotation such that $R_{P,\theta}(A) = A'$, where $P = (h,k)$, then $A' = (x',y')$, where $x' = [(x - h) \cos \theta - (y - k) \sin \theta] + h$ and $y' = [(x - h) \sin \theta + (y - k) \cos \theta] + k$.

Proof. Since we know how to find the image of any point that has been rotated about the origin, we shall first use Theorem 5.4.2 to translate P to the origin,[31] then perform a rotation about the origin using Theorem 5.4.3, and finally translate back (Figure 5.4.3). In particular, we will perform the fol-

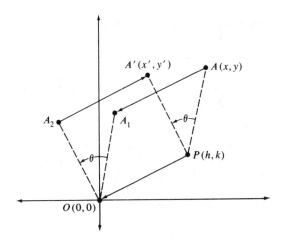

Figure 5.4.3

lowing transformation. $T_{OP} \circ R_{(0,0),\theta} \circ T_{PO}$. We now consider $R_{(h,k),\theta}(x,y) = (T_{OP} \circ R_{(0,0),\theta} \circ T_{PO})(x,y) = (T_{OP} \circ R_{(0,0),\theta})(x - h, y - k) = T_{OP}\{[(x - h) \cos \theta - (y - k) \sin \theta], [(x - h) \sin \theta + (y - k) \cos \theta]\} = \{[(x - h) \cos \theta - (y - k) \sin \theta] + h, [(x - h) \sin \theta + (y - k) \cos \theta + k\}.$

---

[30] As you will see in the proof of Corollary 5.4.4, the technique of translating, performing some isometry, and then translating back, will prove to be a highly effective tool in deriving analytical formulas.

[31] All other points in the plane have been translated similarly, including point A.

Example 5.4.3

(a) $R_{(0,0),30°}$ (2,3) = (2 cos 30° − 3 sin 30°, 2 sin 30° + 3 cos 30°) = [2($\sqrt{3}/2$) − 3($\frac{1}{2}$), 2($\frac{1}{2}$) + 3 ($\sqrt{3}/2$)] = ($\sqrt{3}$ − $\frac{3}{2}$, 1 + 3$\sqrt{3}/2$). (b) $R_{(2,1),90°}$ (5,3) = {[(5 − 2) cos 90° − (3 − 1) sin 90°] + 2, [(5 − 2) sin 90° + (3 − 1) cos 90] + 1} = [3(0) − 2(1) + 2, 3(1) + 2(0) + 1] = (0,4).

We now turn our attention to the analytical description of a line reflection. First, we will consider several special cases, those of reflections over horizontal and vertical lines.

THEOREM 5.4.5.   Let $A = (x,y)$ be any point in the plane. (i) If l is the line $y = 0$, then $R_{y=0}$ $(x,y) = (x,-y)$, and (ii) if l is the line $x = 0$, then $R_{x=0}$ $(x,y) = (-x,y)$. The proof is left as an exercise.

COROLLARY 5.4.6.   Let $A = (x,y)$ be any point in the plane. (i) If l is the line $y = k$, then $R_{y=k}$ $(x,y) = (x, -y + 2k)$, and (ii) if l is the line $x = h$, then $R_{x=h}$ $(x,y) = (-x + 2h,y)$. The proof is left as an exercise.

Example 5.4.4

(a) $R_{y=0}$ (2,4) = (2,−4). (b) $R_{x=3}$ (2,4) = (−2 + 6, 4) = (4,4).

Next we consider the special case of an oblique line $y = mx$ where $m \neq 0$.

THEOREM 5.4.7.   If $A = (x,y)$ is any point in the plane and $R_{y=mx}$ is the reflection about the line $y = mx$ such that $R_{y=mx}$ $(x,y) = (x',y')$, then $x' = x$ cos 2α + y sin 2α and $y' = x$ sin 2α − y cos 2α, where α = arctan m.

Proof. Consider the diagram in Figure 5.4.4 where $R_{y=mx}(A) = A'$. Now observe that $R_{y=mx}(A) = R_{(0,0),2\alpha} \circ R_{y=0}(A) = A'$ and apply the necessary formulas to obtain the result.

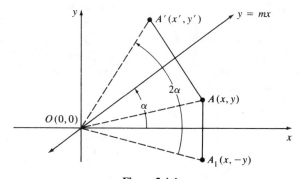

Figure 5.4.4

Finally, we consider the case where l is a general oblique line of the form $y = mx + b$ and $m \neq 0$.[32]

COROLLARY 5.4.8.   If $A = (x,y)$ is any point in the plane and $R_{y=mx+b}$ is the reflection about the line $y = mx + b$ such that $R_{y=mx+b}(A) = A'$, then $A' = (x'y')$, where $x' = x \cos 2\alpha + (y - b) \sin 2\alpha$ and $y' = x \sin 2\alpha - (y - b) \cos 2\alpha + b$, and $\alpha = \arctan m$. The proof is left as an exercise.

Example 5.4.5

(a) Let $R_{y=\sqrt{3}x}(x,y) = (x',y')$. Now since $m = \sqrt{3}$, we know that $\alpha = \arctan \sqrt{3} = 60°$, and $(x',y') = (x \cos 120° + y \sin 120°, x \sin 120° - y \cos 120°) = [-\frac{1}{2}x + (\sqrt{3}/2)y, (\sqrt{3}/2)x + \frac{1}{2}y]$. (b) Let $R_{y=(1/2)x+2}(x,y) = (x',y')$. Since $m = \frac{1}{2}$, we know that $\alpha = \arctan \frac{1}{2}$ which implies that $\sin \alpha = 1/\sqrt{5}$ and $\cos \alpha = 2/\sqrt{5}$. Using familiar trigonometric identities, $\sin 2\alpha = 2 \sin \alpha \cos \alpha = 2(1/\sqrt{5})(2/\sqrt{5}) = \frac{4}{5}$, and $\cos 2\alpha = \cos^2 \alpha - \sin^2 \alpha = (2/\sqrt{5})^2 - (1/\sqrt{5})^2 = \frac{3}{5}$. Now $R_{y=(1/2)x+2}(x,y) = (x',y') = [x(\frac{3}{5}) - (y - 2)(\frac{4}{5}), x(\frac{4}{5}) - (y - 2)(\frac{3}{5}) + 2]$.

Analytical Equations for Similarities

As in the previous section, we will begin our derivation of analytical equations for similarities with a special case that will provide additional insight into future derivations.

THEOREM 5.4.9.   If $A = (x,y)$ is any point in the plane and $H_{O,k}$ is a homothety centered at $O = (0,0)$ such that $H_{O,k}(A) = A'$, then $A' = (x',y')$, where $x' = kx$ and $y' = ky$.

Proof. Let $A = (x,y)$ and $H_{O,k}(A) = A'$, where $A' = (x',y')$ and $k > 1$.[33] By the definition of homothety, O-A-A' (Figure 5.4.5). Now by a previous theorem, since $\overline{AB} \parallel \overline{A'B'}$, $\triangle OBA$ is similar to $\triangle OB'A'$, and therefore $OA/OA' = OB/OB'$. Since we know that $OA'/OA = k$, then $OB'/OB = k$, and thus $x' = OB' = k(OB) = kx$. In similar fashion, $y' = ky$.

Let us now suppose that the homothety in Theorem 5.4.9 was not centered at the origin. As in the previous section, the strategy of first translating the center to the origin, then applying the homothety, and then translating back produces the following result.

---

[32] In this section we chose not to derive general equations for a glide reflection. This decision was based on the fact that a glide reflection is simply the product of a translation and a reflection, each of whose analytic equations are known.

[33] A derivation for $k < 0$ and $0 < k < 1$ are done similarly.

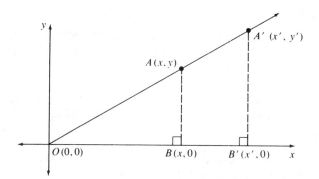

Figure 5.4.5

COROLLARY 5.4.10.   If $A = (x,y)$ is any point in the plane and $H_{P,k}$ $(A) = A'$, where $P = (p,q)$ then $A' = (x',y')$, where $x' = kx + (1 - k)p$ and $y' = ky + (1 - k)q$. The proof is left as an exercise.

Example 5.4.6

(a) $H_{(0,0),2} (3,1) = [2(3), 2(1)] = (6,2)$. (b) $H_{(0,0),-2} = (-6,-2)$. (c) $H_{(-1,4),1/3}$ $(2,5) = \{(\frac{1}{3})(2) + [1-(\frac{1}{3})](-1), (\frac{1}{3})(5) + [1 - (\frac{1}{3})](4)]\} = [(\frac{2}{3} + (-\frac{2}{3}), (\frac{5}{3}) + (\frac{8}{3})] = (0, \frac{13}{3})$.

Now, as we did in the previous section with glide reflections, we will refrain from deriving general equations for similarities, since all similarities are products of isometries and homotheties and the general equations for each of these has been previously discussed. Instead, the remainder of this section will consist of several specific examples.

Example 5.4.7

Find the analytical equations for a similarity composed of a rotation about the point $(0,0)$ through an angle of $30°$ and a homothety of ratio 2 about the point $(0,0)$.

$S_2(x,y) = H_{(0,0),2} \circ R_{(0,0),30°}(x,y) = H_{(0,0),2} [(\sqrt{3}/2)x - \frac{1}{2}y, \frac{1}{2}x + (\sqrt{3}/2)y] = (\sqrt{3}x - y, x + \sqrt{3}y)$.

Example 5.4.8

Find the analytical equations for a similarity composed of a reflection about the line $y = (\sqrt{3}/3)x$ and a homothety of ratio 1 : 3 about the point $(0,0)$.

$S_{1/3} (x,y) = H_{(0,0),1/3} \circ R_{y=(\sqrt{3}/3)x} (x,y) = H_{(0,0),1/3} (x \cos 60° + y \sin 60°, x \sin 60° - y \cos 60°) = H_{(0,0),1/3} [\frac{1}{2}x + (\sqrt{3}/2)y, (\sqrt{3}/2)x - \frac{1}{2}y] = [\frac{1}{6}x + (\sqrt{3}/6)y, (\sqrt{3}/6)x - \frac{1}{6}y]$.

Example 5.4.9

Find the analytical equations for a similarity composed of a reflection about the line $y = \frac{3}{4}x + 1$ [see Exercise Set 5.4.1, Problem 1(j)] and a homothety of ratio -5 about the point $(0,1)$.

$S_{-5}(x,y) = H_{(0,1),-5} \circ R_{y=(3/4)x+1}(x,y) = H_{(0,1),-5}[\frac{7}{25}x + \frac{24}{25}(y-1), \frac{24}{25}x - \frac{7}{25}(y-1) + 1] = \{-5[\frac{7}{25}x + \frac{24}{25}(y-1)] + [1-(-5)](0), (-5)[\frac{24}{25}x - \frac{7}{25}(y-1) + 1] + [1-(-5)](1)\} = [(-\frac{1}{5})(7x + 24y - 24), (-\frac{1}{5})(24x - 7y + 2)].$

Example 5.4.10

Find the image of $\triangle ABC$ where $A = (0,0)$, $B = (1,0)$, and $C = (1,1)$ under the similarity described in Example 5.4.9.

From Example 5.4.9, $S_{-5}(x,y) = [(-\frac{1}{5})(7x + 24y - 24), (-\frac{1}{5})(24x - 7y + 2)]$. Then $S_{-5}(A) = S_{-5}(0,0) = \{(-\frac{1}{5})[7(0) + 24(0) - 24)], (-\frac{1}{5})[24(0) - 7(0) + 2]\} = (\frac{24}{5}, -\frac{2}{5})$, $S_{-5}(B) = S_{-5}(1,0) = \{-\frac{1}{5}[7(0) + 24(1) - 24], (-\frac{1}{5})[24(0) - 7(1) + 2]\} = (\frac{17}{5}, -\frac{26}{5})$, and $S_{-5}(C) = S_{-5}(1,1) = \{(-\frac{1}{5})[7(1) + 24(1) - 24], (-\frac{1}{5})[24(1) - 7(1) + 2]\} = (-\frac{7}{5}, -\frac{19}{5})$.

Applications of Isometries and Similarities Using Analytical Transformations

The purpose of the two previous sections was to assist in the derivation of a variety of analytical equations representing transformations of the plane. With these equations in hand, we are now able to apply analytic transformational techniques to the area of geometric theorem proving. In addition, analytical transformations have many applications in the field of computer graphics.

Our procedure in proving theorems will be a combination of those used in Sections 5.2.5, 5.3.4, and 5.3.5. First, we will coordinatize our figures, then we will determine which isometry or similarity meets our needs, and finally, using analytical equations, we will show that our chosen isometry or similarity actually accomplishes our purpose. The following examples demonstrate the necessary procedure.

THEOREM.   The base angles of an isosceles triangle are congruent.

Proof. Let $\triangle ABC$ be isosceles with $\overline{AC} \cong \overline{BC}$, and let $A = (0,0)$, $B = (2b,0)$, and $C = (b,c)$, where b and c are real numbers with $b \neq 0$ and $c > 0$ (Figure 5.4.6). Let $\overleftrightarrow{CD}$ be the altitude of $\triangle ABC$, where $D = (b,0)$, and show that $\angle CAB$ is the image of $\angle CBA$ under the reflection $R_{\overleftrightarrow{CD}}$. Since $\overleftrightarrow{CD}$ is a vertical line, the equation of $\overleftrightarrow{CD}$ is $x = b$, and $R_{\overleftrightarrow{CD}}(x,y) = R_{x=b}(x,y) = (x',y')$, where $(x',y') = (-x + 2b, y)$. Now $R_{x=b}(A) = R_{x=b}(0,0) = (0 + 2b, 0) = B$, $R_{x=b}(B) = R_{x=b}(2b,0) = (-2b + 2b, 0) = (0,0) = A$, and $R_{x=b}(C) = R_{x=b}(b,c) = (-b + 2b, c) = (b,c) = C$, and the proof is complete.

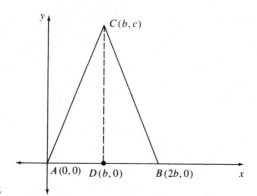

Figure 5.4.6

THEOREM.   The diagonals of a square are congruent.

Proof. Let $\square PQRS$ be a square such that $P = (2p,0)$, $Q = (0,0)$, $R = (0,2p)$, and $S = (2p,2p)$, where p is a real number, and show that $\overline{PR}$ is the image of $\overline{QS}$ under the rotation $R_{M,90°}$, where M is the midpoint of $\overline{QS}$. The coordinates of M are (p,p), and $R_{M,90°}(x,y) = (x',y') = [(x - p) \cos 90° - (y - p) \sin 90° + p, (x - p) \sin 90° + (y - p) \cos 90° + p] = [(x - p)(0) - (y - p)(1) + p, (x - p)(1) + (y - p)(0) + p] = (-y + 2p,x)$. Now $R_{M,90°}(Q) = R_{M,90°}(0,0) = (0 + 2p,0) = (2p,0) = P$, and $R_{M,90°}(S) = R_{M,90°}(2p,2p) = (-2p + 2p, 2p) = (0,2p) = R$. Therefore $R_{M,90°}(\overline{QS}) = \overline{PR}$, hence $\overline{QS} = \overline{PR}$.

THEOREM.   If a line intersects two sides of any triangle in distinct points and is parallel to the third side, then the new triangle formed is similar to the original triangle.

Proof. Let $\triangle ABC$ be coordinatized as $A = (0,0)$, $B = (a,0)$, and $C = (c,d)$, where a, c, d, and f are real numbers and $a \neq 0$, and let $\overrightarrow{DE}$ intersect

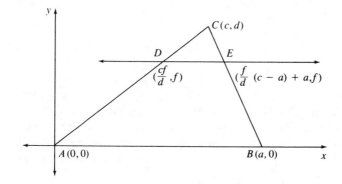

Figure 5.4.7

sides $\overline{AC}$ and $\overline{BC}$ at points $D = (cf/d,f)$ and $E = [(f/d)(c - a) + a, f]$, respectively, where $a \neq 0$, $c > 0$, and $0 < f < d$ (Figure 5.4.7). If we now show that $H_{C,k} (\triangle CDE) = \triangle CAB$, where $k = AC/DC = d/(d - f)$, we are finished. Using the results of Corollary 5.4.10, we find that the analytical representation of $H_{C,k}$ is $H_{C,k} (x,y) = [kx + (1 - k)c, ky + (1 - k)d]$, where $k = d/(d - f)$. Clearly, $H_{C,k} (C) = C$. Now $H_{C,k} (D) = H_{C,k} (cf/d,f) = [k(cf/d) + (1 - k)c, kf + (1 - k)d] = \{[d/(d - f)](cf/d) + [1 - (d/(d - f))]c, [d/(d - f)]f + [1 - d/(d - f)]d\} = [0/(d - f), 0/(d - f)] = (0,0) = A$. In similar fashion, $H_{C,k} (E) = B$, and as a result, $H_{C,k}(\triangle CDE) = \triangle CAB$, hence $\triangle CDE$ is similar to $\triangle CAB$.

Finally, the following problem provides an example of the use of a similarity to prove two triangles similar. In order to simplify algebraic calculations, a specific triangle has been chosen.

Problem

In Figure 5.4.8, $\triangle ABC$ has been coordinatized as follows: $A = (0,0)$, $B = (4,0)$, and $C = (1,\sqrt{3})$, and $\overline{CD}$ is the altitude to side $\overline{AB}$. Show that $\triangle ADC$ is similar to $\triangle ACB$.

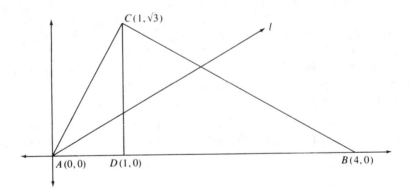

Figure 5.4.8

Solution: We will consider the similarity $S_k = H_{A,k} \circ R_l$, where $k = AC/AD$ and l is the line that bisects $\angle CAB$. First, we must find the analytical representation of R_l. To do so, we will use Theorem 5.4.7 and observe the fact that it is not necessary to know the equation of l explicitly,[34] but that it suffices to recognize that l bisects $\angle CAB$ and that the required angle of 2α is

---

[34] In this special case the coordinates indicate that $m\angle CAD = 60°$ and that the equation of l is $y = (\sqrt{3}/3)x$. But, as the remainder of the proof indicates, neither fact must be specifically stated to complete the proof.

determined by the slope of $\overrightarrow{AC}$. As a result, $\cos 2\alpha = \frac{1}{2}$ and $\sin 2\alpha = \sqrt{3}/2$, and therefore $R_l(x,y) = [\frac{1}{2}x + \sqrt{3}/2y \ (\sqrt{3}/2)x + \frac{1}{2}y]$. Next, as a result of Theorem 5.4.9, $H_{A,k}(x,y) = (kx,ky)$, and since $k = AC/AD = \frac{2}{1}$, $H_{A,k}(x,y) = (2x,2y)$. Finally, by composing R_l and $H_{A,k}$, we find that $S_2(x,y) = (H_{A,k} \circ R_l)(x,y) = (x + \sqrt{3}y, \sqrt{3}x - y)$. We must now show that $S_2(\triangle ADC) = \triangle ACB$. Clearly, $S_2(A) = A$. Now $S_2(D) = S_2(1,0) = (H_{A,k} \circ R_l)(1,0) = [1 + (\sqrt{3})(0), (\sqrt{3})(0)] = (1,\sqrt{3}) = C$, and $S_2(C) = S_2(1,\sqrt{3}) = [1 + (\sqrt{3})(\sqrt{3}), (\sqrt{3})(1) - \sqrt{3}] = (4,0) = B$, and therefore $S_2(\triangle ADC) = \triangle ACB$.

EXERCISE SET 5.4

Analytical Equations for Isometries

1. Find the analytical equations for each of the following:
   (a) A translation through a distance of 3 units at an angle of 30° with the horizontal
   (b) A translation through vector **RS**, where $R = (p,q)$ and $S = (r,s)$
   (c) A rotation of 30° about the point $(2,-1)$
   (d) A rotation of 60° about the point $(2,-1)$
   (e) A rotation of 135° about the point $(2,-1)$
   (f) A half-turn about the point $(2,-1)$
   (g) A reflection about the line $y = (\sqrt{3}/3)x$
   (h) A reflection about the line $y = (\sqrt{3}/3)x + 4$
   (i) A reflection about the line $y = \frac{3}{4}x$
   (j) A reflection about the line $y = \frac{3}{4}x + 1$
   (k) A glide reflection that is the product of $R_{y=(\sqrt{3}/3)x}$ and T_{OP}, where $P = (2\sqrt{3}, 2)$
   (l) A glide reflection using a reflection in the line in part (j) and a translation through a vector whose length is 3

2. Prove Theorem 5.4.2.

3. Prove Theorem 5.4.5.

4. Prove Corollary 5.4.6. (*Hint:* Consider $T_{(0,k)} \circ R_{y=0} \circ T_{(0,-k)}$.)

5. Prove Corollary 5.4.8. (*Hint:* Consider $T_{(0,b)} \circ R_y = mx \circ T_{(0-b)}$.)

Analytical Equations for Similarities

6. Find the analytical equations for each of the following similarities:
   (a) A homothety centered at $(0,0)$ with ratio 3.
   (b) A similarity composed of a rotation about the point $(1,2)$ through an angle of 45° and a homothety of ratio 2 about the point $(1,2)$.
   (c) A similarity composed of a rotation about $(-2,-1)$ through an angle of 45° and a homothety of ratio 2 about $(1,2)$.
   (d) A similarity composed of a reflection over the line $y = (\sqrt{3}/3)x + 4$ and a homothety of ratio $\frac{1}{2}$ about the point $(3,2)$.
   (e) A similarity composed of a reflection over the line $y = (\sqrt{3}/3)x + 4$ and a homothety of ratio -2 about the point $(3,2)$.

7. Find the image of the $\triangle ABC$ where $A = (0,0)$, $B = (3,0)$, and $C = (3,4)$, under each of the transformations in Problem 6.

8. Prove Corollary 5.4.10.

9. Prove: Each isometry can be written analytically in the form, $\alpha\,(x,y) = (x',y')$, where $x' = ax - by + c$, $y' = \pm(bx + ay) + d$, and a, b, c, and d are real numbers with $\sqrt{a^2 + b^2} = 1$.

10. Prove: Each similarity can be written analytically in the form, $S_k(x,y) = (x',y')$, where $x' = ax - by + c$, $y' = \pm(bx + ay) + d$, and a, b, c, and d are real numbers with $\sqrt{a^2 + b^2} = k$.

11. Prove: If a, b, c, and d are real numbers with $a^2 + b^2 \neq 0$ and $\sigma(x,y) = (x',y')$, where $x' = ax - by + c$, $y' = \pm(bx + ay) + d$, then σ is a similarity.

12. Prove, analytically that the composition of two similarities is a similarity.

13. Prove that in a similarity the isometry and the homothety are commutative.

14. Find the equations of two similarities, one direct and one opposite, each of which maps line segment $\overline{AB}$ onto line segment $\overline{A'B'}$, where $A = (1,0)$, $B = (2,3)$, and $A' = (-1,2)$, $B' = (-3,-3)$.

15. Find the equations of the similarity that maps $\triangle ABC$ onto $\triangle A'B'C'$ where $A = (0,0)$, $B = (1,0)$, and $C = (0,2)$ and where (i) $A' = (3,0)$, $B' = (3,2)$, and $C' = (7,0)$, and (ii) $A' = (-5,5)$, $B' = (-6,\frac{7}{2})$, and $C' = (-2,3)$.

Applications of Analytical Transformations

16. Using analytical representations for transformations, show that $\triangle ABC \cong \triangle DEF$ where $A = (1,1)$, $B = (4,2)$, $C = (5,4)$, $D = (5,3)$, $E = (8,4)$, and $F = (9,6)$.

17. Using analytical representations for transformations, show that $\triangle GHI \cong \triangle JKI$ where $G = (0,0)$, $H = (2,2)$, $I = (3,4)$, $J = (6,8)$, and $K = (4,6)$.

18. Using analytical representations for transformations, show that $\triangle LMN \cong \triangle OPQ$ where $L = (1,3)$, $M = (2,7)$, $N = (4,6)$, $O = (3,1)$, $P = (7,2)$, and $Q = (6,4)$.

19. Using analytical representations for transformations, show that $\triangle RST \cong \triangle RUV$ where $R = (0,0)$, $S = (1,0)$, $T = (1,2)$, $U = (3,0)$, and $V = (3,6)$.

20. Using analytical representations for transformations, show that $\triangle WYZ$ is similar to $\triangle ZYX$ where $W = (0,0)$, $X = (4,0)$, $Y = (6,0)$, and $Z = (3,\sqrt{3})$.

21. Using analytical representations for transformations, prove that the diagonals of a parallelogram determine congruent triangles.

22. Using analytical representations for transformations, prove that the diagonals of a rhombus are perpendicular.

23. Using analytical representations for transformations, prove that the composition of two translations is another translation.

24. Using analytical representations for transformations, prove that if $\triangle ABC$ is equilateral, then $R_{P,120°}(\triangle ABC) = \triangle BCA$ and that $R_{P,240°}(\triangle ABC) = \triangle CAB$, where P is the point of intersection of the altitudes (medians) of $\triangle ABC$.

25. Using analytical representations for transformations, prove that if $\triangle ABC$ is equilateral, then
    (a) $R_l(\triangle ABC) = \triangle ACB$, where l is the altitude from A to $\overline{BC}$
    (b) $R_l(\triangle ABC) = \triangle BAC$, where l is the altitude from C to $\overline{AB}$
    (c) $R_l(\triangle ABC) = \triangle CBA$, where l is the altitude from B to $\overline{AC}$

26. Using analytical representations for transformations, prove that the product of two rotations about the same point is another rotation about that point.

27. Using analytical representations for transformations, prove that the product of a reflection about any line l with itself is the identity transformation.

28. Using analytical representations for transformations, prove that the product of reflections about two parallel lines is a translation.
29. Using analytical representations for transformations, prove that the product of reflections about two intersecting lines is a rotation.

5.5 INVERSION

Introduction

In previous sections of this chapter we investigated two specific groups of transformations and the properties they preserve: (1) the group of isometries that preserve the congruence of geometric figures and all of the associated properties such as distance, collinearity, betweeness, and angle congruence, and (2) the group of similarities that preserve the similarity of geometric figures and all of the same associated properties with the notable exception of distance. Lest we be led to believe that all useful transformations preserve either congruence or similarity of geometric figures, we shall investigate a significantly different type of geometric transformation, called *inversion*, which not only fails to preserve the congruence or similarity of figures but in fact transforms certain lines into circles, and vice versa. In addition to studying inversions because of their unique properties, we shall find some interesting connections with previously studied topics and set the stage for applications in Chapter 6.

Inversion in a Circle

We shall begin our investigation of inversion with the following definition:

DEFINITION.   Let O be the center of a fixed circle of radius r in the Euclidean plane. Furthermore, let P be any point in the plane other than O. An *inversion in circle O, $I_{O,r}$*, is a function such that if $I_{O,r}(P) = P'$ then $P' \in \overrightarrow{OP}$ and $(OP)(OP') = r^2$. Here P' is called the *inverse* of P, O is called the *center of inversion*, r is called the *radius of inversion*, and r^2 is called its *power*.

In order that inversion be a transformation of the plane, it is necessary that the center of inversion O have an image under $I_{O,r}$. Clearly, no ordinary Euclidean point can satisfy the definition; therefore it is necessary to append to the Euclidean plane a special point Ω. Point Ω is called an *ideal point* or a *point at infinity*[35] and is defined to be the inverse of the center of inversion. Point Ω is considered to lie on every line in the plane. Also, as a consequence of the definition, it should be obvious that $I_{O,r}(\Omega) = O$.

The addition of an ideal point to the Euclidean plane precipitates the following definition:

DEFINITION.   The Euclidean plane with one ideal point appended to it is referred to as the *inversive plane*.

With the definition of an inversive plane, we now are able to prove the following properties of inversion.

THEOREM 5.5.1.   Inversion is a transformation of the inversive plane. The proof entails showing that an inversion is a one-to-one mapping of the inversive plane onto itself and is left as an exercise.

THEOREM 5.5.2.   Points on a circle of inversion are fixed.

Proof. Let O be the center of a circle of inversion whose radius is r, and let P be a point on the circle. We must now show that $I_{O,r}(P) = P$. Suppose that $I_{O,r}(P) = X$. By definition, $X \in \overrightarrow{OP}$ and $(OP)(OX) = r^2$. Now since $OP = r$, it follows that $OX = r$, and thus $X = P$.

THEOREM 5.5.3.   If a point is in the interior of a circle of inversion, then its image is a point in the exterior of the circle, and conversely. The proof is left as an exercise.

Before we proceed further in our investigation of the properties of inversion, it may prove informative to construct, using the compass and straightedge techniques developed in the last chapter, the image of a point under inversion in a circle. As a previous theorem has indicated, if P is on the circle of inversion, it is fixed, therefore we must consider only the cases where P either lies in the interior of the circle or in its exterior.

Case 1. Let O be the center of a fixed circle of radius r and suppose that P is in the interior of circle O (Figure 5.5.1). First, we will construct a chord perpendicular to $\overrightarrow{OP}$ at P. Let Q be one end of that chord and construct the tangent line $\overleftrightarrow{QS}$ to circle O at point Q. Now $\overleftrightarrow{QS}$ will intersect $\overrightarrow{OP}$ at the desired point P'. The fact that P' is actually the inverse of P in circle O is an immediate consequence of Problem 15b in Exercise Set 4.4, which states that the length of a leg of a right triangle is the geometric mean between the

---

[35] This name seems particularly appropriate, since as P approaches O, OP approaches zero and therefore OP' increases without bound; thus, in the limiting case, the image of O must be a "point at infinity."

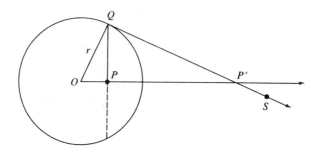

Figure 5.5.1

length of the hypotenuse and the length of that leg's projection on the hypotenuse. Hence $OP/r = r/OP'$, and therefore $(OP)(OP') = r^2$ as required.

Case 2. Let O be the center of a fixed circle of radius r and suppose that P is in the exterior of circle O (Figure 5.5.2). First we locate M, which is the midpoint of $\overline{OP}$. Now using M as its center and $\overline{OM}$ as its radius, we construct a circle that intersects circle O at points Q and R. Then $\overleftrightarrow{QR}$ intersects $\overline{OM}$ at the desired point P'. The proof that P' is the inverse of P is left as an exercise.

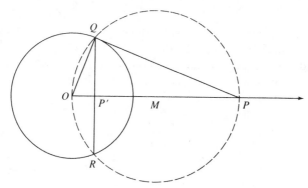

Figure 5.5.2

It should be apparent from its definition and from the previous discussion that inversion does not preserve distance. It is true, however, that points "near" each other before inversion have images that remain "near" each other following the inversion. We find therefore that "neighborhoods" are preserved and, as a consequence, inversion is a continuous transformation, a property that will prove useful in the future.

Our next logical step is to investigate whether or not inversion preserves angle measure. That is, if A', B', and C' are the images of non-collinear points A, B, and C under an inversion, is it true that $m\angle A'B'C' = m\angle ABC$? Before addressing this question, we need the following result:

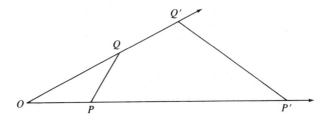

Figure 5.5.3

THEOREM 5.5.4.   Let O be the center of an inversion whose radius is r. If P and Q are distinct points that are not both collinear with O, and if P' and Q' are the inverses of P and Q, respectively, then $\triangle OPQ$ is similar to $\triangle OQ'P'$ (Figure 5.5.3).[36]

Proof. Since P' and Q' are the inverses of P and Q under an inversion whose center is O and whose radius is r, we know that $(OP)(OP') = r^2$ and $(OQ)(OQ') = r^2$, and consequently that $OP'/OQ = OQ'/OP$. Now consider the similarity $S_k = H_{O,k} \circ R_l$, where l is the bisector of $\angle OPQ$ and $k = OQ'/OP$. Since $S_k(\triangle OPQ) = \triangle OQ'P'$ (see Exercise Set 5.5, Problem 6), $\triangle OPQ$ is similar to $\triangle OQ'P'$ as required.

We are now prepared to consider the original question, Does inversion preserve angle measure? Let O be the center of an inversion of radius r, and let A, B, and C be three noncollinear points, no two of which are collinear with O. Now suppose that A', B',C', are the inverses of A, B, and C, respectively (Figure 5.5.4).

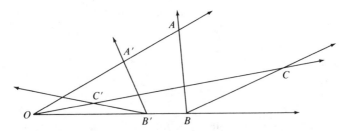

Figure 5.5.4

Using results from elementary Euclidean geometry, we see that $m\angle AOC + m\angle OAB = m\angle ABC + m\angle OCB$, and thus $m\angle ABC = m\angle AOC + m\angle OAB - m\angle OCB$. Now as a consequence of Theorem 5.5.4, $m\angle OAB = m\angle OB'A'$ and $m\angle OCB = m\angle OB'C'$. Substituting, we find that

---

[36] In the interest of simplicity, since the actual circle of inversion is not essential to our discussion, it has been eliminated from our figures.

$m\angle ABC = m\angle AOC + m\angle OB'A' - m\angle OB'C'$. But since $m\angle A'B'C' = m\angle OB'A' - m\angle OB'C'$, we find that $m\angle ABC = m\angle AOC + m\angle A'B'C'$. Therefore unless $m\angle AOC = 0$, that is, unless A, O, and C are collinear, we find that $m\angle ABC \neq m\angle A'B'C'$. Therefore, in general, if $I_{O,r}(A) = A'$, $I_{O,r}(B) = B'$, and $I_{O,r}(C) = C'$, $m\angle ABC \neq m\angle A'B'C'$.

In light of the previous result, the following theorem may seem contradictory:

THEOREM 5.5.5.   The measure of the angle between two curves is invariant under inversion.

Proof. Let O be the center of an inversion of radius r, and let c_1 and c_2 be two curves that intersect at point P ($\neq O$). Now suppose that $\overrightarrow{OQ}$, which is distinct from $\overrightarrow{OP}$, intersects curves c_1 and c_2 at points Q and R, respectively, and furthermore, suppose that P', Q', R', c_1', and c_2' are the images of P, Q, R, c_1, and c_2 under $I_{O,r}$ (Figure 5.5.5).

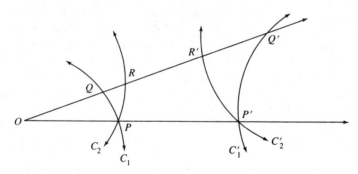

Figure 5.5.5

By Theorem 5.5.4, $\triangle OPQ$ is similar to $\triangle OQ'P'$ and $\triangle OPR$ is similar to $\triangle OR'P'$. Using the corresponding congruent angles, it can easily be shown that $\angle QPR \cong \angle R'PQ'$ (see Exercise Set 5.5, Problem 7). Now in the limiting case,[37] as $m\angle POQ$ approaches zero, each of the secant lines $\overrightarrow{PQ}$, $\overrightarrow{PR}$, $\overrightarrow{P'Q'}$, and $\overrightarrow{P'R'}$ becomes a tangent line and $\angle QPR$ and $\angle R'PQ'$ become the angles between c_1 and c_2 and between c_1' and c_2', respectively, and we are finished.

Why is it then, that the measure of the angle between two curves is preserved while the angle between two straight lines generally is not? The answer, as we will see from the following theorem, lies in the fact that the image of a straight line under an inversion is not always a straight line.

[37] Recall that inversion is a continuous transformation.

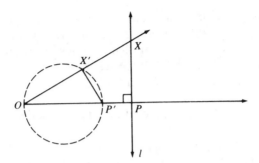

Figure 5.5.6

THEOREM 5.5.6.   If $I_{O,r}$ is an inversion and l is any line that does not contain O, then the image of l under $I_{O,r}$ is a circle that contains O.

Proof. Let O be a point, let l be any line that does not contain O, and let m be a line through O that is perpendicular to l at point P. Now suppose that X is any other ordinary point on l and that $I_{O,r}(P) = P'$ and $I_{O,r}(X) = X'$ (Figure 5.5.6). By Theorem 5.5.4, $\triangle OPX$ is similar to $\triangle OX'P'$, and thus $\angle OPX \cong \angle OX'P'$. Now since $\angle OPX$ is a right angle, $\angle OX'P'$ must be one also, and as a consequence, each X' lies on a circle whose diameter is $\overline{OP'}$. Therefore the image of each ordinary point on l is a point on a circle containing O. Finally, if Ω is the ideal point on l, by definition, $I_{O,r}(\Omega) = O$ and thus lies on the same circle. To complete the proof, it must also be shown that the image of l is the entire circle, which is left as an exercise.

As a result of this theorem, the apparent contradiction created by Theorem 5.5.5 is resolved when we note that, in Figure 5.5.4, $\overleftrightarrow{AB}$ does not contain the center of the inversion and therefore its image is not $\overleftrightarrow{A'B'}$ but is a circle containing A' and B' and thus $I_{O,r}(\angle ABC) \neq \angle A'B'C'$.

The following theorems and corollaries provide additional insight into inversion as a transformation of the inversive plane.

COROLLARY 5.5.7.   If $I_{O,r}$ is an inversion and α is a circle that contains O, then the image of α is a line parallel to the tangent to circle α at O. The proof is left as an exercise.

THEOREM 5.5.8.   If $I_{O,r}$ is an inversion and β is a circle that does not contain O, then the image of β is a circle that does not contain O.

Proof. Let l be the line joining point O and the center of circle β. Then l intersects β at points Q and R, which are endpoints of a diameter. Now let X be any other point on β, and let X', Q', and R' be the images of X, Q, and R under $I_{O,r}$ (Figure 5.5.7). By Theorem 5.5.4, $\triangle OQX$ is similar to $\triangle OX'Q'$

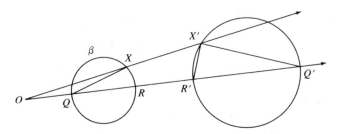

Figure 5.5.7

and $\triangle ORX$ is similar to $\triangle OX'R'$, and as a result, their corresponding angles are congruent. Now using results from elementary Euclidean geometry and the fact that $\angle QXR$ is inscribed in a semicircle, it is easily shown that $\angle Q'X'R'$ is a right angle, and thus the locus of points X' is a circle whose diameter is $\overline{Q'R'}$.

THEOREM 5.5.9.   If $I_{O,r}$ is an inversion and l is a line containing O, then l is invariant (but not pointwise invariant). The proof is left as an exercise.

THEOREM 5.5.10.   Any circle other than the circle of inversion is invariant if and only if it is orthogonal[38] to the circle of inversion.

Proof. Let $I_{O,r}$ be an inversion in circle O of radius r. First, we will show that any circle that is orthogonal to circle O is invariant. Let γ be a circle orthogonal to circle O. If P and Q are the points of intersection of circle O and γ, then $\overrightarrow{OP}$ and $\overrightarrow{OQ}$ are tangent to γ at P and Q, respectively (Figure 5.5.8). Now let X be any other point on γ, and let $\overrightarrow{OX}$ intersect γ at X'. By Theorem 4.6.15, $(OX)(OX') = (OP)^2 = r^2$, and therefore $I_{O,r}(X) = X'$. Now since P and Q remain fixed under $I_{O,r}$ and the image of any other point of γ is another point on γ, then γ is invariant under $I_{O,r}$.[39] Conversely, we must show that if δ is any circle (other than the circle of inversion) that remains invariant under $I_{O,r}$, then δ is orthogonal to circle O. It should be clear that since δ is invariant, it may not lie entirely within the interior of circle O or within its exterior. Therefore circles O and δ must intersect in distinct points R and S, and O is in the exterior of δ. Furthermore, since δ is invariant, if Y is any other point on δ such that $I_{O,r}(Y) = Y'$, then Y' must be on δ, and $(OY)(OY') = (OR)^2$. Now there exists a tangent from O to circle δ at some point R_1, and $(OR_1)^2 = (OY)(OY')$. Substituting, we find that $OR_1 =$

---

[38] Two circles that intersect at distinct points are said to be *orthogonal* if and only if their respective tangents at the points of intersection are perpendicular.

[39] Recall that in Euclidean geometry three noncollinear points uniquely determine a circle.

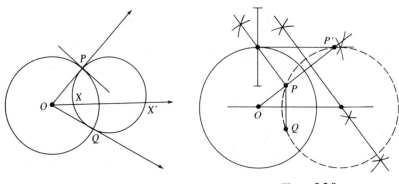

Figure 5.5.8   Figure 5.5.9

OR, and therefore R_1 lies on circle O and R_1 is either R or S; but in either case, circles O and δ have perpendicular radii and are orthogonal.

COROLLARY 5.5.11.   If P and P' are distinct points that are images of each other under an inversion in circle O, then any circle containing both P and P' is orthogonal to circle O. The proof is an application of Theorem 5.5.10 and is left as an exercise.

Problem

Given a circle whose center is the point O and whose radius is r, and two ordinary points P and Q that are not collinear with O, construct a circle containing P and Q that is orthogonal to circle O.

Solution: Theorem 5.5.10 and Corollary 5.5.11 tell us that any circle containing P and Q that is orthogonal to circle O must also contain the images of P and Q under the inversion $I_{O,r}$. Since a circle is uniquely determined by three noncollinear points, it is sufficient to construct $P' = I_{O,r}(P)$ and locate the center of the required circle that lies on the point of intersection of the perpendicular bisectors of $\overline{PP'}$ and $\overline{PQ}$ (Figure 5.5.9).

The following set of exercises provides examples of inversion and its applications. In the subsequent chapter these applications play a significant role in the construction of a model for non-Euclidean geometry.

EXERCISE SET 5.5

1. Prove Theorem 5.5.1.
2. Prove Theorem 5.5.3.
3. Complete the proof that justifies the construction of the image of a point lying in the exterior of the circle of inversion.

4. Let $I_{O,r}$ be an inversion with $r = 2$ units. If P is 4 units from O, construct P', the image of P under $I_{O,r}$.
5. Suppose that $I_{O,r}$ is as described in Problem 4, and let Q be $\frac{1}{2}$ unit from O. Construct Q', the image of Q under $I_{O,r}$.
6. In the proof of Theorem 5.5.4 we stated that $S_k (\triangle OPQ) = \triangle OQ'P'$. Justify this statement.
7. In the proof of Theorem 5.5.5 we claimed that $\angle QPR \cong \angle R'PQ'$. Justify this statement.
8. Prove Corollary 5.5.7.
9. Suppose that $I_{O,r}$ is as described in Problem 4 and that α is a circle whose radius is 1 unit and whose center is 3 units from O. Construct the image of α under $I_{O,r}$.
10. Prove Theorem 5.5.9.
11. Prove Corollary 5.5.11.
12. Prove that the composition of inversions in two concentric circles with center O and with radii r_1 and r_2 is the homothety $H_{O,k}$ where $k = (r_2/r_1)^2$.
13. In Section 5.4 we derived analytical equations for isometries and similarities. Derive the equations for $I_{O,r}$ where O is $(0,0)$ and r is any positive real number.
14. Using the results from Problem 13, derive the equations for $I_{P,r}$ where P is (h,k) and r is any real number.

CHAPTER 5 SUMMARY

5.2 Analytic Geometry

Theorem 5.2.1.   Two intersecting lines determine a unique plane.

Theorem 5.2.2.   If $M = (x_1,y_1)$ and $N = (x_2,y_2)$ are any two points in the plane, then $MN = \sqrt{(x_1 - x_2)^2 + (y_1 - y_2)^2}$.

Theorem 5.2.3.   Given $A = (x_1,y_1)$, $B = (x_2,y_2)$, and r which is any positive real number, if P is on $\overline{AB}$ such that $A\text{-}P\text{-}B$ and $AP/PB = r$, then P has coordinates: $\left(\dfrac{x_1 + rx_2}{1 + r}, \dfrac{y_1 + ry_2}{1 + r} \right)$.

Theorem 5.2.4.   If l is any nonvertical line having slope m, then the coordinates of every point $P = (x,y)$ on l must satisfy an equation of the form $y = mx + k$ where k is a real number.

Theorem 5.2.5.   Parallel lines are lines having the same slope.

Theorem 5.2.6.   Let l_1 and l_2 be two nonvertical lines with slopes m_1 and m_2, respectively. If l_1 and l_2 are perpendicular, then $m_2 = -1/m_1$.

Theorem 5.2.7.   The angle α, measured counterclockwise from line l_2, whose slope is m_2, to line l_1, whose slope is m_1, is

$$\alpha = \arctan\left(\frac{m_1 - m_2}{1 + m_1 m_2}\right)$$

Theorem 5.2.8. Any point, $P = (x,y)$ on a circle whose center is $O = (h,k)$ and whose radius has length r must satisfy an equation of the form $(x - h)^2 + (y - k)^2 = r^2$.

Theorem 5.2.9. The opposite sides of a parallelogram are congruent.

Theorem 5.2.10. The diagonals of a parallelogram bisect each other.

Theorem 5.2.11. The medians of a triangle are concurrent.

Theorem 5.2.12. The base angles of an isosceles trapezoid are congruent.

Theorem 5.2.13. Any three noncollinear points lie on a circle.

5.3 Transformational Geometry

Definition. Let A and B be sets. Then a *mapping* or *function f from A to B* is a rule that assigns to each element x in A exactly one element y in B [denoted $y = f(x)$]. The set A is called the *domain of f*, y is called the *image of x under f*, and x is called the *preimage of y under f*. The set of all images of elements of A under f (which is some subset of B) is called the *range of f*.

Definition. A mapping f from A to B is *onto* B if for any y in B there is at least one x in A for which $f(x) = y$.

Definition. A mapping f from A to B is *one-to-one* if each element of the range has exactly one preimage. For example, if a and b are elements of A such that $f(a) = f(b)$, then $a = b$.

Definition. If $f: A \rightarrow B$ and $g: B \rightarrow C$ are mappings, then the *product* (or composition) of f and g, $g \circ f$, is a mapping $h: A \rightarrow C$ such that $h(x) = g(f(x))$.

Definition. If $f: A \rightarrow B$ is a one-to-one, onto mapping, then $f^{-1}: B \rightarrow A$ is called the *inverse* of f if and only if $(f \circ f^{-1})(x) = x$ for all $x \in A$ and $(f^{-1} \circ f)(y) = y$ for all $y \in B$.

Definition. A *transformation of the plane t* is a one-to-one mapping of points of the plane onto points in the plane.

Definition. A transformation of the plane for which distance between points is invariant is called an *isometry*. For example, if t is a transformation of the plane and P and Q are any points in the plane such that $t(P) = P'$ and $t(Q) = Q'$, and if $PQ = P'Q'$, then t is called an isometry.

Definition. Two figures are said to be *congruent* if and only if one is the image of the other under some isometry.

Definition. If **PQ** is a vector, then a *translation* through vector **PQ**, denoted by T_{PQ}, is a transformation of the plane such that if A is any point in the plane, and $T_{PQ}(A) = A'$, then AA' and PQ are equal vectors.

Definition. A *rotation* $R_{P,\theta}$ about a point P through an angle θ is a transformation of the plane where $R_{P,\theta}(P) = P$, and if $A \ne P$, $R_{P,\theta}(A) = A'$ such that $\overline{PA} \cong \overline{PA'}$ and m$\angle APA' = \theta$. And P is called the center of the rotation.

Definition. A *reflection* R_l in a line l, is a transformation of the plane where if A is not on l, then $R_l(A) = A'$ such that l is the perpendicular bisector of $\overline{AA'}$, and if P is on l, then $R_l(P) = P$.

Definition. Any point that is its own image under a transformation is called an *invariant* or *fixed point*.

Definition. $\triangle ABC$ has a *clockwise* (or *counterclockwise*) *orientation* if when we traverse ABC so as to encounter A,B, and C, in that order, the direction traversed is clockwise (or counterclockwise).

Definition. If, under a transformation, figure orientation is invariant, then the transformation is called a *direct transformation*. Otherwise, the transformation is an *opposite transformation*.

Definition. A *glide reflection* $G_{EH,l}$ is the product of R_l and T_{EH}, where $EH \parallel l$; that is, $G_{EH,l} = R_l \circ T_{EH} = T_{EH} \circ R_l$.

Definition. A *homothety* $H_{O,k}$, where O is a fixed point and k is a nonzero real number, is a transformation of the plane where $H_{O,k}(O) = O$, and if $P \ne O$, then $H_{O,k}(P) = P'$, such that O, P, and P' are collinear and $OP' = k(OP)$. If $k > 1$, then O-P-P'; if $0 < k < 1$, then O-P'-P and if $k < 0$, then P'-O-P. Also, O is called the *center* of the homothety, and k is called the *constant of proportionality*.

Definition. A *similarity* S_k is a transformation of the plane that is the product of an isometry and a homothety with ratio k.

Definition. Two polygons are similar if and only if one is the image of another under some similarity

Theorem 5.3.1.   The product of two transformations of the plane is a transformation of the plane.

Theorem 5.3.2.   The product of two isometries is an isometry.

Theorem 5.3.3.   Collinearity is invariant under an isometry.

Theorem 5.3.4.   Betweeness is invariant under an isometry.

Corollary 5.3.5.   The image of a line segment (or ray, angle, or triangle) under an isometry is a line segment (or ray, angle, or triangle).

Corollary 5.3.6. The image of a line segment under an isometry is a congruent line segment.

Theorem 5.3.7. The image of a triangle under an isometry is a congruent triangle.

Corollary 5.3.8. Angle measure is invariant under an isometry.

Theorem 5.3.9. The image of a circle under an isometry is a congruent circle.

Lemma 5.3.10. An isometry that leaves each of three noncollinear points fixed leaves every point of the plane fixed.

Lemma 5.3.11. Each point in the plane is uniquely determined by its given distances from three noncollinear points.

Theorem 5.3.12. As isometry is uniquely determined by three noncollinear points and their images under that isometry.

Theorem 5.3.13. A translation is an isometry.

Corollary 5.3.14. A translation maps a line segment onto a parallel line segment.

Theorem 5.3.15. A rotation about a point P through an angle θ is an isometry.

Theorem 5.3.16. A reflection is an isometry.

Theorem 5.3.17. A glide reflection is an isometry.

Lemma 5.3.18. Let $R_{P,180°}$ be denoted by σ_P. Then $\sigma_D \circ \sigma_C \circ \sigma_B \circ \sigma_A = I$ if and only if $\square ABCD$ is a parallelogram.

Lemma 5.3.19. The product of three reflections in three concurrent lines is a reflection in a line that passes through the point of concurrency.

Theorem 5.3.20. Betweeness is invariant under a homothety.

Corollary 5.3.21. The image of a line segment (or ray or line) under a homothety is a line segment (or ray or line).

Theorem 5.3.22. If O, A, and B are not collinear, and if $H_{O,k}(\overline{AB}) = \overline{A'B'}$, then $\overline{AB} \parallel \overline{A'B'}$.

Theorem 5.3.23. A homothety maps triangles onto similar triangles.

Corollary 5.3.24. A homothety is a direct and conformal transformation.

Theorem 5.3.25. If two triangles are similar and noncongruent, then there exists a transformation S, which is the product of an isometry and a homothety, that maps one onto the other.

Corollary 5.3.26.   A similarity maps a polygon onto a similar polygon.

Corollary 5.3.27.   Angle measure and collinearity are invariant under a similarity.

5.4 Analytical Transformations

Definition.   If **PQ** is a vector in the plane, P is called the *initial point* and Q is called the *terminal point*. The *magnitude* of **PQ** is equivalent to the length of line segment PQ, and the *direction* of **PQ** is determined by the counterclockwise angle of inclination of PQ.

Definition.   A *position vector* is any vector in the plane whose initial point is the origin.

Theorem 5.4.1.   Any vector **RS** in the plane is equivalent to a position vector **OP**, and if $R = (a,b)$ and $S = (c,d)$, then $P = (c - a, d - b)$.

Theorem 5.4.2.   If $A = (x,y)$ is any point in the plane and if $T_{OP}(A) = A'$ such that $O = (0,0)$ and $P = (h,k)$, then $A' = (x',y')$, where $x' = x + h$ and $y' = y + k$.

Theorem 5.4.3.   If $A = (x,y)$ is any point in the plane and if $R_{0,\theta}(A) = A'$ such that $O = (0,0)$, then $A' = (x',y')$, where $x' = x \cos \theta - y \sin \theta$ and $y' = x \sin \theta + y \cos \theta$.

Corollary 5.4.4.   If $A = (x,y)$ is any point in the plane and $R_{P,\theta}(A) = A'$ such that $P = (h,k)$, then $A' = (x',y')$, where $x' = [(x - h) \cos \theta - (y - k) \sin \theta] + h$ and $y' = [(x - h) \sin \theta + (y - k) \cos \theta] + k$.

Theorem 5.4.5.   Let $A = (x,y)$ be any point in the plane. (i) If l is the line $y = 0$, then $R_{y=0}(x,y) = (x, -y)$, and (ii) if l is the line $x = 0$, then $R_{x=0}(x,y) = (-x,y)$.

Corollary 5.4.6.   Let $A = (x,y)$ be any point in the plane. (i) If l is the line $y = k$, then $R_{y=k}(x,y) = (x, -y + 2k)$, and (ii) if l is the line $x = h$, then $R_{x=h}(x,y) = (-x + 2h, y)$.

Theorem 5.4.7.   If $A = (x,y)$ is any point in the plane and $R_{y=mx}(x,y) = (x',y')$, then $x' = x \cos 2\alpha + y \sin 2\alpha$ and $y' = x \sin 2\alpha - y \cos 2\alpha$, where $\alpha = \arctan m$.

Corollary 5.4.8.   If $A = (x,y)$ is any point in the plane and $R_{y=mx+b}(A) = A'$, then $A' = (x',y')$, where $x' = x \cos 2\alpha + (y - b) \sin 2\alpha$ and $y' = x \sin 2\alpha - (y - b) \cos 2\alpha + b$ and $\alpha = \arctan m$.

Theorem 5.4.9.   If $A = (x,y)$ is any point in the plane and $H_{O,k}$ is a homothety centered at $O = (0,0)$ such that $H_{O,k}(A) = A'$, then $A' = (x',y')$, where $x' = kx$ and $y' = ky$.

Corollary 5.4.10. If $A = (x,y)$ is any point in the plane and $H_{P,k}(A) = A'$, where $P = (p,q)$, then $A' = (x',y')$, where $x' = kx + (1 - k)p$ and $y' = ky + (1 - k)q$.

5.5 Inversion

Definition. Let O be the center of a fixed circle of radius r in the Euclidean plane. Furthermore, let P be any point in the plane other than O. An *inversion in circle* O, $I_{O,r}$, is a function such that if $I_{O,r}(P) = P'$, then $P' \in$ **OP** and $(OP)(OP') = r^2$. Here P' is called the *inverse* of P, O is called the *center of inversion*, r is called the *radius of inversion*, and r^2 is called its *power*.

Definition. The Euclidean plane with one ideal point appended to it is referred to as the *inversive plane*.

Theorem 5.5.1. Inversion is a transformation of the inversive plane.

Theorem 5.5.2. Points on a circle of inversion are fixed.

Theorem 5.5.3. If a point is in the interior of a circle of inversion, then its image is a point in the exterior of the circle, and conversely.

Theorem 5.5.4. Let O be the center of an inversion whose radius is r. If P and Q are distinct points that are not both collinear with O, and if P' and Q' are the inverses of P and Q, respectively, then $\triangle OPQ$ is similar to $\triangle OQ'P'$.

Theorem 5.5.5. The measure of the angle between two curves is invariant under inversion.

Theorem 5.5.6. If $I_{O,r}$ is an inversion and l is any line that does not contain O, then the image of l, under $I_{O,r}$ is a circle that contains O.

Corollary 5.5.7. If $I_{O,r}$ is an inversion and α is a circle that contains O, then the image of α is a line tangent to circle α at O.

Theorem 5.5.8. If $I_{O,r}$ is an inversion and β is a circle that does not contain O, then the image of β is a circle that does not contain O.

Theorem 5.5.9. If $I_{O,r}$ is an inversion and l is a line that contains O, then l is invariant (but not pointwise invariant).

Theorem 5.5.10. Any circle other than the circle of inversion is invariant if and only if it is orthogonal to the circle of inversion.

Corollary 5.5.11. If P and P' are distinct points that are images of each other under an inversion in circle O, then any circle containing both P and P' is orthogonal to circle O.

6

OTHER WAYS TO GO

Non-Euclidean Geometries

6.1 INTRODUCTION

In Chapter 2 we investigated the controversy surrounding Euclid's fifth postulate and implied the existence of both Euclidean and non-Euclidean geometries. In Chapter 3 we avoided the issue by deriving theorems that did not depend on Euclid's controversial fifth postulate. Chapters 4 and 5 extended the axiomatic development of geometry (begun in Chapter 3 with neutral geometry) by positing a form of the Euclidean parallel postulate and investigating its consequences. The result of that extension is what we call Euclidean geometry.

In this chapter we shall return briefly to neutral geometry to extend its results, but this time we will take two "less-traveled" roads by assuming two non-Euclidean parallel postulates. We begin with a detailed look at hyperbolic geometry, since within that domain we can use all of the results derived in neutral geometry. We conclude the chapter with a brief discussion of elliptic geometry, which is substantially different from the Euclidean, hyperbolic, and even neutral geometries that precede it.

The bodies of theorems resulting from the two non-Euclidean parallel postulates that we will investigate are aptly known as non-Euclidean geometries. We caution you that our development of non-Euclidean geometries as consistent axiomatic systems in the spirit of Chapter 1 may produce results that are counterintuitive and may contradict your perceptions of the real world. Recall, however, that Einstein's well-known theories of relativity are based on a non-Euclidean geometry that was useful in explaining some

physical phenomena (e.g., gravity) that had perplexed physicists for centuries. Non-Euclidean geometries have also provided scientists with more
precise models of planetary motion and have even been useful in describing
phenomena on earth.

Before we begin our trip through this mysterious region, we shall in the
next section prove several additional neutral theorems that will be useful in
the study of hyperbolic geometry.

6.2 A RETURN TO NEUTRAL GEOMETRY—THE ANGLE OF PARALLELISM

Suppose that we are given a line l and a point P that is not on l. We know
from Chapter 3 that there is at least one line m containing P that is parallel to
l (Corollary 3.4.2). What we were not able to prove in neutral geometry was
whether or not there are other lines through P that are parallel to l. Consider
Figure 6.2.1 where $\overrightarrow{PR}$ is perpendicular to l at R and $m\angle RPQ = d°.$[1]

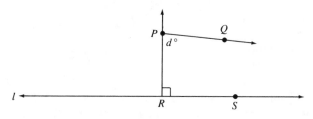

Figure 6.2.1

It should be clear that for some values of d, $\overrightarrow{PQ}$ will intersect l, while
for other values of d it will not. In particular, if $d = m\angle RPS$, then $\overrightarrow{PQ} \cap l =
S$, while if $d = 90°$, then $\overrightarrow{PQ} \cap l = \varnothing$.

Suppose that we now consider the set of all values of d for which $\overrightarrow{PQ}$
intersects line l. To be more precise, let's define a set D in the following
way:

$$D = \{d \mid d = m\angle RPQ \text{ and } \overrightarrow{PQ} \cap \overrightarrow{RS} \neq \varnothing\}$$

By the previous discussion, we know that D is nonempty (since $d =
m\angle RPS$ is in D) and is bounded above by 180°. Therefore, by the complete-
ness property of real numbers,[2] D must have a *least upper bound,* called
LUB(D) [LUB(D) is the smallest number that bounds D from above]. If we

---

[1] Recall that the protractor postulate restricts d so that $0° < d° < 180°$.

[2] The completeness property for real numbers asserts that every nonempty set of real
numbers that is bounded above has a least upper bound (LUB) (i.e., a smallest number that is
greater than or equal to each member of the set).

let $d_0 = \text{LUB}(D)$, then d_0 is greater than or equal to every d in D. This means that d_0 defines a separation between those rays $\overrightarrow{PQ}$ that intersect $\overrightarrow{RS}$ and those that do not. While the existence of such a separation may not be obvious at first, we will shortly see that there must be one. Because of its separation role, d_0 is very important in our discussion of parallel lines. We begin the discussion with the following neutral theorem:

THEOREM 6.2.1.   (i) If $m\angle RPQ = d_0$, then $\overrightarrow{PQ} \cap \overrightarrow{RS} = \varnothing$. (ii) If $m\angle RPQ < d_0$, then $\overrightarrow{PQ} \cap \overrightarrow{RS} \neq \varnothing$.

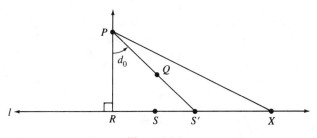

Figure 6.2.2

Proof. (i) Suppose that $m\angle RPQ = d_0$ and that $\overrightarrow{PQ} \cap l = S'$ (Figure 6.2.2). Since R and S' are distinct points on l, there exists a point X such that R-S'-X. Now S' is in the interior of $\angle RPX$, so that $m\angle RPX > m\angle RPS' = d_0$. This means that $m\angle RPX$ is in D and is larger than d_0, which contradicts the notion that d_0 is the least upper bound of D. Hence our assumption that $\overrightarrow{PQ}$ intersects $\overrightarrow{RS}$ is false, so that $\overrightarrow{PQ} \cap \overrightarrow{RS} = \varnothing$.

(ii) Next let us assume that $m\angle RPQ < d_0$ (Figure 6.2.3). So let $m\angle RPQ = k$, where $k < d_0$. Now since k is not an upper bound for D, there must be some $d' \in D$ so that $d' > k$. Let Y be a point on $\overrightarrow{RS}$ such that $m\angle RPY = d'$. (Which neutral postulate allows us to locate such a point?) Since $\overrightarrow{PQ}$ contains a point in the interior of $\triangle RPY$, $\overrightarrow{PQ}$ (hence $\overrightarrow{PQ}$) must intersect $\overrightarrow{RY}$ (and consequently $\overrightarrow{RS}$).

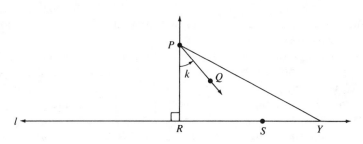

Figure 6.2.3

The previous theorem provides an essential characterization of $\overrightarrow{PQ}$ (and $\overleftrightarrow{PQ}$) with respect to $\overleftrightarrow{RS}$. If $m\angle RPQ = d_0$, $\overrightarrow{PQ}$ is the first nonintersecting line as we move counterclockwise from $\overrightarrow{PR}$, since any value of $m\angle RPQ$ less that d_0 forces $\overrightarrow{PQ}$ to intersect $\overleftrightarrow{RS}$. This result motivates the following definition:

DEFINITION.   If line l and points P, Q, R, and S are as in Figure 6.2.1, and if $m\angle RPQ = d_0$ (with d_0 as defined previously), then $\angle RPQ$ is called the *angle of parallelism* for $\overleftrightarrow{RS}$ and P.

So far we have discussed only those rays that fall on the Q side of $\overleftrightarrow{PR}$. It should be evident, however, that the same arguments apply to those rays that fall on the other side of $\overleftrightarrow{PR}$ as well (Figure 6.2.4).

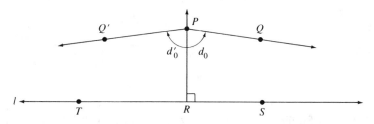

Figure 6.2.4

In particular, let $D' = \{d' \mid m\angle RPQ' = d'$ and $\overrightarrow{PQ'} \cap \overrightarrow{RT} \ne \varnothing\}$, and let $d_0' = \text{LUB}(D')$. Then if $\overrightarrow{PQ'}$ is the first nonintersecting line as we move in a clockwise direction from $\overrightarrow{PR}$, $m\angle RPQ'$ is the angle of parallelism for $\overrightarrow{RT}$ and P. The following theorem concerning this may seem trivial but requires proof:

THEOREM 6.2.2.   If T-R-S and $\angle RPQ$ and $\angle RPQ'$ are the angles of parallelism for P and $\overleftrightarrow{RS}$ and P and $\overrightarrow{RT}$, respectively, then $m\angle RPQ = m\angle RPQ'$.

Proof. In Figure 6.2.5, let $\angle RPQ$ and $\angle RPQ'$ be the angles of parallelism. We will proceed indirectly and assume that these angles do not have the same measure.

Without loss of generality, we may further assume that $m\angle RPQ > m\angle RPQ'$. As a consequence, there exists $\overrightarrow{PX}$ in the interior of $\angle RPQ$ such that $\angle RPX$ is congruent to $\angle RPQ'$. Now since $\overrightarrow{PQ}$ is the first nonintersecting ray on the Q side of $\overrightarrow{RP}$, $\overrightarrow{PX}$ must intersect $\overleftrightarrow{RS}$ at some point Y. If we now let Z be a point on $\overrightarrow{RT}$ such that $\overline{RZ} \cong \overline{RY}$, we find that $\triangle RPY$ and $\triangle RPZ$ are congruent, so that $m\angle RPX = m\angle RPZ$. Therefore $m\angle RPZ = m\angle RPQ'$, so that $\overrightarrow{PQ'} = \overrightarrow{PZ}$, meaning that $\overrightarrow{PQ'} \cap \overrightarrow{RT} = Z$. This, however,

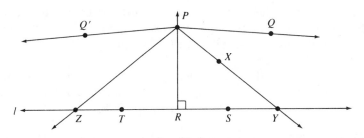

Figure 6.2.5

contradicts the assumption that $\overrightarrow{PQ'}$ is a nonintersecting line, leaving us to conclude that the angles of parallelism are congruent.

The next theorem provides some insight concerning the size of angles of parallelism.

THEOREM 6.2.3.   The angle of parallelism for a given line and point is less than or equal to 90°.

Proof. Let $\angle RPQ$ be the angle of parallelism for $\overrightarrow{RS}$ and P. Then $\overrightarrow{PQ}$ is the first nonintersecting ray as we move in a counterclockwise direction from $\overrightarrow{PR}$. The proof will proceed indirectly by assuming that $m\angle RPQ > 90°$ (Figure 6.2.6). Using the angle construction postulate, we may locate a point Q' in the interior of $\angle RPQ$ such that $m\angle RPQ' = 90°$. By Corollary 3.4.2, $\overrightarrow{PQ'}$ is parallel to $\overrightarrow{RS}$, hence does not intersect $\overrightarrow{RS}$. This, however, contradicts the hypothesis that $\overrightarrow{PQ}$ is the first nonintersecting ray. Therefore $m\angle RPQ$ is not obtuse (i.e., not greater than 90°), and consequently $m\angle RPQ \le 90°$.

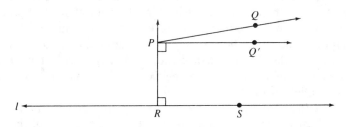

Figure 6.2.6

The following corollary is a direct consequence of Theorem 6.2.3. Its proof is left as an exercise.

COROLLARY 6.2.4.   If $\angle RPQ$ is the angle of parallelism for line l and point P, then $\overleftrightarrow{PQ} \cap l = \varnothing$. (*Note:* This corollary involves line $\overleftrightarrow{PQ}$ rather than $\overrightarrow{PQ}$.)

Next we state a theorem that gets to the heart of the distinction between Euclidean and hyperbolic geometries. It is neutral but sets the stage for our departure from the strictly Euclidean road.

THEOREM 6.2.5.   If the angle of parallelism for a point P and a line $\overleftrightarrow{RS}$ is *less than 90°*,[3] then there exist at least two lines through P that are parallel to $\overleftrightarrow{RS}$.

Although this is a highly significant result, its proof is quite simple and is left as an exercise.

It should by now be evident that Theorem 6.2.5 indicates an essential connection between the measure of the angle of parallelism and the Euclidean versus hyperbolic dichotomy. The next theorem establishes this relationship.

THEOREM 6.2.6.   The Euclidean parallel postulate is equivalent to the following statement: The measure of the angle of parallelism is 90°.

Proof. As usual, establishing the equivalence of two statements such as these involves showing that each is a consequence of the other. We begin by showing that the Euclidean parallel postulate implies that the angle of parallelism is 90°. To do this, we first make note of the fact that the contrapositive of any theorem is still a theorem. In particular, we will state the contrapositive of Theorem 6.2.5:

THEOREM 6.2.5c.   If there exists at most one line through a point P parallel to a given line l, then the angle of parallelism for that point and line is *not* less than 90°.

In this portion of the proof we may assume the Euclidean parallel postulate. By hypothesis then, there is at most (exactly) one line parallel to l through P. Theorem 6.2.5c then assures us that the measure of the angle of parallelism is not less than 90°. In addition, Theorem 6.2.3 tells us that the measure of the angle of parallelism is not greater than 90°. Together these results mean that the measure of the angle of parallelism is 90° *if* we posit the Euclidean parallel postulate.

To prove the converse, we will choose any line l and any point P not on l. Our hypothesis allows us now to assume that the measure of the angle of parallelism is 90°. Let n be the line through P that is perpendicular to l (at point R), and let m be the line through P that is perpendicular to n (Figure 6.2.7). By Corollary 3.4.2, m is parallel to l, and by our assumption, $\angle RPQ$ and $\angle RPQ'$ are the angles of parallelism. Then any other line through P

---

[3] This theorem does not imply that the angle of parallelism *is* less than 90° but rather observes one of the consequences of making such an assumption.

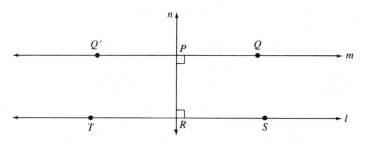

Figure 6.2.7

must intersect l, so that m is the unique line through P that is parallel to l. As a result, we see that there is exactly one line through P that is parallel to l, which indicates that the Euclidean parallel postulate is a consequence of assuming that the angle of parallelism is 90°.

The next five sections of this chapter (Sections 6.3 through 6.7) are devoted to the axiomatic development of hyperbolic geometry. The axiomatic basis for hyperbolic geometry includes all the neutral postulates *and* the assumption that the angle of parallelism is acute.

Following the detailed discussion of hyperbolic geometry, Sections 6.8 and 6.9 survey ideas from elliptic geometry (where parallel lines do not exist) and summarize the relationships among Euclidean, hyperbolic, and elliptic geometries.

EXERCISE SET 6.2

1. Refer to Figure 6.2.1 and prove, using the neutral axioms, that if $\angle RPQ$ is the angle of parallelism for P and $\overrightarrow{RS}$, $\angle R'P'Q'$ is the angle of parallelism for some other point P' and some other line $\overrightarrow{R'S'}$, and if $\overline{PR} \cong \overline{P'R'}$, then $m\angle RPQ = m\angle R'P'Q'$.

2. Prove, using the neutral axioms, that if R-P_1-P_2 (or $RP_1 < RP_2$), and $\angle RP_1Q_1$ is the angle of parallelism for P_1 and $\overrightarrow{RS}$, and $\angle RP_2Q_2$ is the angle of parallelism for P_2 and $\overrightarrow{RS}$ (see the accompanying figure), then $m\angle RP_2Q_2 \leq m\angle RP_1Q_1$.

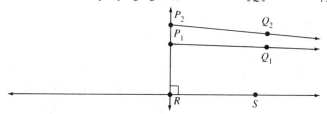

3. Prove Corollary 6.2.4.

4. Prove Theorem 6.2.5.

5. In the proof of Theorem 6.2.2 we assumed, without loss of generality, that $m\angle RPQ > \angle RPQ'$. Show that the theorem is generally true by proving it under the hypothesis that $m\angle RPQ < m\angle RPQ'$.

6.3 THE HYPERBOLIC PARALLEL POSTULATE

You probably recall that in Chapters 2 and 3 we encountered a variety of statements that were equivalent to Euclid's fifth postulate. In particular, we demonstrated that Euclid's fifth postulate, the Euclidean parallel postulate (Playfair's postulate), and the angle sum theorem for triangles (Theorem 4.2.1) are all equivalent in the sense that each can be proven if we assume any one of the others. In fact, there are many other statements that are equivalent to the Euclidean parallel postulate, any of which could be postulated in place of the Euclidean parallel postulate without changing the theorems that constitute the geometry. Our choice of a Euclidean parallel postulate is then mostly a matter of taste or, perhaps, convenience.

The same type of decision confronts us as we begin our journey through hyperbolic geometry. We could simply negate Playfair's postulate and assume that given a line and a point not on the line, there are at least two lines through the point that are parallel to the line. In fact, many developments of hyperbolic geometry do just that. However, in order to proceed a bit more efficiently by using some of the previous results, we shall instead choose to postulate the negation of a statement we have already shown to be equivalent to the Euclidean parallel postulate, Theorem 6.2.6.

Strictly speaking, the negation of "The measure of the angle of parallelism is 90°" is "The measure of the angle of parallelism is *not* 90°." However, since Theorem 6.2.3 assures us that the angle of parallelism is not obtuse, we will complete our set of hyperbolic postulates by positing the hyperbolic parallel Postulate in the following form:

THE HYPERBOLIC PARALLEL POSTULATE.   The measure of the angle of parallelism for a line l and a point P not on l is less than 90°.

An immediate consequence of assuming this statement is a negation of Playfair's postulate. This can now be proved as our first theorem in hyperbolic geometry.

THEOREM 6.3.1.   Given a line l and a point P not on l, there are at least two lines through P that are parallel to l.

Theorem 6.3.1 is an immediate consequence of a previous result, Theorem 6.2.5, which establishes that *if* the angle of parallelism is less than 90°, then there are at least two lines through P that are parallel to l. Of course, now that we have assumed the Hyperbolic parallel postulate, Theorem 6.2.5 implies Theorem 6.3.1. The formal details of the proof are left as an exercise.

Our next theorem will allow us to extend some of the ideas of Chapter 3 that were first discussed by Girolamo Saccheri nearly 300 years ago. Much of Saccheri's work centered around attempts to show that certain four-sided figures, called Saccheri quadrilaterals, were rectangles. The following theorem indicates that in hyperbolic geometry they are not.

THEOREM 6.3.2.   The hyperbolic parallel postulate is equivalent to the following statement. The summit angles of a Saccheri quadrilateral are acute.

Proof. We begin by showing that the summit angles of a Saccheri quadrilateral are acute in hyperbolic geometry.

We will proceed indirectly. Assume that the summit angles of a Saccheri quadrilateral are *not* acute. This hypothesis, in conjunction with Theorems 3.6.2 and 3.6.3, means that the summit angles are both right angles (Figure 6.3.1). From this we may deduce that we have found a rectangle. (Remember our search for a rectangle in Chapter 3?) Next we will recall two other theorems established in Chapter 3. In particular, Theorem 3.6.14 tells us that if one rectangle can be found, then every triangle has an angle sum of 180°. Later, in neutral geometry (Theorem 3.6.18) we proved that the statement "Every triangle has an angle sum of 180°" is equivalent to the Euclidean parallel postulate. Together these results contradict the hyperbolic parallel postulate and imply that in hyperbolic geometry the summit angles of a Saccheri quadrilateral must be acute.

Figure 6.3.1

To complete Theorem 6.3.2 we must also prove that if the summit angles of a Saccheri quadrilateral are acute, then the angle of parallelism is acute. This follows directly from the definition of the angle of parallelism and the fact that the summit and base of a Saccheri quadrilateral are parallel (Corollary 3.6.5). The details of this portion of the proof are left as an exercise.

Embedded within the preceding proof is a proof for the following theorem which was suggested in Chapter 3:

THEOREM 6.3.3.   Rectangles do not exist in hyperbolic geometry.

Now that we have established, to a large extent, the nature of Saccheri quadrilaterals in hyperbolic geometry, we will briefly turn our attention to a second, related, figure discussed in Chapter 3—the Lambert quadrilateral. You should recall that a Lambert quadrilateral is a four-sided polygon with right angles at three of the vertices. John Lambert (1728–1777) investigated

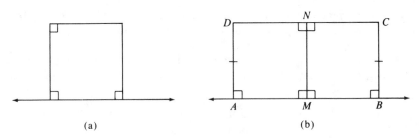

Figure 6.3.2

results concerning parallels in a way similar to the approach used by Saccheri about 30 years earlier. In neutral geometry a Lambert quadrilateral can be constructed by choosing a line segment, constructing perpendiculars at each endpoint, and then constructing another perpendicular at an arbitrary point on either of the perpendiculars from the endpoints (Figure 6.3.2a). A second way to produce a Lambert quadrilateral is to begin with a Saccheri quadrilateral and then connect the midpoints of the summit and base. Since this segment is perpendicular to both the summit and the base (Theorem 3.6.4), it will partition the Saccheri quadrilateral into a pair of Lambert quadrilaterals. This phenomenon is illustrated in Figure 6.3.2b, where Saccheri quadrilateral $\square ABCD$ is partitioned into Lambert quadrilaterals, $\square AMND$ and $\square BMNC$.

Since the fourth angle of a Lambert quadrilateral constructed in this manner serves also as the summit angle of a Saccheri quadrilateral, one may be tempted to conclude that the fourth angle of any Lambert Quadrilateral is acute. Although this statement is true in hyperbolic geometry, proving this property for all Lambert quadrilaterals requires that we begin without assuming anything specific about the configuration of the points. We establish this property in the following theorem.

THEOREM 6.3.4.   The fourth angle of a Lambert quadrilateral is acute.

Proof. Consider Lambert quadrilateral $\square ABCD$ shown in Figure 6.3.3a. We will begin by asking how the length of side $\overline{DC}$ compares with the length of $\overline{AB}$. If $DC = AB$, then $\square ABCD$ is a Saccheri quadrilateral (why?), so that the summit angles (which must be congruent, by Theorem 3.6.2) are both right angles. But if both summit angles are right angles, $\square ABCD$ is a rectangle, which contradicts Theorem 6.3.3. Therefore the assumption that $DC = AB$ must be rejected, leaving us with the following two alternatives: $DC > AB$ and $DC < AB$. We will consider these remaining possibilities separately.

First, suppose that $DC > AB$ (Figure 6.3.3b). Under this assumption we can locate A' on $\overline{DC}$ so that D-A'-C and $AB = A'C$. Since the angles at B and C are both right angles, $\square ABCA'$ is a Saccheri quadrilateral, so that

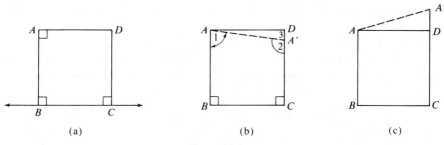

Figure 6.3.3

$m\angle 1 = m\angle 2$. Now since $\overrightarrow{AA'}$ is in the interior of $\angle DAB$, we may conclude that $m\angle 1 < m\angle DAB = 90°$ and since $m\angle 1 = m\angle 2$ (Theorem 3.6.2), we know that $m\angle 2 < 90°$. Next note that $\angle 2$ is an exterior angle for triangle DAA' and, as such, has a measure that is greater than either remote interior angle (Theorem 3.2.6). This implies that $m\angle 2 > m\angle 3$, so that we conclude $m\angle 3 < m\angle 2 < 90°$. Therefore $\angle 3$, the fourth angle of the Lambert quadrilateral, is acute under the assumption that $DC > AB$.

The other possibility we must consider is that $AB > DC$ (Figure 6.3.3c). Under this premise we can locate A' on $\overrightarrow{CD}$ so that A'-D-C and $AB = A'C$. Once again we have a Saccheri quadrilateral ($\square ABCA'$), but in this case $\angle A'AB$ is obtuse (why?), a consequence that contradicts Theorem 6.3.2. From this, we may conclude that $AB \not< DC$, and having already dismissed the possibility that $AB = DC$, we conclude that $DC > AB$, so that the fourth angle of a Lambert quadrilateral is acute.

Contained within the preceding proof is, in essence, a proof for the following theorem. Details of the proof are left as an exercise.

THEOREM 6.3.5.   In a Lambert quadrilateral, the sides contained in the sides of the acute angle are longer than the sides that they are opposite.

Next we will discuss some relationships concerning lengths in hyperbolic geometry. As we proceed, we will need to make use of the distance between a point and a line. We will use the standard definition given in Chapter 4, which is repeated here for convenience.

DEFINITION.   The distance between a line and a point not on the line is the length of the (unique) perpendicular from the point to the line. (The distance between a point and any line containing the point is, by definition, zero.)

The following idea, while counterintuitive, is a theorem in hyperbolic geometry.

THEOREM 6.3.6.   Parallel lines are *not* everywhere equidistant.

To prove this, we must show that if we are given any pair of parallel lines *l* and *m*, we can find at least one pair of points on line *l* for which the distances to line *m* are not the same. To prove this, we will begin with any three points on *l* and proceed in the following way.

Proof. Choose any pair of parallel lines *l* and *m* and any three distinct points *A*, *B*, and *C* on *l* such that *A*-*B*-*C* (Figure 6.3.4). *D*, *E*, and *F* are the feet of the perpendiculars drawn, respectively from *A*, *B*, and *C* to *m*.

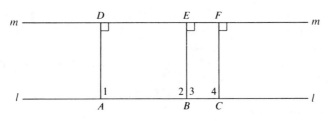

Figure 6.3.4

We wish to show that the lengths *AD, BE,* and *CF* are not all the same, so we shall proceed indirectly and assume otherwise, that is, *AD* = *BE* = *CF*. Under this assumption, Figure 6.3.4 displays three "inverted" Saccheri quadrilaterals: □*DABE*, □*EBCF*, and □*DACF*. Therefore m∠1 = m∠2, m∠3 = m∠4 and m∠1 = m∠4 (each equality involves the summit angles of a Saccheri quadrilateral). From these equalities we see that m∠2 = m∠3, and since ∠2 and ∠3 form a linear pair, each has a measure of 90°. Consequently,

$$\text{m}\angle 1 = \text{m}\angle 2 = \text{m}\angle 3 = \text{m}\angle 4 = 90°$$

so that each of the Saccheri quadrilaterals is a rectangle. But this is in direct contradiction to Theorem 6.3.3, so that our assumption that *AD* = *BE* = *CF* must be rejected and we may conclude that parallel lines are *not* everywhere equidistant.

It should not come as a surprise that Saccheri quadrilaterals play a major role in many of our proofs in hyperbolic geometry, since Saccheri used this idea extensively in *Euclides Vindicatus*. We are now in a position to make some of our neutral theorems concerning Saccheri quadrilaterals more specific. For example,

THEOREM 6.3.7.   The summit of a Saccheri quadrilateral is longer than the base.

Proof. The proof of this theorem is an easy consequence of Theorem 6.3.5. To see this, begin with Saccheri quadrilateral $\square ABCD$ (Figure 6.3.5) in which the midpoints of $\overline{AB}$ and $\overline{CD}$ are labeled M and N, respectively.

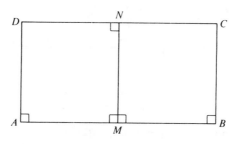

Figure 6.3.5

Theorem 3.6.4 assures us that $\overleftrightarrow{MN}$ is perpendicular to both $\overleftrightarrow{AB}$ and $\overleftrightarrow{CD}$, so that $\square ADNM$ and $\square BCNM$ are both Lambert quadrilaterals. Theorem 6.3.5 implies that $DN > AM$ and $NC > MB$. Adding these lengths, we see that $DC > AB$, which completes the proof.

Next, we will consider a property that pertains to parallels that share a common perpendicular (Figure 6.3.6).[4] If t is any transversal that contains the midpoint M of $\overline{AB}$, it is easy to show (Exercise Set 6.3, Problem 9) that $\triangle MBD \cong \triangle MAC$, so that $\angle 1 = \angle 2$; that is, the alternate interior angles are congruent. Conversely related to this matter are the following two questions.

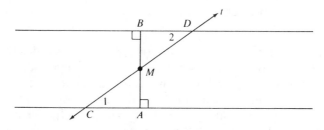

Figure 6.3.6

If two parallel lines are crossed by a transversal in such a way as to make the alternate interior angles congruent,

---

[4] At this point we know that such lines exist, namely, lines that contain the summit and base of a Saccheri quadrilateral. One should not assume, however, that all pairs of parallel lines share a common perpendicular, for they do not.

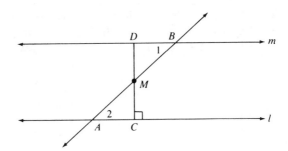

Figure 6.3.7

1. Do the lines have a common perpendicular?
2. If so, does the transversal contain the midpoint of the common perpendicular?

To see this, let l and m be lines crossed by transversal t in such a way as to make $m\angle 1 = m\angle 2$ (Figure 6.3.7). Let M be the midpoint of $\overleftrightarrow{AB}$, and let C be the foot of the perpendicular from M to l. Extend $\overleftrightarrow{CM}$, and let $\overleftrightarrow{CM} \cap m = D$. (How do we know that $\overleftrightarrow{CM} \cap m \neq \varnothing$?) We can show that $\triangle CMA = \triangle DMB$ (ASA), so that $m\angle BDM = 90°$, making $\overline{CD}$ a common perpendicular. In addition, $DM = MC$, so that M is not only the midpoint of $\overline{AB}$ but also of the common perpendicular $\overline{CD}$. Consequently, the answer to each of the questions posed above is yes. This leads to our next hyperbolic theorem.

THEOREM 6.3.8. Two parallel lines that are crossed by a transversal have congruent alternate interior angles if and only if the transversal contains the midpoint of a segment that is perpendicular to both lines.

To conclude this section, we will look at one further idea that helps to characterize hyperbolic geometry and has a clear, but contradictory, Euclidean counterpart. In Chapter 4 we showed that in Euclidean geometry every triangle can be circumscribed, and that the center of the circumscribing circle can be located by finding the point where the perpendicular bisectors of two sides intersect. In hyperbolic geometry this is not always the case. As we shall soon see, certain triangles cannot be circumscribed.

To illustrate this we will examine a Saccheri quadrilateral, $\square ABCD$, in which M and N are the midpoints of the summit and base, respectively, as shown in Figure 6.3.8.

Theorem 6.3.5 assures us that $AB > MN$, so that we can locate point E on $\overline{NM}$ in such a way that $\overline{NE} \cong \overline{AB}$ (also, $\overline{NE} \cong \overline{CD}$). Since N, M, and E are collinear and A, M, and D are collinear, we can conclude that A, E, and D

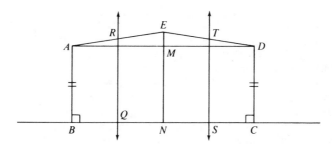

Figure 6.3.8

are *not* collinear (why?), so that we may talk of $\triangle AED$. Is there a circle that contains all three vertices of $\triangle AED$? We will show that the answer is no.[5]

To do this, note that $\square ABNE$ is itself a Saccheri quadrilateral. (Verify this.) Next we can locate points Q and R, the midpoints of the base and summit, respectively, of $\square ABNE$. Theorem 3.6.4 assures us that $\overleftrightarrow{QR}$ is perpendicular to $\overleftrightarrow{AE}$ (and to $\overleftrightarrow{BN}$), so that $\overleftrightarrow{QR}$ is the perpendicular bisector of $\overline{AE}$. Similarly, since $\square DCNE$ is a Saccheri quadrilateral, $\overleftrightarrow{ST}$, which connects the midpoints of $\overline{NC}$ and $\overline{ED}$, is the perpendicular bisector of $\overline{ED}$. Furthermore, since both $\overleftrightarrow{QR}$ and $\overleftrightarrow{ST}$ are perpendicular to $\overleftrightarrow{BC}$, they are parallel. Now if there is a circle that contains A, E, and D, its center would have to be on the perpendicular bisectors of sides $\overline{AE}$ and $\overline{ED}$, that is $\overleftrightarrow{QR}$ and $\overleftrightarrow{ST}$. However, since $\overleftrightarrow{QR} \parallel \overleftrightarrow{ST}$, there is no such point, and consequently, there can be no circumcircle. This result is formalized as Theorem 6.3.9.

THEOREM 6.3.9.   In hyperbolic geometry there exist triangles that cannot be circumscribed.

This theorem does *not* say that no triangles can be circumscribed, since it can be shown that certain triangles do have circumcircles. Rather, Theorem 6.3.9 says that if we look hard enough, we can find examples of circles that cannot be circumscribed. This result is in clear contradiction to the results of Euclidean geometry, so that it is *strictly hyperbolic.*

EXERCISE SET 6.3

In the proofs involved in the following exercises, you may use all of the neutral postulates, all of the results from neutral geometry, and the hyperbolic parallel postulate and its consequences thus far derived.

1. Supply the details for the proof of Theorem 6.3.1.

2. Complete the proof of Theorem 6.3.2.

---

[5] In hyperbolic geometry we can think of circles in the usual way: a set of points, all of which are equidistant from a given point.

3. Supply the details of the proof that rectangles do not exist in hyperbolic geometry (Theorem 6.3.3).

4. Suppose lines l and m are parallel and that points A and B are on line l. (a) Is it possible that $d(A,m) = d(B,m)$? (b) Is it necessary that $d(A,m) = d(B,m)$? Explain why or why not in each.

5. Explain why lines with a common perpendicular cannot have a second common perpendicular.

6. A Lambert quadrilateral is, by definition, a quadrilateral with three right angles.
   (a) Explain how a Lambert quadrilateral can be constructed using compass and straightedge alone.
   (b) Do the constructions employed in the solution to part (a) depend on a form of the Euclidean Parallel postulate? Explain.
   (c) Which constructions from Section 4.9 do you think are valid in hyperbolic geometry? Explain.
   (d) Prove that the sides of a Lambert quadrilateral contained in the acute angle are longer than their opposite sides.

7. Prove that if two lines share a common perpendicular, the distance between the lines is least on that segment.

8. In Figure 6.3.5 assume that $AB = CD$ and prove, under this hypothesis, that $\square AMND$ is congruent to $\square CMNB$. What does this congruence imply about the measures of $\angle C$ and $\angle D$? Does this result in a contradiction? What does this imply about the assumption that $AB = CD$?

9. Prove that if a transversal crossing two parallel lines contains the midpoint of a common perpendicular to those lines, then the alternate interior angles are congruent.

10. Theorem 6.3.1 asserts that through a point P there are at least two lines parallel to any line l that does not contain P. Can there be exactly two such parallels? Explain why or why not.

11. In Chapter 4 we defined a parallelogram as a quadrilateral with opposite sides parallel. We then proved (Theorem 4.2.3) that the opposite sides of a parallelogram are congruent.
    (a) Do parallelograms exist in hyperbolic geometry? Explain how you know.
    (b) Is Theorem 4.2.3 true in hyperbolic geometry? Explain why or why not.
    (c) Do you think Theorem 4.2.3 could be proved in neutral geometry? Explain why or why not.

12. Since rectangles do not exist in hyperbolic geometry, squares do not exist either. Do you think rhombuses can exist in hyperbolic geometry? If so, explain how you could construct one. If not, explain why not.

13. Suppose that $\square ABCD$ and $\square EFGH$ are Saccheri quadrilaterals (right angles at B, C, F, and G) in which $\overline{AB} \cong \overline{EF}$, $\overline{BC} \cong \overline{FG}$ and $\overline{CD} \cong \overline{GH}$. Prove that $m\angle CDA = m\angle HEF$.

14. Suppose that lines l and m are parallel and that t is a transversal crossing both of them. What can you say about the measures of the alternate interior angles? Is there a pair of points P and Q (one on each of l and m) such that the distance from P to Q is smallest? Is there a pair of points, P and Q such that the distance from P to Q is greatest? Explain.

15. In the proof of Theorem 6.3.9 we showed that, in at least one case, the perpendicular bisectors of two sides of a triangle are parallel (refer to Figure 6.3.8). Show that the third perpendicular bisector of that triangle is parallel to both of the others.

16. Theorem 6.3.9 provides a hyperbolic characteristic that is clearly not Euclidean, since it contradicts a Euclidean theorem.

  (a) Which Euclidean theorem is contradicted by Theorem 6.3.9?

  (b) Where in the proof of the theorem is the hyperbolic parallel postulate (or an equivalent form) invoked?

6.4 SOME HYPERBOLIC RESULTS CONCERNING POLYGONS

At the end of the previous section we proved a result concerning triangles that is in direct conflict with a Euclidean theorem established in Chapter 4. In order to emphasize the contrast, each theorem is restated here.

THEOREM 4.7.2 *(Euclidean).*   The three perpendicular bisectors of the sides of any triangle are concurrent (at a point that serves as the center of a circle that circumscribes the triangle).

THEOREM 6.3.9 *(Hyperbolic).*   There exist triangles that cannot be circumscribed.

We see then that a hyperbolic theorem can directly contradict a Euclidean theorem. This is possible because the parallel postulates for the two geometries are contradictory. The proof of each theorem listed above relies in some way on a form of the parallel postulate for the geometry in which the statement was proved. Of course, not all statements about triangles in hyperbolic geometry contradict results in Euclidean geometry. For example, in Chapter 4 we proved (Theorem 4.7.3) that we can inscribe a circle within any given triangle and that this circle is centered at the point where the bisectors of the interior angles of the triangle intersect. The proof of this does not require the use of the Euclidean parallel postulate or any equivalent form. (Verify this.) This means that Theorem 4.7.3 could have been included in Chapter 3 as a neutral theorem[6] and so is valid in hyperbolic geometry as well.

In this section we investigate some results concerning triangles in hyperbolic geometry. Most of these results are strictly hyperbolic, meaning that they cannot be proved in neutral geometry. A few, however, do not require use of the hyperbolic parallel postulate. Keep this in mind as you study the proofs of the theorems in this section. Several of the exercises center around this distinction.

We begin the discussion by recalling a highly significant theorem from Chapter 3—the Saccheri-Legendre theorem.

---

[6] Keep in mind that every neutral theorem is both Euclidean *and* hyperbolic. Conversely, some Euclidean theorems in Chapter 4 are also neutral. Distinguishing between neutral, strictly Euclidean, and strictly hyperbolic theorems is discussed later in Section 6.7.

THEOREM 3.5.1.   The angle sum of any triangle is less than or equal to 180°.

For the sake of convenience, we will use the notation $S(\triangle ABC)$ to represent the sum of the measures of the interior angles of $\triangle ABC$. Using this convention, we can express Theorem 3.5.1 as follows:

$$S(\triangle ABC) \leq 180° \qquad (1)$$

Equation (1) has several equivalent forms. First, we can convert the inequality to an equality by adding a nonnegative value k to the lesser quantity, giving

$$S(\triangle ABC) + k° = 180° \qquad k \geq 0 \qquad (2)$$

Performing some arithmetic on Equation (2) yields two other equivalent statements:

$$S(\triangle ABC) = 180° - k° \qquad k \geq 0 \qquad (3)$$

and

$$k = 180° - S(\triangle ABC) \qquad (4)$$

From Equation (4) we see that k represents the difference between 180° (the Euclidean angle sum for triangles) and the angle sum for $\triangle ABC$ in hyperbolic geometry. This difference k is generally referred to as the triangle's *defect,* and because of the way in which it is derived, it is clearly nonnegative. The notation $d(\triangle ABC)$ is used to denote k for a specific $\triangle ABC$. The first theorem of this section specifies the range of the defects for hyperbolic triangles.

THEOREM 6.4.1.   If $\triangle ABC$ is a triangle, then its defect $d(\triangle ABC)$ is positive [i.e., $d(\triangle ABC) > 0$ or, equivalently, $S(\triangle ABC) < 180°$].

Proof. Theorem 6.4.1 in effect asserts that the angle sum for hyperbolic triangles is *strictly less than 180°*. To show this, we need only show that $d(\triangle ABC) \neq 0$, since the Saccheri-Legendre theorem guarantees that $S(\triangle ABC) \leq 180°$ so that $d(\triangle ABC) \geq 0$.

We will proceed indirectly and assume that we have a found a triangle $\triangle ABC$ whose defect $d(\triangle ABC)$ is zero. This means that $S(\triangle ABC) = 180°$. But in Chapter 3 we deduced (Theorem 3.6.14) that if any triangle has an angle sum of 180°, then all triangles have that angle sum. Further, we saw (Theorem 3.6.18) that the statement "All triangles have angle sum 180°" was equivalent to the Euclidean parallel postulate. Combining these results, we see that the existence of even one triangle with $S(\triangle ABC) = 180°$ [or with $d(\triangle ABC) = 0$] implies the Euclidean parallel postulate, which clearly contradicts the hyperbolic parallel postulate. Consequently, in hyperbolic geometry there can exist no triangles with $S(\triangle ABC) = 180°$, from which we may conclude that every triangle has a positive defect.

An immediate consequence of Theorem 6.4.1 is the following corollary which asserts that quadrilaterals also have positive defects.[7]

THEOREM 6.4.2 *(Corollary).*   The sum of the measures of the interior angles of a convex[8] quadrilateral is less than 360°.

Proof. The proof of this corollary involves using a diagonal to partition the quadrilateral into a pair of triangles and then applying the previous theorem. The details are elementary and have been left as an exercise.

Now that we know that every hyperbolic triangle has a positive defect, a reasonable question to ask is, How big is the defect? Is the defect a constant, such as 1°, making the angle sum a constant for every triangle? Is the defect more than 1°? Less than 1°? (How much less could it be?)

It turns out that the defect varies among triangles. In fact, it is not difficult to prove that if the defect is the same for all triangles, it must be zero, making the angle sum for triangles a constant 180°,[9] which would contradict Theorem 6.4.1. From this we must conclude that in hyperbolic geometry defects vary among triangles.

To illustrate how defects vary in size in hyperbolic geometry, consider $\triangle ABC$ in Figure 6.4.1.

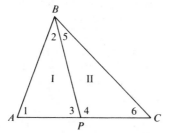

Figure 6.4.1

Point P is any Menalaus point on side $\overline{AC}$, and the Cevian line $\overleftrightarrow{BP}$ partitions $\triangle ABC$ into two other triangles that we will call $\triangle\,$I and $\triangle\,$II. From Theorem 6.4.1, we know that

---

[7] The defect of a convex quadrilateral is the difference between the sum of the measures of its interior angles and 360°. In general, the defect for any convex n-gon is $(n - 2)(180 - \Sigma m\angle i)$, $i = 1$ to n, where $m\angle i$ represents the measure of the ith interior angle of the n-gon.

[8] Nonconvex polygons present an annoying problem here since they involve at least one interior angle whose measure is not between 0° and 180°. This result can be extended to nonconvex quadrilaterals if we are willing to deal with angles of this type.

[9] In Section 2.2 there is a proof that all triangles have an angle sum of 180°. An exercise at the end of that section asked you to find the hidden assumption in the proof. The unstated assumption was that all triangles have the same angle sum.

$$d(\triangle \text{I}) = 180° - S(\triangle \text{I}) = 180° - (\text{m}\angle 1 + \text{m}\angle 2 + \text{m}\angle 3)$$

$$d(\triangle \text{II}) = 180° - S(\triangle \text{II}) = 180° - (\text{m}\angle 4 + \text{m}\angle 5 + \text{m}\angle 6)$$

$$d(\triangle ABC) = 180° - S(\triangle ABC) = 180° - (\text{m}\angle 1 + \text{m}\angle 2 + \text{m}\angle 5 + \text{m}\angle 6)$$

where $d(\triangle \text{I})$, $d(\triangle \text{II})$, and $d(\triangle ABC)$ are the defects for $\triangle \text{I}$, $\triangle \text{II}$, and $\triangle ABC$, respectively. Now consider the sum $d(\triangle \text{I}) + d(\triangle \text{II})$.

$$d(\triangle \text{I}) + d(\triangle \text{II})$$
$$= [180° - (\text{m}\angle 1 + \text{m}\angle 2 + \text{m}\angle 3)] + [180° - (\text{m}\angle 4 + \text{m}\angle 5 + \text{m}\angle 6)]$$

or

$$d(\triangle \text{I}) + d(\triangle \text{II}) = 360 - \text{m}\angle 1 - \text{m}\angle 2 - \text{m}\angle 3 - \text{m}\angle 4 - \text{m}\angle 5 - \text{m}\angle 6$$

However, if we use the fact that $\angle 3$ and $\angle 4$ form a linear pair, the second expression reduces to

$$d(\triangle \text{I}) + d(\triangle \text{II}) = 180° - (\text{m}\angle 1 + \text{m}\angle 2 + \text{m}\angle 5 + \text{m}\angle 6) = d(\triangle ABC),$$

so that

$$d(\triangle ABC) = d(\triangle \text{I}) + d(\triangle \text{II})$$

Since neither $d(\triangle \text{I})$ nor $d(\triangle \text{II})$ can be zero, we know that $d(\triangle ABC) \neq d(\triangle \text{I})$ and that $d(\triangle ABC) \neq d(\triangle \text{II})$, so that $d(\triangle ABC)$ is clearly different from both $d(\triangle \text{I})$ and $d(\triangle \text{II})$. Additionally, we see that the defect for hyperbolic angle sums is additive in the sense that if a triangle is partitioned into component triangles by means of a Cevian line, the defect of the whole is equal to the sum of the defects of the parts. This results in our next theorem.

THEOREM 6.4.3.   If $\triangle ABC$ is partitioned into a pair of component triangles by a Cevian line, the defect of $\triangle ABC$ is equal to the sum of the defects of the component triangles.

A consequence of Theorem 6.4.3 is that triangles with arbitrarily small defects can be generated. To see this, note that any time a triangle is partitioned by a Cevian line, the result is a pair of triangles, one of which has a defect smaller than or equal to one-half the defect of the triangle with which we began. Repeating this process produces a triangle with a defect less than or equal to one-quarter of the original triangle's defect. By repeating this process, triangles with defects arbitrarily close to zero can be generated.

Theorem 6.4.3 established the additivity of the defect of a triangle partitioned by a Cevian line. In fact, the defect is additive even if the partitioning line does not contain a vertex of the triangle. The proof proceeds in much the same way.

In Figure 6.4.2, $\triangle ABC$ is partitioned into $\triangle \text{I}$ and $\square \text{II}$ by $\overrightarrow{PQ}$.

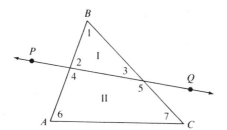

Figure 6.4.2

The quantities $d(\triangle\,\mathrm{I})$, $d(\square\,\mathrm{II})$, and $d(\triangle ABC)$ measure the defects of $\triangle\,\mathrm{I}$, $\square\,\mathrm{II}$, and $\triangle ABC$ in the following ways:

$$d(\triangle\,\mathrm{I}) = 180° - (\mathrm{m}\angle\,1 + \mathrm{m}\angle\,2 + \mathrm{m}\angle\,3)$$

$$d(\square\,\mathrm{II}) = 360° - (\mathrm{m}\angle\,4 + \mathrm{m}\angle\,5 + \mathrm{m}\angle\,6 + \angle\,7)$$

$$d(\triangle ABC) = 180° - (\mathrm{m}\angle\,1 + \mathrm{m}\angle\,6 + \mathrm{m}\angle\,7)$$

A bit of arithmetic (and the fact that $\angle\,2$ and $\angle\,4$, and $\angle\,3$ and $\angle\,5$, are linear pairs) can be used to show that $d(\triangle ABC) = d(\triangle\,\mathrm{I}) + d(\square\,\mathrm{II})$, leading to a companion theorem for Theorem 6.4.3.

THEOREM 6.4.4.   If $\triangle ABC$ is partitioned into a triangle and a quadrilateral by a line, the defect of $\triangle ABC$ is equal to the sum of the defects of the component pieces (i.e., the triangle and the quadrilateral).

Before generalizing further on the additivity of defects, we will take the opportunity to use the preceding theorems to establish an interesting distinction between Euclidean and hyperbolic geometries concerning congruence conditions. This distinction was mentioned in Chapters 2 and 3 and involves the absence of similar, noncongruent triangles in hyperbolic geometry. We will show that in hyperbolic geometry any two triangles with corresponding angles of equal measure are congruent. The argument proceeds as follows:

Suppose that $\triangle ABC$ and $\triangle DEF$ have three pairs of congruent angles, $\angle A \cong \angle D$, $\angle B \cong \angle E$, and $\angle C \cong \angle F$, as shown in Figure 6.4.3.

If any two corresponding sides are congruent, we can conclude that the triangles are congruent using ASA. Therefore let's assume that no two corresponding sides have the same length. From this assumption, we know that one of the triangles must have at least two sides that are longer than the corresponding sides of the other triangle. (Why?) We proceed by assuming that $\triangle ABC$ has two longer sides and that $\overline{AB}$ and $\overline{AC}$ are those sides. Under this hypothesis, we can locate E' between A and B in such a way as to make $AE' = DE$. Similarly, we can locate F' between A and C so that $AF' = DF$.

Having done this, we can show that $\triangle AE'F' = \triangle DEF$ (SAS), so that

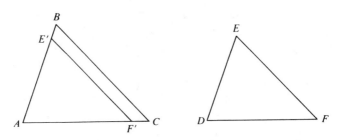

Figure 6.4.3

their corresponding angles are congruent, implying that $S(\triangle AE'F') = S(\triangle DEF)$ and $d(\triangle AE'F') = d(\triangle DEF)$. Also, since $\triangle ABC$ and $\triangle DEF$ have corresponding angles congruent, their angle sums and defects must be equal. Together these facts indicate that

$$d(\triangle ABC) = d(\triangle AE'F')$$

However, from this and Theorem 6.4.4, we see that

$$d(\triangle ABC) = d(\triangle AE'F') + d(\square E'BCF')$$

so that

$$d(\square E'BCF') = 0$$

But this statement is a direct contradiction of Theorem 6.4.2, which assures us that the defect of every quadrilateral is positive. Therefore we must reject the assumption that the sides are of different lengths and conclude that AAA is a *congruence* condition in hyperbolic geometry. This is stated as our next theorem.

THEOREM 6.4.5. *AAA Congruence Condition.*   Two triangles are congruent if their corresponding angles have the same measure.

Together Theorems 6.4.3 and 6.4.4 make explicit the additivity of defects for the two configurations that are possible when a triangle is parti-

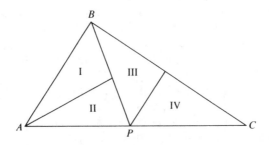

Figure 6.4.4

tioned by a single line. However, it will be useful to extend this idea of additivity to partitions that are more complex than this. For example, consider the partition shown in Figure 6.4.4, where $\triangle ABC$ is partitioned into four component triangles. We would like to show that the defect of $\triangle ABC$, $d(\triangle ABC)$, equals the sum of $d(\triangle I)$, $d(\triangle II)$, $d(\triangle III)$, and $d(\triangle IV)$, the defects of the component triangles. This is easily done for specific cases such as this. In Figure 6.4.4 we need only apply Theorem 6.4.3 to see that

$$d(\triangle ABC) = d(\triangle ABP) + d(\triangle PBC)$$

$$d(\triangle ABP) = d(\triangle I) + d(\triangle II)$$

$$d(\triangle PBC) = d(\triangle III) + d(\triangle IV)$$

so that

$$d(\triangle ABC) = d(\triangle I) + d(\triangle II) + d(\triangle III) + d(\triangle IV)$$

indicating that, at least in this case, defects are additive in complex partitions of triangles. In fact, it can be proved that the defect of a polygon is additive relative to any partition using triangles, but the proof is tedious and not particularly enlightening. For those interested, the details can be found in David Hilbert's *Grundlagen der Geometrie*. We state this result as Theorem 6.4.6.

THEOREM 6.4.6.   If a convex polygon is partitioned into triangles in any manner, the defect of the polygon is equal to the sum of the defects of the component triangles.

Next we turn our attention to pairs of polygons that have defects that are numerically equal. In particular, we can show that every triangle has associated with it a Saccheri quadrilateral with a defect of the same magnitude. To see this, we will perform a procedure that may seem familiar, since it was used during the discussion of similarity in Chapter 4.

The procedure involves constructing a line that contains the midpoints of two of the sides of the triangle. We then draw perpendiculars to this line from each vertex of the triangle, labeling the feet of the perpendiculars P, Q, and R as shown in Figure 6.4.5.

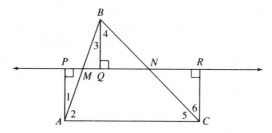

Figure 6.4.5

We claim the following:
1. $\square APRC$ is a Saccheri quadrilateral
2. $d(\triangle ABC) = d(\square APRC)$

To show that $\square APRC$ is a Saccheri quadrilateral, we need only verify that $AP = CR$. We note first that $\triangle MAP \cong \triangle MBQ$ and $\triangle NBQ \cong \triangle NCR$ (both by AAS), so that $\overline{AP} \cong \overline{BQ}$ and $\overline{CR} \cong \overline{BQ}$. From this, it follows that $\overline{AP} \cong \overline{CR}$, so $\square APRC$ is in fact a Saccheri quadrilateral. $\square APRC$ is called the *associated Saccheri quadrilateral* with respect to $\triangle ABC$.

From the congruence of $\triangle MAP$ and $\triangle MBQ$, it follows that m∠1 = m∠3, and from the congruence of $\triangle NBQ$ and $\triangle NCR$, it follows that m∠6 = m∠4. These relationships will allow us to compare $d(\triangle ABC)$ and $d(\square APRC)$ as follows:

$$d(\triangle ABC) = 180° - (m\angle 2 + m\angle 3 + m\angle 4 + m\angle 5) \tag{5}$$

$$d(\square APRC) = 360° - (90° + m\angle 1 + m\angle 2 + m\angle 5 + m\angle 6 + 90°) \tag{6}$$

But Equation 6 simplifies to

$$d(\square APRC) = 180° - (m\angle 1 + m\angle 2 + m\angle 5 + m\angle 6)$$

and by using the congruence of ∠1 and ∠3 and of ∠6 and ∠4, we find that

$$d(\square APRC) = 180° - (m\angle 3 + m\angle 2 + m\angle 5 + m\angle 4) = d(\triangle ABC)$$

so that the defect of $\triangle ABC$ is equal to the defect of its associated Saccheri quadrilateral $\square APRC$.

Actually, we can make a slightly stronger statement about the relationship between a triangle and its associated Saccheri quadrilateral. If we refer to Figure 6.4.5, we see that $\square APRC$ can be cut up into three pieces ($\triangle MAP$, $\square AMNC$, and $\triangle NCR$) and pasted back together to form a figure congruent to $\triangle ABC$ by appropriately placing the triangles $\triangle MAP$ and $\triangle NCR$ on $\overline{MN}$. In fact, both $\triangle ABC$ and $\square APRC$ can be viewed as being comprised of three pieces that are, respectively, congruent. If we go a step further and partition $\square AMNC$ into two triangles, we see that these two polygons can be partitioned into equivalent sets[10] of triangles that are, correspondingly, congruent. This relationship motivates the following definition:

DEFINITION. *Equivalent Polygons.* Two polygons P_1 and P_2 are said to be equivalent if and only if each polygon can be partitioned into a finite set of triangles in such a way that (i) the two sets of triangles can be placed into a one-to-one correspondence, and (ii) the one-to-one correspondence associates triangles that are congruent.

---

[10] In set theory we say that two sets are *equivalent* if and only if their elements can be put into a one-to-one correspondence.

This definition, along with the discussion that preceded it, leads to the next theorem.

THEOREM 6.4.7.   Every triangle is equivalent to its associated Saccheri quadrilateral.

Clearly, any pair of congruent triangles is also equivalent, since the partitions required by the definition of equivalence can be the triangles themselves which are, by hypothesis, congruent. It is almost as obvious that if two quadrilaterals are congruent, they are equivalent, since a diagonal can be drawn in each quadrilateral that produces a partition involving pairs of congruent triangles. We can proceed, inductively, to convex polygons with more than four sides by arguing that any n-gon can be partitioned into a triangle and an $(n - 1)$-gon. A proof using induction on n can then be used to justify the following theorem.

THEOREM 6.4.8.   Any two congruent convex polygons are equivalent.

At this point one might be tempted to speculate that equivalence and congruence of polygons are essentially the same relation. This is not, however, the case. Figure 6.4.6 illustrates this. Figure 6.4.6a is partitioned into $\triangle$ I and $\triangle$ II which are congruent to $\triangle$ III and $\triangle$ IV, respectively, in Figure 6.4.6b. Clearly, however, $\square ABCD$ is *not* congruent to $\square EFGH$, so that the quadrilaterals are equivalent although not congruent. We see then that equivalence is a slightly more general relationship than congruence. Nonetheless, several important properties we associate with congruence carry over to equivalence. For example, equivalence of triangles is clearly both reflexive and symmetric (see Exercise Set 6.4, Problems 10 and 11).

Equivalence of polygons is also transitive, although the proof of this is a bit more difficult. To illustrate how this works, consider $\square ABCD$ and

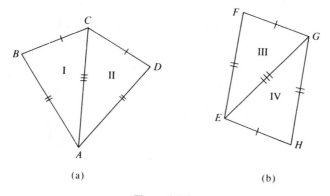

(a)   (b)

Figure 6.4.6

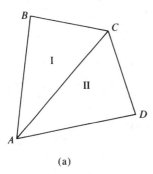

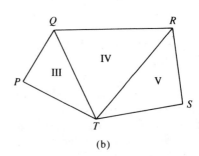

(a) (b)

Figure 6.4.7

⬠ $PQRST$ in Figure 6.4.7. Each of these is equivalent to a single quadrilateral $\square WXYZ$, as indicated by the two partitions shown in Figure 6.4.8.

To be more explicit, we can indicate the correspondences between pairs of congruent triangles using a table.

$\square ABCD \leftrightarrow \square WXYZ$	⬠ $PQRST \leftrightarrow \square WXYZ$
$\triangle ABC \cong \triangle WXY$	$\triangle PQT \cong \triangle XVY$
$\triangle ACD \cong \triangle YWZ$	$\triangle TQR \cong \triangle VYZ$
	$\triangle TRS \cong \triangle ZVW$

We see from this that each of $\square ABCD$ and $\square PQRST$ is equivalent to $\square WXYZ$. To show that $\square ABCD$ is equivalent to ⬠ $PQRST$, we need to show that each can be partitioned into a set of triangles and that these sets can be matched in a one-to-one fashion involving congruent pairs. To accomplish this, we will superimpose the two previous partitions on a single copy of $\square WXYZ$, as shown in Figure 6.4.9. $\square WXYZ$ is now partitioned into five triangles numbered I through V. These five triangles can be reassem-

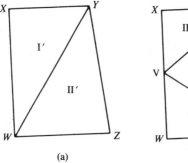

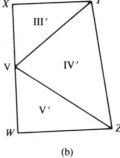

(a) (b)

Figure 6.4.8

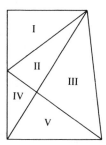

Figure 6.4.9

bled as either □$ABCD$ or as ⌐$PQRST$, indicating that each of the polygons can be partitioned into five triangles that are congruent, in a one-to-one fashion, to △I through △V in Figure 6.4.9. This means that □$ABCD$ and $PQRST$ are, as we expected, equivalent.

The general argument in favor of the transitivity of equivalence takes into account certain other possible configurations of the polygons involved but proceeds in essentially the same manner.[11] These properties concerning the equivalence of polygons are summarized in the following theorem:

THEOREM 6.4.9.   The equivalence relation as defined for polygons is reflexive, symmetric, and transitive.

Theorem 6.4.9 allows us to conclude that equivalence of polygons is an equivalence relation. (Is this a redundant statement?) This will be useful later when we define the idea of area in hyperbolic geometry.

With these results in place, we can establish the relationship between defect and equivalence for triangles. In fact, these terms are very closely related, as we will soon see. However, before we can deduce this relationship, we will need a preliminary proof, or lemma.

THEOREM 6.4.10 *(Lemma)*.   Two Saccheri quadrilaterals are congruent if their summits and summit angles are congruent.

Proof. To see this, assume that □$ABCD$ and □$EFGH$ are quadrilaterals with summits and summit angles congruent (Figure 6.4.10). If $AB = EF$ (so that $DC = HG$ also—why?), then the two quadrilaterals are congruent by SASAS. Assume then that $AB \neq EF$ and, without loss of generality, assume that $AB < EF$. Under these assumptions, we can locate B' on $\overline{EF}$ so that $EB' = AB$. In like fashion, we can locate C' on $\overline{HG}$ so that $HC' = DC$ (see Exercise Set 6.4, Problem 12). Having done this, we may conclude that □$ABCD = $ □$EB'C'H$ (SASAS), so that $m\angle 1 = m\angle 2$ and $m\angle 3 = m\angle 4$.

[11] For a more complete discussion of the transitivity of equivalence, see Henry George Forder, *The Foundations of Euclidean Geometry* (New York: Dover Publications, Inc., 1958).

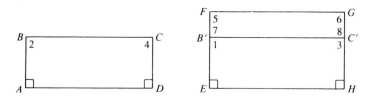

Figure 6.4.10

But since, by hypothesis, the summit angles of $\square ABCD$ and $\square EFGH$ are congruent, we know also that $m\angle 2 = m\angle 5$ and $m\angle 4 = m\angle 6$. Now consider the sum of the measures of the interior angles of $\square B'FGC'$:

$$S(\square B'FGC') = m\angle 5 + m\angle 7 + m\angle 6 + m\angle 8$$

Substituting, we obtain

$$S(\square B'FGC') = m\angle 1 + m\angle 7 + m\angle 3 + m\angle 8$$

But since $\angle 1$ and $\angle 7$, and $\angle 3$ and $\angle 8$, form linear pairs, we have

$$S(\square B'FGC') = 360°$$

which contradicts Theorem 6.4.2. Thus we must reject the assumption that $AB \neq EF$, and conclude that the quadrilaterals are congruent.

We now proceed with a sequence of four theorems that will establish the relationship between defect and equivalence for triangles and will set the stage for our discussion of area in hyperbolic geometry in the next section.

THEOREM 6.4.11.   If two triangles have the same defect and a pair of congruent sides, then they are equivalent.

Proof. Suppose that a pair of triangles $\triangle ABC$ and $\triangle DEF$ have the same defect $[d(\triangle ABC) = d(\triangle DEF)]$ and a pair of corresponding sides that are congruent. (See Figure 6.4.11 where we assume that $\overline{AC} \cong \overline{DF}$.)

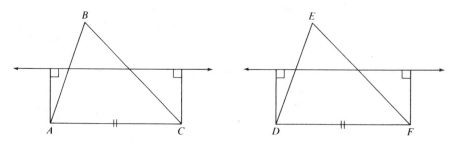

Figure 6.4.11

If we construct the Saccheri quadrilateral associated with the triangles upon the congruent bases, we can show that these quadrilaterals are congruent (see Exercise Set 6.4, Problem 14). Next recall from Theorem 6.4.7 that each of the triangles is equivalent to its associated Saccheri quadrilateral. Since the two Saccheri quadrilaterals are congruent (hence equivalent), we may apply the transitivity of equivalence (Theorem 6.4.9) to conclude that $\triangle ABC$ and $\triangle DEF$ are equivalent.

Of course, we can't assume that all triangles with equal defects have a pair of congruent sides, for this is not always the case. Our next task is to show the equivalence of pairs of triangles with equal defects but without a pair of congruent sides.

THEOREM 6.4.12.   If two triangles have the same defect, then they are equivalent.

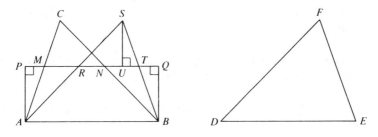

Figure 6.4.12

Proof. In Figure 6.4.12 we assume that $\triangle ABC$ and $\triangle DEF$ have equal defects but do not share a pair of congruent sides. In particular (and without loss of generality), we may assume that $DF > AC$. Upon $\overline{AB}$ we will construct $\square APQB$, the Saccheri quadrilateral associated with $\triangle ABC$, using M and N to represent the midpoints of $\overline{AC}$ and $\overline{BC}$. Now since $DF > AC$, we know that $\frac{1}{2}DF > AM$. With this in mind, we construct the circle $C(A, \frac{1}{2}DF)$[12] and use R to denote the intersection of this circle and $\overleftrightarrow{MN}$. Next we extend $\overrightarrow{AR}$ and locate point S on $\overrightarrow{AR}$ so that R is the midpoint of $\overline{AS}$. Finally, we complete $\triangle ASB$ by constructing $\overline{SB}$. Since S and B are on opposite sides of $\overleftrightarrow{PQ}$ (why?), we know that $\overline{SB}$ intersects $\overleftrightarrow{PQ}$ at some point which we will call T.

Our first claim is that $\square APQB$, which was constructed to be the Saccheri quadrilateral associated with $\triangle ABC$, is also the Saccheri quadrilateral associated with $\triangle ASB$. In order to substantiate this claim, we need to show that T is the midpoint of $\overline{SB}$. To see this, we drop a perpendicular from S to $\overleftrightarrow{PQ}$ and use U to label the foot of the perpendicular. From this, we see that $\triangle APR \cong \triangle SUR$ (AAS), so that $\overline{AP} \cong \overline{SU}$. This congruence allows us to

[12] Recall from Chapter 4 that $C(P, r)$ denotes a circle centered at P with radius r.

establish the congruence of $\triangle TUS$ and $\triangle TQB$,[13] so that $\overline{TS} \cong \overline{TB}$, which means that T is the midpoint of $\overline{SB}$ and $\square APQB$ is the Saccheri quadrilateral associated with both $\triangle ABS$ and $\triangle ABC$.

By Theorem 6.4.7 and the transitivity of equivalence (Theorem 6.4.6), we may conclude that $\triangle ABC$ and $\triangle ABS$ are equivalent. From an earlier discussion, we can show (Exercise Set 6.4, Problem 19) that each triangle has an angle sum equal to the sum of the measures of the summit angles of its associated Saccheri quadrilateral. Consequently, $\triangle ABS$ and $\triangle ABC$ have the same angle sum, hence the same defect, and since our hypothesis is that $\triangle ABC$ and $\triangle DEF$ have the same defect, we may conclude that $\triangle ABC$, $\triangle ABS$, and $\triangle DEF$ all have the same defect.

To complete the argument that $\triangle ABC$ and $\triangle DEF$ are equivalent, we need only note that $\triangle ABS$ and $\triangle DEF$ are triangles with equal defects and a pair of congruent sides (Remember how we constructed AS?), so that by Theorem 6.4.11 $\triangle ABS$ and $\triangle DEF$ are equivalent. If we once again apply the transitivity of equivalence, we may conclude that $\triangle ABC$ is equivalent to $\triangle DEF$, which finally completes the proof.

We now know that pairs of triangles are equivalent providing they have the same defect. The next theorem establishes the converse of this idea.

THEOREM 6.4.13.   If two triangles are equivalent, then they have the same defect.

Proof. Assume that two triangles $\triangle ABC$ and $\triangle DEF$ are equivalent. By definition, we can partition each into a set of triangles, and these sets can be paired in a one-to-one way involving congruent pairs. Let T_1, T_2, T_3, . . . , T_n represent the partition for $\triangle ABC$, and R_1, R_2, R_3, . . . , R_n represent the partition for $\triangle DEF$. Then presume that we have ordered these sets of triangles so that $T_i = R_i$, where $i = 1$ to n.

Using Theorem 6.4.6, we know that

$$d(\triangle ABC) = d(T_1) + d(T_2) + d(T_3) + \cdots + d(T_n)$$

and

$$d(\triangle DEF) = d(R_1) + d(R_2) + d(R_3) + \cdots + d(R_n)$$

But since $T_i \cong R_i$ for all i, $d(T_i) = d(R_i)$ for each i, hence

$$d(\triangle ABC) = d(\triangle DEF)$$

Combining Theorems 6.4.12 and 6.4.13, we obtain the final result for this section, which puts forth the close relationship of defect and equivalence for triangles and, by implication, polygons in general.

---

[13] These triangles are congruent by AAS also—but which is the pair of congruent sides? See Exercise Set 6.4, Problem 17.

THEOREM 6.4.14.   Two triangles have the same defect if and only if they are equivalent.

EXERCISE SET 6.4

1. Explain how the statements below are all equivalent to the following statement: The sum of the measures of the interior angles of a triangle $\triangle ABC$ is less than or equal to $180°$.
   (a) $S(\triangle ABC) + k° = 180°$
   (b) $k° = 180° - S(\triangle ABC)$
   (c) $S(\triangle ABC) = 180° - k°$

2. Reread the proof of Theorem 6.4.1. Specifically, consider the reference to Theorem 3.6.18. Does Theorem 3.6.18 imply that if $S(\triangle ABC) = 180°$ for every triangle, then the Euclidean parallel postulate is valid? Explain.

3. Prove that the sum of the measures of the interior angles of a quadrilateral must be less than $360°$.

4. Theorem 6.4.1 and Corollary 6.4.2 can be generalized to convex polygons with any number of sides. Give the general statement concerning the angle sum for polygons and provide a proof for the statement using mathematical induction.

5. Verify that the proof of Theorem 4.7.3 does not rely on any form of the Euclidean parallel postulate.

6. Prove Theorem 6.4.4.

7. Suppose that $\triangle ABC$ and $\triangle DEF$ are equilateral triangles with $AB = 2DE$. Show that the corresponding angles are *not* congruent and that the larger triangle has the smaller interior angles.

8. In the proof of Theorem 6.4.5 why can we assume that one of the triangles has at least two sides longer than the corresponding sides of the other triangle?

9. Let $\triangle ABC$ be an obtuse triangle with the obtuse angle at A. Let M and N be the midpoints of sides $\overline{AB}$ and $\overline{BC}$, respectively. Can you construct an associated Saccheri quadrilateral for $\triangle ABC$ in which the base is contained in $\overleftrightarrow{MN}$ and the summit is $\overline{AC}$? If so, do so and verify that the vertical sides are congruent. If not, explain why not.

10. Give a proof for the fact that the equivalence of polygons is a reflexive relation.

11. Give a proof for the fact that the equivalence of polygons is a symmetric relation.

12. In the proof of Theorem 6.4.10 a point C' was located on $\overleftrightarrow{HG}$ so that $HC' = DC$. Explain how you can be sure that C' is between H and G.

13. Restate Theorem 6.4.10 as a biconditional. Is the statement still true? If so, prove it. If not, explain why not.

14. Prove that if $\triangle ABC$ and $\triangle DEF$ are triangles with equal defects and a pair of congruent sides, their associated Saccheri quadrilaterals are congruent. (*Hint:* Recall that the angle sum for a triangle is equal to the sum of the measures of the summit angles of its associated Saccheri quadrilateral and that the two summit angles of a Saccheri quadrilateral are congruent.)

15. In Figure 6.4.12a how do we know that S and B are on opposite sides of $\overleftrightarrow{PQ}$?

16. What is the defect for every triangle in Euclidean geometry? With this in mind, is Theorem 6.4.14 valid in Euclidean geometry? Explain why or why not.

17. In the proof of Theorem 6.4.10 we used the AAS congruence condition to show

that $\triangle TUS$ and $\triangle TQB$ are congruent (see Figure 6.4.12). Which pairs of angles and which pair of sides are congruent? Justify your choices.

18. Theorem 6.4.14 states the relationship between defect and equivalence for triangles. Can this relationship be generalized to polygons in general? Explain why or why not.

19. Prove that the sum of the measures of the interior angles of any triangle is equal to the sum of the measures of the summit angles of the associated Saccheri quadrilateral. (*Note:* This means that the defect of any triangle is equal to the difference between 180° and the sum of the measures of the summit angles of its associated Saccheri quadrilateral.)

6.5 AREA IN HYPERBOLIC GEOMETRY

In Section 4.3 we discussed the idea of area for polygonal regions in Euclidean geometry. This discussion was greatly simplified by the use of SMSG Postulates 17 through 20, each of which addresses a property we normally associate with area measure. Development of the concept of area in hyperbolic geometry must proceed differently than in Euclidean geometry because of the difference in the postulate sets. To begin the discussion, we will take a second look at SMSG Postulate 20, which for convenience is restated here.

POSTULATE 20.   The area of a rectangle is the product of the length of its base and the length of its altitude.

The use of this postulate makes computations of areas of polygonal regions quite easy, since every polygon can be partitioned into triangles and every triangle can be shown to have an area equal to one-half that of an associated rectangle. We see then that the area of any polygonal region can ultimately be determined using Postulate 20 or consequences of it. It should be clear, however, that this postulate is not applicable in hyperbolic geometry, since we saw earlier that rectangles do not exist in hyperbolic geometry. This leaves us, at least temporarily, with no means of assigning areas to polygonal regions. Our current goal then is to formulate an alternative means of associating a number (i.e., an area measure) with the interior of each simple polygon.

As we formulate a method for assigning areas, we must take care to preserve as many as possible of the Euclidean properties of area so that the results are consistent with our intuitive ideas concerning area. In particular, we would like to preserve Postulates 17 through 19 from the SMSG set, which posit the following properties for area measure:

1. Every polygonal region has one and only one area (SMSG Postulate 17).
2. Congruent triangles have equal areas (SMSG Postulate 18).

3. If a polygonal region is partitioned into a pair of subregions, the area of
    the region will equal the sum of the areas of the two subregions (SMSG
    Postulate 19).

One approach to defining area in hyperbolic geometry would be to try
to define a unit of area measure that is not based on rectangles. For example, we might try to use a unit of measure based on the interior of equilateral
triangles using some arbitrary unit of length for the sides of the unit triangle.
Using this approach, we might measure area in terms of "triangular units"
rather than square units and assign to each polygonal region an area equal to
the number of "unit triangles" into which the region could be partitioned.

In a sense, this is what we will do. However, in order to proceed more
efficiently we will recall that discussions of the area of a region (whether
Euclidean, hyperbolic, or other) always seem to involve counting pieces of a
special[14] partition of the region. From this, it follows that regions of equal
area ought to have partitions that involve congruent pairs of polygons (e.g.,
in Euclidean geometry the pairs involve unit squares or portions thereof). If
we allow the congruent pairs to be pairs of triangles, we may then say that
polygonal regions with equal area are those that are equivalent, using the
idea of equivalence defined in the preceding section.

The preceding comments provide for us a clue to a formal definition of
area in hyperbolic geometry. First, recall that Theorem 6.4.14 provides us
with a means of assigning a single number (the defect) to all triangles that are
equivalent to one another. Second, we want to define area in such a way
that two polygonal regions have the same area providing they are equivalent
(i.e., can be partitioned into triangles that are, in pairs, congruent). Taken
together, these ideas suggest that it is reasonable to say that regions have
equal areas as long as they have the same defect.

The idea we will pursue is this. In order to assign an area measure to
triangles (hence to polygonal regions in general), can we simply compute the
defect and, by definition, say that the region's area is equal to its defect?
This might work, but there are two items left to consider:

1. Will this interpretation of area satisfy SMSG Postulates 17 through 19?
2. Is this approach to the definition of area sufficiently general to allow for
    variations in systems of measure, for example, conversion between
    standard and metric units in Euclidean geometry?

In order to generalize to alternative units of measure, we need only
require that the area of a polygon be *proportional* to its defect. For example,
if our unit of angular measurement is degrees, and we wish to express the

---

[14] The special nature of the partition is that it is comprised of "unit pieces," that is,
pieces with area of 1 unit.

area of a polygon in terms of radian measure, we could multiply the polygon's "degree defect" by the constant $\pi/180$ in order to make the conversion[15] to a defect in radians. Using this idea as motivation, we state the following definition for the area of a polygon in hyperbolic geometry:

DEFINITION.   The *area* A(P) for a simple polygon P in the hyperbolic plane is directly proportional to the polygon's defect,[16] so that

$$A(P) = k \times d(P)$$

In order to ensure that A(P) is unique and positive, we will require that k be a fixed, positive constant. Other than this restriction, the value of k is arbitrary and in effect defines a measurement scale.

With this definition in place, we are prepared to argue that the properties of area as given by SMSG Postulates 17 through 19 are valid. These properties can now be stated as theorems.

THEOREM 6.5.1.   To every polygonal region there corresponds a unique positive area (SMSG Postulate 17).

THEOREM 6.5.2.   If two polygons are congruent, then the polygonal regions have the same area (SMSG Postulate 18).

These two theorems follow directly from the definition of area and some of the hyperbolic properties of polygons discussed previously in this chapter. Their proofs are left as exercises.

THEOREM 6.5.3.   If a polygonal region P is partitioned into two subregions, R_1 and R_2, then

$$A(P) = A(R_1) + A(R_2)$$

(SMSG Postulate 19).

Proof. Suppose that a polygonal region P is partitioned into regions R_1 and R_2 (Figure 6.5.1). We know that the areas of P, R_1, and R_2 are, by definition, proportional to their defects. Therefore, in order to show that

$$A(P) = A(R_1) + A(R_2)$$

[15] Clearly, if our angular measurements are in radians and the area measure is to be in terms of degrees, the constant would be $180/\pi$. Any other constant could be used to define an alternative measurement schema.

[16] Recall from Section 6.4 that the defect for a polygon P with n sides is given by

$$d(P) = (n - 2)(180° - \Sigma m\angle i) \qquad i = 1 \text{ to } n$$

where $m\angle i$ represents the measure of the ith interior angle of the polygon.

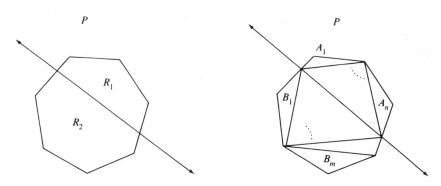

Figure 6.5.1

it will suffice to show that

$$k \times d(P) = k \times d(R_1) + k \times d(R_2)$$

To do this, we will first partition R_1 into a set of triangles A_1, A_2, A_3, $\ldots$, A_n by drawing all possible diagonals from a single vertex A. Likewise, we will partition R_2 into triangles B_1, B_2, B_3, $\ldots$, B_m. Note that these two partitions taken together partition P into $A_1, \ldots, A_n, B_1, \ldots, B_m$. Theorem 6.4.6 tells us that

$$d(P) = d(A_1) + \cdots + d(A_n) + d(B_1) + \cdots + d(B_m)$$

Since k and all of the defects are numbers, we may apply the multiplicative property of equality to show that

$$k \times d(P) = k[d(A_1) + \cdots + d(A_n) + d(B_1) + \cdots + d(B_m)]$$

Distributing, we find that

$$k \times d(P) = k \times d(A_1) + \cdots + k \times d(A_n) + k \times d(B_1) + \cdots + k \times d(B_m)$$

but since

$$k \times d(R_1) = k \times d(A_1) + \cdots k \times d(A_n)$$

and

$$k \times d(R_2) = k \times d(B_1) + \cdots k \times d(B_m)$$

we see that

$$k \times d(P) = k \times d(R_1) + k \times d(R_2)$$

so that

$$A(P) = A(R_1) + A(R_2)$$

This means that area, as defined in the hyperbolic plane is additive, and concludes the proof of the theorem.

Our definition for area has been given for polygons in general so that it may be applied to any specific type of polygon. In particular, we may say that the area of a triangle is proportional to its defect, so that

$$A(\triangle ABC) = k \times d(\triangle ABC)$$

for any $\triangle ABC$. We are now in a position to explore a surprising property of hyperbolic area.

THEOREM 6.5.4.   If $\triangle ABC$ is a hyperbolic triangle, then

$$A(\triangle ABC) < k \times 180°$$

where degrees are used as the unit of angular measure.

Proof. Consider $\angle A$, $\angle B$, and $\angle C$ of $\triangle ABC$. Each of these angles has a measure between $0°$ and $180°$ (SMSG Postulate 11), so that using $S(\triangle ABC)$ to represent the sum of the measures of the interior angles of $\triangle ABC$, we have

$$S(\triangle ABC) > 0 \tag{7}$$

If we subtract $180°$ from each side of Equation (7), we will see that

$$S(\triangle ABC) - 180° > -180° \tag{8}$$

Multiplying each side of Equation (8) by -1, we obtain

$$180° - S(\triangle ABC) < 180° \tag{9}$$

But since

$$180° - S(\triangle ABC) = d(\triangle ABC)$$

Equation (9) becomes

$$d(\triangle ABC) < 180°$$

so that

$$A(\triangle ABC) = k \times d(\triangle ABC) < k \times 180° \tag{10}$$

Theorem 6.5.4 tells us in effect that there is an upper bound, $k \times 180°$, for the area of triangles [and, more generally, $k(n - 2)(180°)$ for n-gons] in hyperbolic geometry. This is true even though lengths are not limited by any upper bound, so that the lengths of the sides of triangles can be arbitrarily large. This idea is counterintuitive but is consistent with the list of equivalences of the parallel postulate listed in Chapter 2. In fact, it can be shown that the Euclidean parallel postulate is equivalent to the following statement: There is no upper bound for the areas of triangles.

EXERCISE SET 6.5

1. We defined the area of a triangle to be proportional to the defect of the triangle, so that for $\triangle ABC$, we have $A(\triangle ABC) = k \times d(\triangle ABC)$. Since there is a close relationship between $d(\triangle ABC)$ and the angle sum for the triangle [recalling that $S(\triangle ABC) = 180° - d(\triangle ABC)$], could we have defined $A(\triangle ABC) = k \times S(\triangle ABC)$? Explain why or why not.

2. Prove Theorem 6.5.1, being careful to show that the area must be *positive*.

3. Prove Theorem 6.5.2.

4. Determine the area of the hyperbolic triangles having the following angle measures if $k = \pi/180$:
   - **(a)** $m\angle A = 60°$
     $m\angle B = 60°$
     $m\angle C = 58.5°$
   - **(b)** $m\angle A = 40°$
     $m\angle B = 40°$
     $m\angle C = 70°$
   - **(c)** $m\angle A = 20°$
     $m\angle B = 30°$
     $m\angle C = 25°$

5. Repeat Problem 4 using $k = 2$ and $k = \frac{1}{3}$.

6. What would be the angle sum (in degrees) of a triangle with an area of 1 unit in a system of area measurement in which $k =$
   - **(a)** 2   **(b)** $\frac{1}{2}$ **(c)** 6   **(d)** $\frac{1}{4}$ **(e)** $\pi/180$

7. Repeat Problem 6 using a radian defect rather than a degree defect.

8. Restate Theorem 6.5.4 using radians as the unit of angular measure.

9. Suppose that the measure of a summit angle of a Saccheri quadrilateral is 2°. What is the area of the quadrilateral? Derive a formula for the area of a Saccheri quadrilateral in terms of the measure of its summit angles.

10. Suppose that the measure of the acute angle of a Lambert quadrilateral is 78°. What is the area of the quadrilateral? Derive a formula for the area of a Lambert quadrilateral in terms of the measure of its acute angle.

11. Suppose that the measure of each interior angle of an equilateral triangle is 55°. What is the area of the triangle? Derive a formula for the area of an equilateral triangle with interior angles measuring $x°$.

6.6 SHOWING CONSISTENCY—A MODEL FOR HYPERBOLIC GEOMETRY

In the first five sections of this chapter we stated and proved a number of theorems using the postulates of hyperbolic geometry (namely, the neutral postulates and the hyperbolic parallel postulate). Many of the hyperbolic theorems seem strange, since they contradict theorems from Euclidean geometry. In fact, these results were strange enough to convince many pre-nineteenth century mathematicians that non-Euclidean geometries were inconsistent with the real world, hence not viable geometries at all. You may recall from discussions in Chapters 2 and 3 that Girolamo Saccheri (the eighteenth century geometer who in a very real sense developed hyperbolic geometry) rejected the hyperbolic theorems he had proved because they were "repugnant to the nature of a straight line."

At about the same time, Emmanuel Kant, perhaps the leading western philosopher of the eighteenth century, argued that worldly truths do not come from experience only—that if truths exist they must be a priori (deductive) judgments. Kant reasoned that the human mind was endowed (by God or by nature) with the capacity to make sense of the universe without relying on empirical data. Consequently, it was necessary that the results of reason be consistent with worldly observations, since the two were necessarily linked. Using this idea as a premise, Kant argued that since we could perceive no geometry other than Euclid's, there could be no geometry other than Euclid's. This led many mathematicians and philosophers to believe that somewhere in the realm of each non-Euclidean geometry there could be found at least one inconsistency which, when unearthed, would confirm the notion that no geometry but Euclid's is consistent.

For example, hyperbolic geometry could be dismissed immediately if one could find within the list of hyperbolic theorems even one pair of contradictory statements. But try as they[17] might, geometers could derive no contradictions from the postulates for hyperbolic geometry. Nevertheless, the academic climate was not yet right for the acceptance of non-Euclidean geometries. Thus when two rather obscure Eastern European mathematicians, Nicholai Lobachevsky (1830) and Janos Bolyai (1832), published independent (and logically sound) developments of hyperbolic geometry, their works were subjected to ridicule and were, at least initially, ignored.

Ironically, the greatest mathematician of the time, Karl Friedrich Gauss, had been silently engaged in the same research and had achieved essentially the same results. It was only after Gauss's death (1855), when this work became public, that the mathematical community began to seriously consider hyperbolic geometry as a viable alternative to Euclidean geometry.

Still the question remained, How do we show that a non-Euclidean geometry (e.g., hyperbolic geometry) is consistent? The failure of mathematicians to derive a pair of contradictory theorems is not sufficient, since regardless of how many consistent theorems are derived, there may still be an undiscovered contradictory pair. A different approach was clearly needed.

By the second half of the nineteenth century, with the posthumous blessing of Gauss, mathematicians took on the task of establishing the consistency of non-Euclidean geometries, including hyperbolic geometry. In order to do this, an alternative strategy was applied—model building.

You will recall from Chapter 1 that one means of establishing the consistency of an axiomatic system is to build a model for the system that

---

[17] Besides Saccheri, other, better known, mathematicians such as J. H. Lambert (1728–1777) and A. Legendre (1752–1833) pursued the goal of showing that Euclidean geometry alone is consistent.

demonstrates all of its postulates simultaneously. Models can be either concrete or abstract. A concrete model for hyperbolic geometry would interpret the terms and relationships of the system (point, line, incidence, and so on) as entities in the real world. But this runs counter to our intuition (the hyperbolic parallel postulate being, for most people, nonintuitive) and would be difficult since even Euclidean geometry is interpreted as an idealized model of the real world. It becomes clear then that if a model for hyperbolic geometry is to be constructed, it will necessarily be an abstract model, that is, one in which the interpretations are taken from another axiomatic system whose consistency is assumed. One problem with constructing an abstract model is that if successful, one can only conclude the relative consistency of the axiomatic systems. However, if we choose to interpret hyperbolic geometry as a model within Euclidean geometry, and if we are successful in building the model in this way, we will at least be able to conclude that hyperbolic geometry is consistent providing Euclidean geometry is. Since this is as much as we can hope for, we'll pursue this avenue.

Over the last century and a half several abstract models for hyperbolic geometry have been developed. One of the most elegant of these was offered by the French mathematician Henri Poincaré (1854–1912). We will use this model, in a slightly modified form, to illustrate the relative consistency of hyperbolic and Euclidean geometries.

We will restrict our discussion to "plane" hyperbolic geometry and embed a model of hyperbolic geometry within a Euclidean plane. In this model the Euclidean plane will be coordinatized, making it in effect a piece of graph paper with origin O.

In this model hyperbolic points will be interpreted as only those points (x, y) of the Euclidean plane having the property that $x^2 + y^2 < 1$.

In other words, hyperbolic points are only those Euclidean points in the interior of the unit circle $\mathcal{C}$ centered at the origin. Since the term "point" is undefined, we are free to interpret it in this way. We shall show later that this interpretation of "point" is in keeping with the hyperbolic axioms within Poincaré's model.

The second undefined term, "line", will appear within the model in two forms. First, every diameter[18] of $\mathcal{C}$ will serve as a line in the hyperbolic model (Figure 6.6.1a). Second, if $\mathcal{C}'$ is a Euclidean circle *orthogonal* to $\mathcal{C}$, then the points of $\mathcal{C}'$ in the interior of $\mathcal{C}$ will also constitute a hyperbolic line (Figure 6.6.1b).

Obviously, "lines" interpreted in this way don't look like lines. Keep in mind, however, that "line," like "point," is undefined and will acquire

---

[18] The endpoints of each diameter must be excluded from the line since these endpoints are not interior points of C and therefore are not points of our hyperbolic plane. When we use the term "diameter" to describe a hyperbolic line we will by implication mean a diameter of C excluding its endpoints.

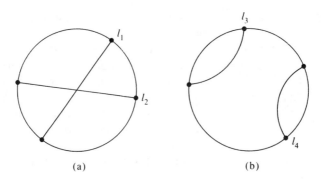

Figure 6.61

meaning implicitly from the postulates. As long as these interpretations of the undefined terms demonstrate the properties required of them by the postulates, they will serve as valid interpretations. The trick is to choose interpretations for "point," "line," and the other geometric terms that are consistent with the postulates of hyperbolic geometry.

Recall that the set of hyperbolic postulates consists of all of the neutral postulates (SMSG Postulates 1 through 15)[19] and the hyperbolic parallel postulate stated in Section 6.3. Since our model is planar, the postulates concerning space relations (SMSG Postulates 5 through 8) need not be considered. In addition, the postulates concerning separation (SMSG Postulates 9 and 10) follow from the fact that the Poincaré model is embedded in a Euclidean plane. Also, recall that Postulates 16 through 20 (those concerning area) were disposed of in Section 6.5. Our task then is to show that the remaining neutral postulates and the hyperbolic parallel postulate are valid in Poincaré's model of the hyperbolic plane.

To begin, consider neutral Postulate 1:

POSTULATE 1.   Given any two different points, there is exactly one line that contains both of them.

Suppose that we choose a pair of hyperbolic points from our model (i.e., any two points in the interior of circle $\mathscr{C}$). Is it necessarily true in the model that there is exactly one hyperbolic line containing both points as required by Postulate 1? The answer is yes, and the verification proceeds as follows:

Suppose A and B are two distinct points. Suppose also that, by chance, A, B, and C (the origin and center of circle $\mathscr{C}$) are collinear (Figure 6.6.2). A and B then define a diameter of $\mathscr{C}$, and this diameter is, according

<hr/>

[19] From here on, references to the neutral postulates will mean SMSG Postulates 1 through 15.

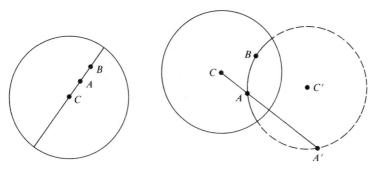

Figure 6.6.2   **Figure 6.6.3**

to the model, a line. Furthermore, since there is exactly one (Euclidean) line that contains both A and B, there can be only one diameter for $\mathscr{C}$ that contains both A and B, so that the line containing A and B is unique.

Of course, it is quite likely that A, B, and $\mathscr{C}$ will *not* be collinear. We now consider this possibility. To do so, we will make use of several results from Chapters 4 and 5. First, recall from Section 5.5 that every point in the interior of circle $\mathscr{C}$ is mapped to a point in the exterior by the transformation of inversion. In particular, the image of A under inversion in circle $\mathscr{C}$ is some exterior point A' (Figure 6.6.3). We know that A, A', and B are not collinear. (Why?) Therefore, by Theorem 4.5.1, there is a unique circle $\mathscr{C}'$ that contains A, A', and B. Corollary 5.5.11 guarantees that circle $\mathscr{C}'$ is orthogonal to $\mathscr{C}$, the circle of inversion. Consequently, the portion of $\mathscr{C}'$ that is in the interior of $\mathscr{C}$ is, according to the model, a line. It can be shown (see Exercise Set 6.6, Problem 28) that this line is unique. From this we see that Postulate 1 is satisfied for all choices of points A and B.

It is interesting and important to note that in verifying neutral Postulate 1 we used several results that are strictly Euclidean. This is legitimate, since our model for hyperbolic geometry is embedded within a Euclidean plane. Because of this, all Euclidean results apply to the model from this global perspective. The interior of circle $\mathscr{C}$ is the domain in which we hope to establish the validity of hyperbolic postulates using the interpretations from the model.

Neutral Postulates 2 through 4 are related to one another, since each involves some notion of distance. In order to verify that these postulates apply to our model, we will need a means of assigning distance to pairs of hyperbolic points. This is more easily said than done.

We cannot measure distance in the usual (Euclidean) manner. If we did, the distance between points A and B would be limited[20] to a finite

---

[20] If measured in a Euclidean fashion, $d(A,B)$ would always be less than 2π, the circumference of circle C.

value. This would contradict the ruler postulate (see Exercise Set 6.6, Problem 2). Therefore an alternate method of measurement is needed. It is a credit to the genius of Poincaré that he was able to construct a distance formula applicable within the model. Poincaré chose to measure the distance between hyperbolic points A and B using the following *hyperbolic distance formula* (Figure 6.6.4).

$$d(A,B) = \left| \ln\left(\frac{AQ}{BQ} \times \frac{BP}{AP}\right)\right| \qquad (11)$$

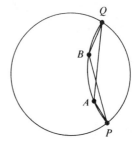

Figure 6.6.4

In this formula, P and Q represent the points of intersection of $\overrightarrow{AB}$ and circle $\mathscr{C}$, ln represents the natural logarithm function, and a notation such as AQ represents (as usual) the Euclidean length of $\overline{AQ}$. Note that this expression yields a unique value for any two hyperbolic points A and B, thereby satisfying neutral Postulate 2.

The choice of the product

$$\frac{AQ}{BQ} \times \frac{BP}{AP}$$

was not arbitrary. This product is known as the *cross ratio* of the range of points A, B, Q, and P (taken in that order) for any four concyclic or collinear points. The notation (AB,QP) is often used to represent this cross ratio. Cross ratios are often computed using directed segments (i.e., signed values), but since we will restrict ourselves to positive distances, this aspect of the ratio will not be of concern to us.

The cross ratio (AB,QP) appears to be a good candidate for use as part of a distance formula, since $(AB,QP) = (BA,PQ)$ (see Exercise Set 6.6, Problem 4), so that $d(A,B) = d(B,A)$, a property required by the ruler postulate. However, it is not obvious that distance as defined by Equation (11) will satisfy all requirements of neutral Postulates 3 and 4. Verifying that it does will require some effort.

To begin, we will consider the simplest type of hyperbolic line—a diameter. Our initial goal will be to show that the hyperbolic distance formula given earlier satisfies the requirements of the ruler postulate (neutral Postulate 3), which for convenience is restated here.

POSTULATE 3.   *The Ruler Postulate.* The points of a line can be placed in correspondence with the real numbers in such a way that

(i) To every point of the line there corresponds exactly one real number.
(ii) To every real number there corresponds exactly one point of the line.
(iii) The distance between two points is the absolute value of the difference of the corresponding numbers.

As a starting point, we will investigate distance along $\overline{PQ}$ shown in Figure 6.6.5. We will illustrate neutral Postulate 3 by showing how this hyperbolic line (open segment $\overline{PQ}$) can be converted into a hyperbolic number line consistent with the ruler postulate.

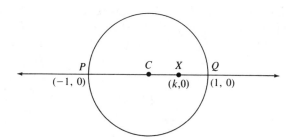

Figure 6.6.5

To accomplish this, we must establish a one-to-one correspondence between the points of $\overline{PQ}$ and the real numbers in such a way as to make $d(A,B) = |a - b|$ for any points A and B on $\overline{PQ}$ with corresponding hyperbolic coordinates a and b. A reasonable starting point in this process is to assign the coordinate 0 (zero) to the origin C and allow each point X with Euclidean coordinates $(k,0)$ on diameter $\overline{PQ}$ to correspond to its signed distance from 0.[21] Using this convention, we find that the hyperbolic coordinate x for X can be found using the formula

$$x = d(C,X) = \ln\left(\frac{CQ}{XQ} \times \frac{XP}{CP}\right)$$

But from Figure 6.6.5 we can see that $CQ = 1$, $XQ = 1 - k$, $XP = 1 + k$, and $CP = 1$, so that the hyperbolic coordinate x for point X is given by

$$x = d(C,X) = \ln\left(\frac{1}{1 - k} \times \frac{k + 1}{1}\right)$$

From this we see that

---

[21] Note that by using signed values here, we allow the correspondence to include negative values as well as positive ones for the coordinates of points.

$$x = \ln\left(\frac{1 + k}{1 - k}\right)$$

This means that for every point X on $\overline{PQ}$ (i.e., for every value of k between -1 and 1), there corresponds exactly one real number x. For example, if point X has Euclidean coordinates $(0.6,0)$, then its hyperbolic coordinate x on the $\overrightarrow{PQ}$ number line is given by

$$x = \ln\left(\frac{1 + 0.6}{1 - 0.6}\right)$$

so that

$$x = \ln 4 \approx 1.38629$$

Less obvious is the converse, that is, whether each real number corresponds to a unique point on hyperbolic number line $\overrightarrow{PQ}$. For example, is there exactly one point on $\overrightarrow{PQ}$ with coordinate 5? In other words, does the equation

$$5 = \ln\left(\frac{1 + k}{1 - k}\right)$$

have a unique solution k?

Written exponentially, the equation given above is

$$e^5 = \frac{1 + k}{1 - k}$$

Solving for k gives

$$k = \frac{e^5 - 1}{e^5 + 1}$$

or

$$k \approx 0.9866143$$

This means that the Euclidean point ($\approx 0.966143,0$) corresponds to 5 (or has coordinate 5) on the $\overrightarrow{PQ}$ number line.

Of course, there is nothing special about the choice of 5 as the coordinate in the preceding example. In fact, for every real number x there is a unique Euclidean point $(k,0)$ on $\overrightarrow{PQ}$ with hyperbolic coordinate x. The value of k, which clearly depends on x, can be found using the function

$$k(x) = \frac{e^x - 1}{e^x + 1}$$

Careful examination of function k will show that it is a one-to-one function that maps all real values of x to the open set $(-1,1)$. This shows that, at least for $\overleftrightarrow{PQ}$, there is a one-to-one correspondence between the points of $\overline{PQ}$ and the set of real numbers, as required by parts (i) and (ii) of the ruler postulate.

We now consider part (iii) of the ruler postulate, which dictates the manner in which distance is computed. Specifically, $d(A,B)$ is to be computed as $|a - b|$, where a and b are the coordinates assigned to A and B by the correspondence discussed in the preceding paragraphs. We will illustrate this idea using points A and B, with Euclidean coordinates $(k_1,0)$ and $(k_2,0)$, on the hyperbolic line $\overleftrightarrow{PQ}$ as shown in Figure 6.6.6. Using the hyperbolic distance formula, we represent the distance between points A and B in the following way:

$$d(A,B) = \left|\ln\left(\frac{AQ}{BQ} \times \frac{BP}{AP}\right)\right| = \left|\ln\left(\frac{1 - k_1}{1 - k_2} \times \frac{1 + k_2}{1 + k_1}\right)\right|$$

We need to show that $d(A,B)$ is equivalent to $|a - b|$, where a and b represent the hyperbolic coordinates of A and B, respectively. Since $|a - b| = |b - a|$ for all real values of a and b, it will suffice to show that $d(A,B) = |b - a|$.

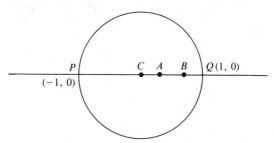

Figure 6.6.6

By our earlier discussion, a and b can be computed as follows:

$$a = d(C,A) = \ln\left(\frac{1 + k_1}{1 - k_1}\right)$$

$$b = d(C,B) = \ln\left(\frac{1 + k_2}{1 - k_2}\right)$$

From this we see that

$$|b - a| = \left|\ln\left(\frac{1 + k_2}{1 - k_2}\right) - \ln\left(\frac{1 + k_1}{1 - k_1}\right)\right|$$

but using a fundamental property of logarithms and some algebra, we see that

$$|b - a| = \left|\ln\left(\frac{1 + k_2}{1 - k_2} \div \frac{1 + k_1}{1 - k_1}\right)\right| = \left|\ln\left(\frac{1 + k_2}{1 - k_2} \times \frac{1 - k_1}{1 + k_1}\right)\right| = d(A,B)$$

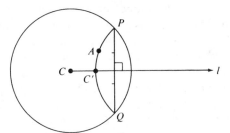

Figure 6.6.7

This shows that, at least for $\overleftrightarrow{PQ}$, part (iii) of the ruler postulate applies within this model of hyperbolic geometry.

In order to illustrate neutral Postulate 3 for lines that are not diameters, we can take a slightly different approach. For example, suppose open arc PAQ in Figure 6.6.7 represents a line in the model.

Our aim is to coordinatize $\overleftrightarrow{PQ}$ in much the same way as we coordinatized diameter $\overleftrightarrow{PQ}$ earlier. In order to locate a zero point on $\overleftrightarrow{PQ}$ we can construct line l, the perpendicular bisector of chord $\overline{PQ}$, and use C' to denote the intersection of l and PAQ. Then C' will correspond to zero as we coordinatize $\overleftrightarrow{PQ}$. Let A be any point on $\overleftrightarrow{PQ}$. The hyperbolic coordinate a for A will be determined in the following way:

$$a = d(C',A) = \ln\left(\frac{C'P}{AP} \times \frac{AQ}{C'Q}\right)$$

Because of the way C' was chosen, we know that $C'P = C'Q$ (see Exercise Set 6.6, Problem 7), and this expression reduces to

$$a = \ln\left(\frac{AQ}{AP}\right)$$

If A and C' happen to coincide, then $AQ = AP$, so that a is $\ln 1 = 0$; that is, the coordinate of C' is zero (this was, of course, by design). Suppose, however, that A is closer to P than to Q. In this case $AQ/AP > 1$, so that $a = \ln(AQ/AP) > 0$, which means that all points on the P side of C' have positive coordinates. Now if we think of point A as moving toward P, we will see that the length AP approaches zero, while the length AQ approaches the length of chord $\overline{PQ}$. This means that the ratio AQ/AP increases without bound and that $\ln(AQ/AP)$ tends toward infinity (see Exercise Set 6.6, Problem 8). Consequently, every positive real number r will serve as the coordinate of some point between C' and P on $\overleftrightarrow{PQ}$. In much the same way, we can argue that each negative real number serves as the coordinate for some point on $\overleftrightarrow{PQ}$ that lies between C' and Q (see Exercise Set 6.6, Problem 9). Taken together, this means that there is a one-to-one correspondence between the points on $\overleftrightarrow{PQ}$ and the set of real numbers, as is required by parts (i) and (ii) of the ruler postulate.[22]

To complete our verification of the ruler postulate, we now need to establish the relationship

$$|a - b| = d(A,B)$$

where a and b represent the coordinates for hyperbolic points A and B on $\overleftrightarrow{PQ}$. If C' has coordinate 0 (zero) in Figure 6.6.8, then the coordinates a and b for points A and B are positive and can be expressed as

$$a = d(C',A) = \ln\left(\frac{C'P}{AP} \times \frac{AQ}{C'Q}\right)$$

$$b = d(C',B) = \ln\left(\frac{C'P}{BP} \times \frac{BQ}{C'Q}\right)$$

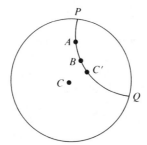

Figure 6.6.8

Then

$$|a - b| = |b - a| = \left|\left|\ln\left(\frac{C'P}{BP} \times \frac{BQ}{C'Q}\right)\right| - \left|\ln\left(\frac{C'P}{AP} \times \frac{AQ}{C'Q}\right)\right|\right|$$

For simplicity, we will assume that each logarithm has a positive value. Then, if we express the difference of the logarithms as the logarithm of a quotient, we see that

$$|a - b| = \left|\ln\left(\frac{\dfrac{C'P}{BP} \times \dfrac{BQ}{C'Q}}{\dfrac{C'P}{AP} \times \dfrac{AQ}{C'Q}}\right)\right|$$

Some algebra will show that this expression is equivalent to

---

[22] Neutral Postulate 4 forces us to generalize this discussion by allowing C' to be any point on PQ. This problem is addressed in Exercise Set 6.6, Problem 10.

$$|a - b| = \left| \ln\left(\frac{AP}{BP} \times \frac{BQ}{AQ}\right) \right| = d(A,B)$$

which verifies part (iii) of the ruler postulate. Demonstrations of this property for other cases proceed similarly.

We have now verified the validity of neutral postulates 1 through 3 in the hyperbolic model. Postulate 4 is closely related to the ruler postulate, and a discussion of it is included in Exercise Set 6.6, Problem 10. Postulates 5 through 8 apply to three-dimensional geometries and not to the two-dimensional model. Therefore we next need to discuss Postulate 11, the angular measurement postulate.

In order to discuss the angular measurement postulate, we must first define what is meant by the term "angle" in the model.

DEFINITION. *Angle.* If two lines intersect, then the angles between them are the Euclidean angles contained in the lines on which they lie.

If both lines are diameters, the angle is simply the angle of intersection of the two diameters, and the measure of the angle is the Euclidean measure of the central angle formed by the diameters (Figure 6.6.9a). If one (or both) of the lines is an arc of a Euclidean circle, the Euclidean angle between the lines will have as one (or both) of its sides a tangent to the Euclidean circle at the point of intersection of the lines (Figure 6.6.9b).

With this interpretation of the undefined term "angle," the measure of any hyperbolic angle has a Euclidean counterpart with the same measure. Because of this, Postulates 11 through 14 are all satisfied, since these properties are assumed to apply to angles in the Euclidean plane.

This brings us to the most difficult of the neutral postulates to verify: the SAS congruence condition for triangles. In order to illustrate this postulate and to demonstrate how one may work within this model of the hyperbolic plane, we will construct a triangle congruent to a given triangle using the SAS conditions.

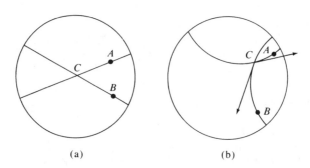

(a)   (b)

Figure 6.6.9

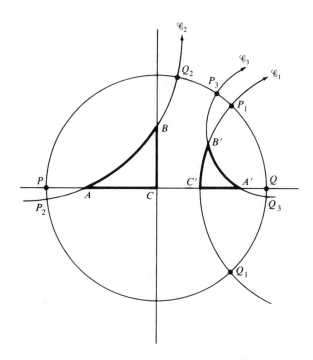

Figure 6.6.10 $\triangle ABC \cong A'B'C'$ (SAS).

To begin, we consider three noncollinear hyperbolic points, C, A, and B, with Euclidean coordinates $(0,0)$, $(-0.6,0)$, and $(0,0.5)$, respectively. These three points define the triangle $\triangle CAB$ shown in Figure 6.6.10. $\triangle CAB$ is a right triangle since $\overline{CA} \perp \overline{CB}$, and $d(A,C)$ and $d(C,B)$ are relatively easy to compute since P and Q (the points of intersection of circle C and the x axis) are $(\pm 1,0)$ and $(0,\pm 1)$, respectively. In particular,

$$d(A,C) = \left| \ln \left(\frac{AP}{CP} \times \frac{CQ}{AQ} \right) \right| = \left| \ln \left(\frac{0.4}{1} \times \frac{1}{1.6} \right) \right| = \left| \ln \left(\frac{1}{4} \right) \right|$$

Since $\ln(\frac{1}{4})$ is negative and since $\ln(4) = -\ln(\frac{1}{4})$, we may say that $d(A,C) = \ln(4)$. In a similar fashion, we can show that $d(C,B) = \ln(3)$ (see Exercise Set 6.6, Problem 13).

To illustrate the SAS postulate in the hyperbolic model, we will construct a second right triangle $\triangle C'A'B'$ having legs of length $\ln(4)$ and $\ln(3)$.[23] We will then show that the hypotenuse and remaining angles of $\triangle C'A'B'$ are congruent to the corresponding parts of $\triangle CAB$. Placement of

[23] Since each of the legs of a right triangle is adjacent to the right angle, the SAS condition is satisfied.

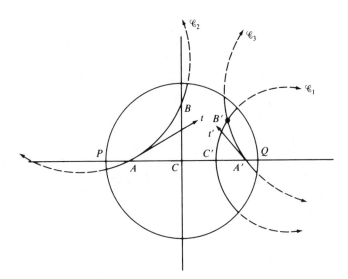

Figure 6.6.11

$\triangle C'A'B'$ is arbitrary, but for convenience we will place C' at the Euclidean point $(0.4,0)$ and locate A' on $\overrightarrow{C'Q}$. Then B' will be located on the Euclidean circle orthogonal to C and containing C' (Figure 6.6.11).

Our first task is to locate A' on $C'Q$ so that $d(C',A') = \ln(4)$. Since A' is located on the x axis, we may denote the coordinates of A' by $(a',0)$. From this, we have

$$d(C'A') = \left| \ln \left(\frac{C'Q}{A'Q} \times \frac{A'P}{C'P} \right) \right| = \left| \ln \left(\frac{0.6}{1 - a'} \times \frac{1 + a'}{1.4} \right) \right|$$

$$= \left| \ln \left(\frac{3}{7} \times \frac{1 + a'}{1 - a'} \right) \right|$$

We wish to find the values of a' for which

$$d(C'A') = \left| \ln \left(\frac{3}{7} \times \frac{1 + a'}{1 - a'} \right) \right| = \ln(4)$$

This equation has two solutions, but we will focus on the solution to

$$\ln \left(\frac{3}{7} \times \frac{1 + a'}{1 - a'} \right) = \ln(4)$$

Since the function $y = \ln(x)$ is one-to-one (from $\mathbb{R}^+ \rightarrow \mathbb{R}$), we may say that

$$\frac{3(1 + a')}{7(1 - a')} = 4$$

which when solved gives $a' = \frac{25}{31}$ (or approximately 0.80645). This means that point A' has Euclidean coordinates $(\frac{25}{31},0)$ and that $d(C',A') = d(C,A) = \ln (4)$, so that A' is appropriately placed.

To complete $\triangle C'A'B'$ we must locate point B' so that $d(C',B') = \ln (3)$ and $\overline{C'B'} \perp \overline{C'A'}$. We will consider the second of these conditions first.

Although there are many hyperbolic lines containing C', only one of them, $\overleftrightarrow{C'B'}$, is perpendicular to $\overleftrightarrow{C'A'}$ (see Exercise Set 3.2, Problem 9). The center C_1 of the Euclidean circle containing $\overleftrightarrow{C'B'}$ is contained in the x axis and has coordinates $(\frac{29}{20},0)$ since it is the midpoint of the Euclidean line segment that joins point C' and its image under inversion (see Exercise Set 6.6, Problem 16) (Note: we will use $\mathscr{C}_1$ to denote both the circle and its center. The meaning will be clear from the context.). Putting these pieces together, we find that the equation

$$(x - \tfrac{29}{20})^2 + y^2 = \tfrac{441}{400} \tag{12}$$

defines circle $\mathscr{C}_1$. [Verify that Equation (12) contains C' and is orthogonal to circle $\mathscr{C}$.]

We now need to locate point B' on the circle defined by Equation (12) so that $d(C',B') = \ln (3)$. Using the hyperbolic distance formula, we see that

$$d(C',B') = \left| \ln \left(\frac{B'P_1}{C'P_1} \times \frac{C'Q_1}{B'Q_1} \right) \right| = \ln(3)$$

But since $C'Q_1 = C'P_1$, we may write

$$d(C',B') = \left| \ln \left(\frac{B'P_1}{B'Q_1} \right) \right| = \ln(3)$$

Since B' is in quadrant I, we know $B'Q_1 > B'P_1$. This means $B'P_1/B'Q_1 < 1$, so that

$$d(C',B') = \left| \ln \left(\frac{B'P_1}{B'Q_1} \right) \right| = \ln \left(\frac{B'Q_1}{B'P_1} \right) = \ln(3)$$

Again, since $y = \ln(x)$ is one-to-one, we may say that

$$\frac{B'Q_1}{B'P_1} = 3$$

so that

$$B'Q_1 = 3(B'P_1) \tag{13}$$

In order to proceed from here we will need to determine the coordinates of P_1 and Q_1, the points where circles $\mathscr{C}_1$ and $\mathscr{C}$ intersect. This means we need to solve the system consisting of Equation (12) and $x^2 + y^2 = 1$. Some algebra (Exercise Set 6.6, Problem 17) can be used to show that $P_1 = (\frac{20}{29}, \frac{21}{29})$ and $Q_1 = (\frac{20}{29}, -\frac{21}{29})$.

Using these coordinates and the Euclidean distance formula, Equation (13) becomes

$$\sqrt{(x - \tfrac{20}{29})^2 + (y - \tfrac{21}{29})^2} = 3\sqrt{(x - \tfrac{20}{29})^2 + (y - \tfrac{21}{29})^2} \qquad (14)$$

where (x, y) represents the Euclidean coordinates of point B'. Squaring both sides of Equation (14) and collecting terms gives the equation

$$(y + \tfrac{21}{29})^2 - 9(y - \tfrac{21}{29})^2 = 8(x - \tfrac{20}{29})^2 \qquad (15)$$

Together Equations (15) and (12) constitute a system that defines the two points on $\overrightarrow{C'B'}$ that are ln (3) hyperbolic units from C'. The algebraic solution to the system is tedious[24] (Exercise Set 6.6, Problem 18) but has approximate solutions $(0.48077, \pm0.40385)$, so that the first-quadrant solution for point B' is approximately $(0.48077, 0.40385)$.

Taking inventory, we see that we have now located points C', A', and B' so that $m\angle A'C'B' = m\angle ACB = 90°$, $C'A' = CA = \ln(4)$, and $C'B' = CB = \ln(3)$. This means that these triangles satisfy the SAS congruence condition. If the hyperbolic model is a valid one, $\triangle ACB$ and $\triangle A'C'B'$ will be congruent, meaning that $\overline{A'B'} \cong \overline{AB}$, $m\angle C'A'B' = m\angle CAB$, and $m\angle C'B'A' = m\angle CBA$.

We will first show that $\overline{A'B'} \cong \overline{AB}$. In order to determine the length of $\overline{AB}$ we must construct the equation for the Euclidean circle that is orthogonal to circle $\mathscr{C}$ and contains points A and B, use this equation to determine the coordinates of P_2 and Q_2 (see Figure 6.6.10), and then apply the hyperbolic distance formula to determine $d(A,B)$.

The center of $\mathscr{C}_2$ (the orthogonal circle containing A and B) will lie on the perpendicular bisector of the Euclidean segment joining $A(-0.6,0)$, and its inversion image $A^*(-\tfrac{5}{3},0)$ (see Exercise Set 6.6, Problem 19). This implies that the first component of point C_2 is $-\tfrac{17}{15}$. In a similar fashion, using point B, we can show that the second component of C_2 is $\tfrac{5}{4}$. Application of the Euclidean distance formula shows that r_2 (the radius of $\mathscr{C}_2$) ≈ 1.35902. From this, we see that the equation that defines the circle C_2 [or more formally $C(C_2, r_2)$] is

$$(x + \tfrac{17}{15})^2 + (y - \tfrac{5}{4})^2 \approx (1.35902)^2 \approx 1.8469$$

P_2 and Q_2 represent the points of intersection of this circle and the circle $x^2 + y^2 = 1$. An algebraic solution of this system gives the points $P_2(-0.994790, -0.101943)$ and $Q_2(0.198615, 0.980078)$. We may then compute $d(A,B)$ as follows:

$$d(A,B) = \left| \ln \left(\frac{AP_2}{BP_2} \times \frac{BQ_2}{AQ_2} \right) \right|$$

The lengths AP_2, AQ_2, BP_2, and BQ_2 can all be computed using the Euclidean distance formula (see Exercise Set 6.6, Problem 20), so that

[24] The approximate solutions given here were determined using the SOLVR function on a HP-28c scientific calculator.

$$d(A,B) \approx \left| \ln \left(\frac{0.407739}{1.162730} \times \frac{0.519540}{1.26453} \right) \right| \approx |\ln 0.144108| \approx 1.93719$$

Next we need to show that $d(A',B')$ is also approximately 1.93719. Once again, a great deal of computational work is involved in this process. First, we must determine an equation for the Euclidean circle containing the line $\overline{A'B'}$ [i.e., the circle orthogonal to $x^2 + y^2 = 1$ that contains $A'(\frac{25}{31}, 0)$ and $B'(0.480769, 0.403846)$]. Some algebra (Exercise Set 6.6, Problem 21) can be used to show that such a circle is centered at (1.023225, 0.50864) with radius of length 0.552377. Some additional computation (Exercise Set 6.6, Problem 22) will show that the circle containing $\overline{A'B'}$ has the equation

$$(x - 1.023226)^2 + (y - 0.508064)^2 = 0.552377^2 \tag{16}$$

We can also show (Exercise Set 6.6, Problem 23) that Equation (16) and the circle $x^2 + y^2 = 1$ intersect at points P_3 (0.568976, 0.822354) and Q_3 (0.999041, -0.043783).

After doing this, $d(A',B')$ can be computed in the usual way:

$$d(A',B') = \left| \ln \left(\frac{A'P_3}{B'P_3} \times \frac{B'Q_3}{A'Q_3} \right) \right|$$

Once again, each of the required lengths can be computed using the Euclidean distance formula (Exercise Set 6.6, Problem 23), so that

$$d(A',B') \approx |\ln(6.939225)| \approx 1.937190$$

from which we see that $d(A,B) = d(A',B')$.

The implication here is that by copying from a triangle two sides and the included angle, we have constructed a second triangle whose third side is congruent to the third side of the triangle with which we began. This result is quite obvious in the Euclidean plane but is not nearly so obvious in the hyperbolic plane. In fact, in Figure 6.6.10 $\triangle ABC$ and $\triangle A'B'C'$ do not appear to be congruent in the usual sense, since $\overline{AB}$ and $\overline{A'B'}$ are not congruent in the Euclidean plane. However, using the interpretations from the model, we see that $\overline{A'B'}$ is congruent to segment $\overline{AB}$, a result that is consistent with the SAS congruence postulate.

The other implication of the SAS congruence postulate is the congruence of corresponding angles. $\angle B'C'A'$ is congruent to $\angle BCA$ by design. We now need to show that the other corresponding angle pairs are also congruent. We will first consider m$\angle BAC$.

Since $\overline{AB}$ is an arc of a Euclidean circle, m$\angle BAC$ is computed as the angle between the x axis and line t, the tangent to circle C_2 at point A (Figure 6.6.11). In view of this we may compute m$\angle BAC$ using the formula m$\angle BAC = \tan^{-1} m$, where m is the slope of line t. Similarly, m$\angle B'A'C' = \tan^{-1} m'$, where m' is the slope of line t', the tangent to C_3 at A'.

Computing m and m' requires that we evaluate the derivative of the equations that define $\mathcal{C}_2$ and $\mathcal{C}_3$. In general, circles are defined by a relation of the form

$$(x - h)^2 + (y - k)^2 = r^2$$

Differentiating implicitly, we find that

$$2(x - h) + 2(y - k) \left(\frac{dy}{dx}\right) = 0$$

Solving for dy/dx, we see that

$$\frac{dy}{dx} = \frac{h - x}{y - k}$$

This means that in order to determine the slope of a tangent to a circle, we need only identify the center (h, k) of the circle and the point on the circle at which the tangent is to be drawn. In particular, slope m is

$$m = \frac{dy}{dx} = \frac{-\frac{17}{15} - 0.6}{0 - \frac{5}{4}} \approx 0.42667$$

so that

$$m\angle BAC \approx \tan^{-1} 0.42667 \approx 23.106°$$

Likewise,

$$m' = \frac{dy}{dx} \approx \frac{1.0232 - \frac{25}{31}}{0 - 0.508064} \approx -0.42667$$

so that

$$m\angle B'A'C' \approx \tan^{-1} -0.42667 \approx -23.106$$

From these calculations we may conclude that $\angle BAC \approx \angle B'A'C'$. In a similar fashion, we can show that $\angle ABC \approx \angle A'B'C'$ (Exercise Set 6.6, Problem 25), so that $\triangle ABC$ and $\triangle A'B'C'$ are, as expected, congruent.

Although we have looked only at a special case, the SAS congruence condition is generally applicable,[25] so that the last of the neutral postulates, Postulate 15, is satisfied by the model. Because of this, all theorems from Chapter 3 are valid in this model of the hyperbolic plane. For example, in Chapter 3 we proved, using only the neutral postulates and their consequences, that ASA, SSS, and AAS (Theorems 3.3.1, 3.3.3, and 3.3.4) are all valid congruence conditions. Since all the neutral postulates apply within the hyperbolic model, these congruence conditions must be valid in this model of the hyperbolic plane.

As a second example, recall the Saccheri-Legendre theorem (Theorem

[25] For a proof of the SAS congruence condition in the model of the hyperbolic plane, see Howard Eves, *A Survey of Geometry*, rev. ed. (Boston: Allyn and Bacon, 1972).

3.5.1) which states that the angle sum for any triangle is always less than or equal to 180°. This theorem sounded strange when stated in Chapter 3 and perhaps still does. But consider as an example △*ABC* in Figure 6.6.10. The angle sum for this triangle is about 146° 36′, considerably less than 180° but consistent with Theorem 3.5.1. In fact, all triangles in the model will have angle sums less than 180°, as was suggested in Chapter 3.

Other results from Chapter 3 can also be clearly illustrated using the model. For example, Theorem 3.6.3 states that the summit angles of a Saccheri quadrilateral are either both acute or both right. To illustrate the hyperbolic interpretation of this theorem, we will consider the Saccheri quadrilateral in Figure 6.6.12. □*RSTU* was constructed by drawing perpendiculars at points *R* and *S* [with Euclidean coordinates (−0.6,0) and (0.6,0), respectively] and measuring two hyperbolic units along each to locate points *T* and *U*. Line $\overline{TU}$ joins these points to complete the Saccheri quadrilateral. Visually it is clear, and formally it can be shown (see Exercise Set 6.6, Problem 26), that the angles between the tangents at points *T* and *U* are acute. (We know they are congruent because of neutral Theorem 3.6.2.) In a similar fashion we can demonstrate, within our model of hyperbolic geometry, that the fourth angle of a Lambert quadrilateral is acute, that the summit of a Saccheri quadrilateral is longer than the base, and various other results from Chapters 3 and 6 that until now may have been hard to visualize.

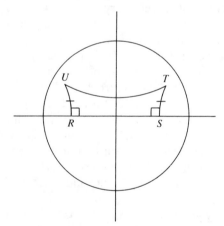

Figure 6.6.12

Since all the neutral postulates are valid within the model of the hyperbolic plane, the results described above that are clearly non-Euclidean must be consequences of the one difference between the SMSG postulate set and the postulate set for hyperbolic geometry—the hyperbolic parallel postulate—an equivalent form of which is as follows:

THEOREM 6.3.1.   Given a line *l* and a point *P* not on *l*, there are at least two lines through *P* that are parallel to *l*.

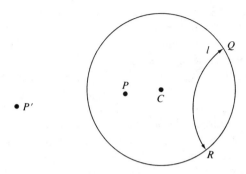

Figure 6.6.13

Before continuing further, we must establish that this property is valid within the model. Consider point P and line l as shown in Figure 6.6.13.

Q and R represent the points of intersection of line l and circle C, and P' represents the inversion image of P using $\mathscr{C}$ as the circle of inversion. Let $\mathscr{C}_1$ represent the unique Euclidean circle containing P, P', and Q, and let $\mathscr{C}_2$ represent the Euclidean circle containing P, P', and R. The portions of $\mathscr{C}_1$ and $\mathscr{C}_2$ that lie in the interior of circle $\mathscr{C}$ constitute hyperbolic lines (why ?), both of which are parallel to l (Exercise Set 6.6, Problem 27).

From the preceding discussion we may conclude that our model satisfies all the axioms of neutral geometry as well as the hyperbolic parallel postulate, so that it is a valid model of hyperbolic geometry. Beyond that, since we have embedded this model in the Euclidean plane, if contradictory statements had been derived in hyperbolic geometry (i.e., if hyperbolic geometry had been found to be inconsistent), a corresponding inconsistency would be found in the interpretations from the Euclidean plane.

Consequently, we may conclude that Euclidean geometry is consistent only if hyperbolic geometry is. Conversely, it can be shown that hyperbolic geometry is consistent only if Euclidean Geometry is. Together this means that Euclidean geometry is consistent if and only if hyperbolic geometry is consistent. This conclusion had enormous historical significance, since it put to rest the idea, championed by Kant, that there could be no consistent geometry except Euclid's.

The model for hyperbolic geometry has a second important consequence: establishment of the independence of the Euclidean parallel postulate. To see this, suppose for a moment that the Euclidean parallel postulate is a consequence of the other postulates of Euclidean geometry (i.e., is not independent of the others). Under this hypothesis, every geometric model that satisfies all the Euclidean assumptions but the Euclidean parallel postulate would have to satisfy the parallel postulate as well, since it could be proved from the others as a theorem (see the definition of independence in Chapter 1).

However, in this section we have constructed a geometry that satisfies all the neutral postulates but fails to satisfy the Euclidean parallel postulate. This forces us to conclude that the parallel postulate is not a consequence of the others, so that, by definition, it is independent of the others. This of course means that Saccheri, Lambert, Legendre, and the others were doomed to failure in their attempts to prove the fifth postulate using the other postulates, since the others are not sufficient to do so.

An important consequence of the acceptance of non-Euclidean geometries was that mathematicians and physicists were at last free to pursue investigations in a variety of non-Euclidean geometries. In fact, it was less than 50 years after Poincaré's work that Albert Einstein used non-Euclidean geometries as the basis for his theories of special and general relativity, the ramifications of which are still being explored.

EXERCISE SET 6.6

1. Suppose A and B are hyperbolic points and that A' is the image of A under an inversion in circle $\mathscr{C}$. Under what circumstances can we be sure that A, B, and A' are *not* collinear? Explain your response.

2. Explain why the ruler postulate disallows the use of Euclidean distance in the Poincaré model of hyperbolic geometry. (*Hint*: The length of an arc of any Euclidean circle is bounded above by the circumference of the circle.)

3. Could distance in the Poincaré model have been defined using base 10 logarithms? Base 2 logarithms? Explain why or why not.

4. Show that, in Figure 6.6.4, the cross ratios (AB,QP) and (BA,PQ) are equal.

5. Determine the hyperbolic coordinate on the $\overrightarrow{PQ}$ number line (see Figure 6.6.5) for each of the following Euclidean points:
   (a) $(0,0)$ **(b)** $(0.3,0)$ **(c)** $(-0.8,0)$ **(d)** $(-0.1,0)$ **(e)** $(0.99,0)$

6. Determine the Euclidean points having the following hyperbolic coordinates on the $\overrightarrow{PQ}$ number line (see Figure 6.6.5).
   (a) 1   **(b)** -8 **(c)** $\frac{1}{2}$ **(d)** $-\frac{2}{3}$ **(e)** 99

7. Refer to Figure 6.6.7. Prove that if l, the perpendicular bisector of chord $\overline{PQ}$, intersects PAQ at C', then chord $\overline{C'P}$ is congruent to chord $\overline{C'Q}$. (*Note*: These are Euclidean lines, segments, and arcs, so the proof should be Euclidean.)

8. Evaluate the following limit: $\lim_{a\to\infty} \ln a$. Explain how this limit relates to the idea that there is a one-to-one correspondence between points on a hyperbolic line and the real numbers.

9. Evaluate the following limit: $\lim_{a\to 0^+} \ln a$. Explain how this limit relates to the idea that there is a one-to-one correspondence between the points on a hyperbolic line and the real numbers. (*Hint*: See Figure 6.6.7 and the discussion that follows it.)

10. In the verification of the ruler postulate, we said that the coordinate a for any point A on $\overrightarrow{PQ}$ was given by

$$a = d(C',A) = \left| \ln\left(\frac{C'P}{AP} \times \frac{AQ}{C'Q} \right) \right|$$

which could be simplified to

$$a = \left| \ln \left(\frac{AQ}{AP} \right) \right|$$

because $C'P = C'Q$ (see Figure 6.6.7). Of course, the reason that $C'P$ and $C'Q$ have the same length is that C' is the midpoint of PQ. Neutral Postulate 4 forces us to generalize this idea and to allow any point of $\overrightarrow{PQ}$ to serve as the origin. With this in mind, the first equation above would not simplify to the second, but could be written as

$$a = \left| \ln \left(\frac{AQ}{AP} \times \frac{C'P}{C'Q} \right) \right| = \left| \ln \left(\frac{AQ}{AP} \right) + \ln \left(\frac{C'P}{C'Q} \right) \right|$$

But for any $\overrightarrow{PQ}$ and for any choice of C', all three of the Euclidean points P, Q, and C' are fixed, making $\ln(C'P/C'Q)$ a constant. Show how this implies a one-to-one correspondence between $\overrightarrow{PQ}$ and the real numbers regardless of the choice of C'

11. Suppose that points A, B and C' all lie on $\overrightarrow{PQ}$ and that C' has coordinate zero (see Figure 6.6.8). Suppose also that the coordinates of A and B (a and b, respectively) are

$$a = -\left| \ln \left(\frac{C'P}{AP} \times \frac{AQ}{C'Q} \right) \right| \, b = -\left| \ln \left(\frac{C'P}{BP} \times \frac{BQ}{C'Q} \right) \right|$$

Show that for these values of a and b

$$|a - b| = d(A,B)$$

12. Suppose $R = (0.8, 0)$ and $S = (0. - 0.7)$. Use the hyperbolic distance formula to determine $d(R, S)$.

13. Show that, in Figure 6.6.10, $d(C,B) = \ln 3$.

14. Use the hyperbolic distance formula to verify that, in Figure 6.6.10, $d(C', A') = \ln 4$.

15. In locating point A' in Figure 6.6.10, we solved the equation

$$\ln \left(\frac{3}{7} \times \frac{1 + a'}{1 - a'} \right) = \ln 4$$

A second point A' can be found by solving the related equation

$$\ln \left(\frac{3}{7} \times \frac{1 + a'}{1 - a'} \right) = \ln \left(\frac{1}{4} \right)$$

Show that the solution to the second equation is $-\frac{5}{19}$ and provide a geometric interpretation for this solution.

16. Show that, in Figure 6.6.10, the Euclidean circle containing $(0.4,0)$ that is perpendicular to $\overrightarrow{PQ}$ has center $(\frac{29}{20}, 0)$.

17. Show that the solutions to the system

$$x^2 + y^2 = 1$$

$$(x - \tfrac{29}{20})^2 + y^2 = (\tfrac{21}{20})^2$$

are $(\frac{20}{29}, \frac{21}{29})$ and $(\frac{20}{29}, -\frac{21}{29})$.

18. Show that the ordered pairs $(0.48077, 0.40385)$ and $(0.48077, -0.40385)$ are (approximate) solutions to the system

$$(y + \tfrac{21}{29})^2 - 9(y - \tfrac{21}{29})^2 = 8(x - \tfrac{20}{29})^2$$

$$(x - \tfrac{29}{20})^2 + y^2 = (\tfrac{21}{20})^2$$

19. (a) In Figure 6.6.10 the Euclidean coordinates of point A are $(-0.6, 0)$. Since A is on the x axis, its inversion image A^* will have coordinates $(k, 0)$, where $-0.6 \times k = 1$ (see the definition of inversion in Section 5.5), so that $k = -\tfrac{5}{3}$. Explain how this implies that the center of circle C_2 lies on the Euclidean line $x = -\tfrac{17}{15}$.

   (b) Show that, in Figure 6.6.10, $\mathscr{C}_2$ must also lie on the line $y = \tfrac{5}{4}$.

   (c) Use the Euclidean distance formula to show that, in Figure 6.6.10, $d(C_2, A) = d(C_2, B) \approx 1.35902$.

   (d) Write the equation for the Euclidean circle that contains the hyperbolic line $\overline{AB}$ in Figure 6.6.10.

20. In Figure 6.6.10 the coordinates of A, B, P_2, and Q_2 are $A(-0.6, 0)$, $B(0, 0.5)$, $P_2(-0.994790, -0.101943)$, $Q_2(0.198615, 0.980078)$. Use the Euclidean distance formula to determine the lengths AP_2, AQ_2, BP_2, and BQ_2.

21. In order to determine an equation for the circle that is orthogonal to $x^2 + y^2 = 1$ and contains A' and B' (see Figure 6.6.10), we first locate the inversion image A'^* for A'. Since A' is on the x axis, A'^* will have coordinates $(k, 0)$ such that $\tfrac{25}{31} \times k = 1$.

   (a) Show that this implies that C_3 (the center of $\mathscr{C}_3$) lies on the line $x = \tfrac{793}{775} \approx 1.023226$.

   (b) C_3 will also lie on the perpendicular bisector of Euclidean segment $\overline{A'B'}$. Show that an (approximate) equation for the perpendicular bisector of $A'B'$ is $y = 0.80645x - 0.317118$.

   (c) Show that the intersection of $x = \tfrac{793}{775}$ and $y = 0.80645x - 0.317118$ is the point $C_3(1.023226, 0.508065)$.

   (d) Show that the radius of the Euclidean circle containing A' and B' is approximately 0.552377.

   (e) Write an equation for the Euclidean circle that is orthogonal to $x^2 + y^2 = 1$ and contains $\overline{A'B'}$.

22. Show that the system defined by the circle discussed in Problem 21 and the circle $x^2 + y^2 = 1$ has as (approximate) solutions the points $P_3(0.568976, 0.822354)$ and $Q_3(.999041, -0.043783)$.

23. Use the Euclidean distance formula to verify the following lengths in Figure 6.6.10. $A'P_3 \approx 0.855956$, $A'Q_3 \approx 0.197503$, $B'P_3 \approx 0.427702$, $B'Q_3 \approx 0.684819$.

24. Suppose point R in the hyperbolic plane has Euclidean coordinates $(0.2, 0)$.

   (a) Locate point S on the positive x axis so that the hyperbolic distance $d(R, S) = 2$.

   (b) Determine an equation for the Euclidean circle $\mathscr{C}^*$ that is orthogonal to $x^2 + y^2 = 1$ and contains point R.

   (c) Determine the first quadrant point T, on $\mathscr{C}^*$ such that $d(R, T) = 3$.

   (d) Determine an equation for the Euclidean circle $\mathscr{C}$ such that $\mathscr{C}$ is orthogonal to $x^2 + y^2 = 1$ and $\mathscr{C}$ contains the hyperbolic line $\overline{ST}$.

   (e) Determine the hyperbolic distance $d(S, T)$.

25. In Figure 6.6.11 we showed that $\angle BAC \approx \angle B'A'C'$. In order to complete the discussion of the SAS congruence condition, we also need to show that $\angle ABC \approx \angle A'B'C'$. In doing this, we will refer to Figure 6.6.14.

   (a) $m\angle ABC = 90° - \tan^{-1} m_2$, where m_2 is the slope of line T_2, the tangent to circle $\mathscr{C}_2$ at point B. Compute $m\angle ABC$ and explain why this formula works.

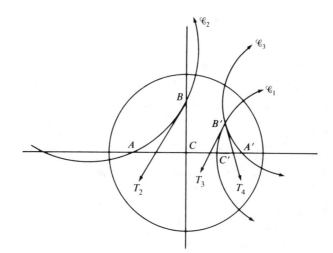

Figure 6.6.14

(b) $m\angle A'B'C' = \tan^{-1}[(m_3 - m_4)/(1 + m_3m_4)]$, where m_3 and m_4 represent the slopes of T_3 and T_4 in Figure 6.6.12. Compute $m\angle A'B'C'$ and explain why this formula works.

(c) Are $\angle ABC$ and $\angle A'B'C'$ congruent? Are the angle sums of $\triangle ABC$ and $\triangle A'B'C'$ equal? Do these triangles have the same area? Explain how you know.

26. Determine the measures of the summit angles of the Saccheri quadrilateral in Figure 6.6.12. (*Note*: To do this you must determine the equations for Euclidean circles containing lines RU, ST, and UT and then determine the measures of the summit angles.)

27. In Figure 6.6.13 explain why circle $\mathscr{C}_2$ (which contains P, P' and Q) can have no hyperbolic points in common with circle $\mathscr{C}^*$ (the circle that contains line l). (*Hint*: Both $\mathscr{C}_2$ and $\mathscr{C}^*$ are orthogonal to $\mathscr{C}$, so that each is invariant under an inversion in circle $\mathscr{C}$.) Repeat for $\mathscr{C}_3$ and $\mathscr{C}^*$.

28. Suppose that A and B are hyperbolic points that do not lie on the same diameter of circle $\mathscr{C}$. Suppose also that A' and B' are the inversion images of A and B, respectively. Show that the Euclidean circle that contains A, B, and A' is the same circle that contains A, B, and B'.

6.7 CLASSIFYING THEOREMS

At this point in our study of geometries we have proved a large number of theorems, and we have grouped them, in a rather loose way, into categories according to the chapters in which they were introduced. In particular, Chapter 3 included theorems that were proved without using either the Euclidean or the hyperbolic parallel postulate. These theorems are known

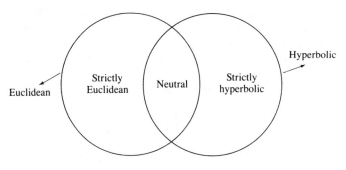

Figure 6.7.1

as neutral theorems and are valid in both Euclidean *and* hyperbolic geometry.

In Chapter 4 we introduced the Euclidean parallel postulate and discussed a number of theorems whose proofs made use of the parallel postulate or results derived from it. Theorems that *require* the use of a form of the Euclidean parallel postulate in their proof are called *strictly Euclidean*.

Thus far in this chapter we have encountered a number of theorems that have made use of the hyperbolic parallel postulate in their proof. Theorems that require use of the hyperbolic parallel postulate (or a consequence of it) in their proof are called *strictly hyperbolic*. Figure 6.7.1 may help clarify the relationships among these three classifications of theorems.

With this distinction in mind, we now pose the following question: Does the use of the Euclidean (or hyperbolic) parallel postulate in the proof of a theorem make the theorem strictly Euclidean (or hyperbolic)? The answer, as we will see, is no, and the purpose of this section is to discuss a technique that can help to classify theorems as strictly Euclidean, strictly hyperbolic, or neutral.

To begin, recall the following theorem, a proof for which was given in Chapter 3.

THEOREM 3.3.1.   *AAS Congruence Theorem.* If the vertices of two triangles are in one-to-one correspondence such that two angles and the side opposite one of them in one triangle are congruent, respectively, to the corresponding parts of the second triangle, then the triangles are congruent.

The standard proof for this theorem found in most elementary Euclidean geometry textbooks proceeds by using the Euclidean angle sum theorem for triangles (Theorem 4.2.1) to show that if two pairs of corresponding angles are congruent, then the third pair must also be congruent. (why?) We may then apply the ASA congruence condition (Theorem 3.3.3) to show that the triangles are congruent (the details of the proof are elementary and have been left as an exercise).

But Theorem 4.2.1, on which the preceding proof was based (and which states that the angle sum for any triangle is 180°), is equivalent to the Euclidean parallel postulate (Theorem 3.6.18) and is therefore strictly Euclidean. Because of this, we might be tempted to believe that the AAS congruence condition is strictly Euclidean, since its proof makes use of a strictly Euclidean result. This conclusion, of course, would be erroneous, since the AAS congruence theorem was proved in Chapter 3 without the use of any form of a parallel postulate. This means that, in general, we have no assurance that theorems proved using the parallel postulate (or its consequences) necessarily depend on it for their proof. Is it not possible that theorems we have proved using the Euclidean parallel postulate (or an equivalent form) can be proved in an alternative manner that makes no use of the parallel postulate? How can we be sure that the proof of a theorem actually *requires* the use of the parallel postulate? This is the issue we pursue in this section.

To illustrate this dilemma, recall Theorem 4.4.8—the well-known Pythagorean theorem. The proof given in Chapter 4 for this theorem relies heavily on the use of similar (but noncongruent) triangles, the existence of which implies Euclidean geometry (see Theorem 3.4.8). This alone, however, doesn't guarantee that an alternate proof isn't possible that makes no use of a statement equivalent to the Euclidean parallel postulate. The Pythagorean theorem is remarkable because of (among other reasons) the large number of proofs that have been given for it. One book alone, *The Pythagorean Proposition,* by Elisha Scott Loomis,[26] presents hundreds of different proofs for the theorem. Could not one of these proofs be neutral? And if not one of these, is it not possible that some other proof exists for the Pythagorean theorem that is neutral? And if such a proof has not yet been constructed, is it not possible that it will be discovered tomorrow? Or the day after? Or . . . ?

One way to address this question is to determine whether or not the Pythagorean theorem is independent of the neutral axioms. If we succeed in showing this independence, we may logically conclude that there is no chance of deducing the Pythagorean theorem from the neutral axioms. From this it follows that any proof we can give for the theorem must depend, at least in part, on some form of a parallel postulate (remember that the only difference between the neutral postulates of Chapter 3 and the Euclidean postulates of Chapter 4 is the Euclidean parallel postulate). From Chapter 1, we know that one way to show the independence of a statement (relative to the neutral axioms) is to construct a model in which all the neutral axioms

---

[26] Elisha Scott Loomis, *The Pythagorean Proposition* (Reston, Va.: National Council of Teachers of Mathematics, 1968).

are valid but in which the statement itself is not. There is, of course, no guarantee that we can find such a model. However, if we can find one, we will have the evidence needed to conclude that the Pythagorean Theorem is independent of the neutral axioms, so that its proof must rely in some way on the only Euclidean postulate not available to us in neutral geometry—the Euclidean parallel postulate. This then would mean that the Pythagorean theorem is strictly Euclidean.

The task at hand, then, is to find a model in which all of the neutral axioms are valid but in which, we hope,[27] the Pythagorean theorem is not. Of course, our model must not satisfy the Euclidean parallel postulate, since then the Pythagorean theorem would necessarily be valid (after all it exists as a theorem in Euclidean geometry). Very clearly, then, we need a geometric model that satisfies all the neutral postulates but is not Euclidean. Fortunately (but not surprisingly), we have one available—the Poincaré model of the hyperbolic plane that we discussed in detail in Section 6.6.

In Figure 6.7.2 we see $\triangle ABC$ with vertices $A(-0.6,0)$, $B(0,0.5)$, and $C(0,0)$. This is, as you may notice, the same triangle used during the discussion of the SAS congruence condition in Section 6.6 (see Figure 6.6.10). From the work in that section, we know the following lengths: $a = d(B,C) = \ln 3$; $b = d(A,C) = \ln 4$; and $c = d(A,B) \approx 1.937190$.

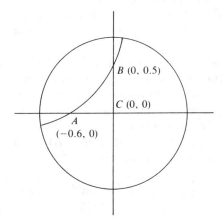

Figure 6.7.2

We now ask the question, Is it true that $a^2 + b^2 = c^2$? Some arithmetic is needed.

$$\ln (4)^2 + \ln (3)^2 \overset{?}{=} (1.937190)^2$$

---

[27] There *is* reason for us to hope that the Pythagorean theorem is not valid within the model we construct, since then we can conclude the theorem is strictly Euclidean. Otherwise no conclusion can be drawn, and we must proceed in some alternative manner. (See Exercise Set 6.7, Problem 2.)

or

$$1.921812 + 1.206949 \overset{?}{=} 3.752705$$

But

$$3.128761 \neq 3.752705$$

even if we allow for some rounding error, so that $\triangle ABC$ provides a counterexample of the Pythagorean Theorem.

This means that in the hyperbolic model the Pythagorean theorem *is not* valid. From this, we may conclude that the validity of the Pythagorean theorem in Euclidean geometry must be due, at least in part, to the parallel postulate, so that every proof of it must make use of it in some form or other. In other words, the Pythagorean theorem is strictly Euclidean.

As a second example, consider the midpoint connection theorem.

THEOREM 4.2.12.   If a line segment has as endpoints the midpoints of two sides of a triangle, then the segment is contained in a line that is parallel to the third side and is one-half the length of the third side.

This theorem is actually comprised of two separate statements. One states that the line segment joining the midpoints of the sides is contained in a line parallel to the third side. The other states that the segment is one-half the length of the third side. We will consider the statement concerning length first.

Once again we will use $\triangle ABC$ with coordinates $A(-0.6,0)$, $B(0,0.5)$, and $C(0,0)$ (see Figure 6.7.2 or 6.6.10). This is a convenient choice, since we have already computed the lengths of the sides of this triangle and can make use of those values as we proceed. In particular, we know that $d(A,B) \approx 1.937190$. We will compare this length to $d(M_1,M_2)$ (Figure 6.7.3).

M_1 represents the midpoint of $\overline{AC}$ so that $d(C,M_1) = \frac{1}{2} \ln 4 = \ln 2$. Using $(x,0)$ as the coordinates for M_1 we see that

$$d(C,M_1) = \left| \ln\left(\frac{1+x}{1} \times \frac{1}{1-x} \right) \right| = \left| \ln\left(\frac{1+x}{1-x} \right) \right|$$

But, since $x < 0$, we may write

$$\left| \ln\left(\frac{1+x}{1-x} \right) \right| = \ln\left(\frac{1-x}{1+x} \right) = \ln 2$$

Solving this equation we find that $x = -\frac{1}{3}$, so that the coordinates of M_1 are $(-\frac{1}{3},0)$. Similarly, we can show that the coordinates of M_2 are $(0, 2 - \sqrt{3})$ (see Exercise Set 6.7, Problem 4).

To compute $d(M_1,M_2)$ we will need to determine an equation for the circle, C', that contains M_1 and M_2 and is orthogonal to circle C. To do this we recall that M_1' and M_2' (the inversion images of M_1 and M_2 using circle C)

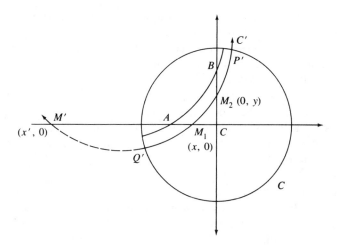

Figure 6.7.3

both lie on C'. If we use $(x',0)$ to represent the coordinates of M_1', we may write $x' \times (-\tfrac{1}{3}) = 1$, so that $x' = -3$. Since M_1 and M_1' define a chord for C', the center of C' must lie on the perpendicular bisector of $\overline{M_1M_1'}$, so that the first coordinate for the center (h,k) of C', is $-\tfrac{5}{3}$. Similarly, we can show that $k = 2$ (see Exercise Set 6.7, Problem 5). Using the distance formula, we find that the radius of circle C' is $\sqrt{52/9}$, so that the equation for C' is

$$(x + \tfrac{5}{3})^2 + (y - 2)^2 = \tfrac{52}{9} \tag{16}$$

Solving Equation (16) with $x^2 + y^2 = 1$ (see Exercise Set 6.7, Problem 6) gives the points $P'(0.463387, 0.886155)$ and $Q'(-0.95519, -0.29599)$ as the points of intersection of C' and C.

The length of $\overline{M_1M_2}$ can then be determined using the hyperbolic distance formula, so that

$$d(M_1,M_2) = \left| \ln\!\left(\frac{M_1Q'}{M_2Q'} \times \frac{M_2P'}{M_1P'} \right) \right| = \left| \ln\!\left(\frac{0.688706}{1.10924} \times \frac{0.772597}{1.191652} \right) \right|$$

This means that $d(M_1,M_2) \approx |\ln[(0.620880)(0.648341)]| \approx |\ln 0.402542| \approx 0.909955$.

But, $\tfrac{1}{2}d(A,B) = \tfrac{1}{2}(1.937190) = 0.968595$. From all of this it is clear that $d(M_1,M_2) \neq \tfrac{1}{2}d(A,B)$. Therefore we can conclude that the portion of Theorem 4.2.12 concerning the lengths of segments is not valid within the model of the hyperbolic plane. This means that the property is strictly Euclidean.

Next we consider the second portion of Theorem 4.2.12. In Figure 6.7.3 we see that the line containing M_1 and M_2 fails to intersect the line containing A and B. Is this observation sufficient to conclude that the following statement is neutral? The line containing the midpoints of two sides of a triangle is parallel to the third side.

The answer, of course, is no. The fact that a single triangle (in this case $\triangle ABC$) demonstrates a property is not sufficient to allow us to conclude that *all* triangles will do likewise. There may be a counterexample elsewhere. In this sense our method for classification is limited. As it happens, the second portion of Theorem 4.2.12 *is* neutral (see Exercise Set 6.7, Problem 7) but we cannot make that judgment based on the example illustrated in Figure 6.7.3.

In summary, we have three means of classifying theorems as neutral, Euclidean, or hyperbolic.

1. A theorem can be classified as neutral if its proof makes no use of the Euclidean parallel postulate or a consequence of it. For example, the ASA congruence condition was proved in Chapter 3 without any use of a parallel postulate.

2. A theorem can be classified as strictly Euclidean if a counterexample to it can be constructed within the model of the hyperbolic plane. For example, the Pythagorean theorem is strictly Euclidean since we were able to find, within the model of the hyperbolic plane, a triangle in which $a^2 + b^2 \neq c^2$.

3. A theorem can be classified as strictly hyperbolic if a counterexample to it can be found within the Euclidean plane. For example Theorem 6.3.4, which states that parallel lines are *not* everywhere equidistant, is strictly hyperbolic since, in the Euclidean plane, we can prove (see Exercise Set 6.7, Problem 8) that any pair of parallel lines are everywhere equidistant.

We see then that the theorems we have proved thus far can be classified according to the parallel postulates that characterize those geometries. This is so because the only difference between the two geometries we have investigated so far is the parallel postulate that was posited. Other geometries, however, involve more and more subtle differences in the postulate sets that are used. We will investigate one of these, elliptic geometry, in the following section.

EXERCISE SET 6.7

1. Provide the details of the standard Euclidean proof for the ASA congruence theorem.

2. We used $\triangle ABC$ (See Figure 6.7.2) to provide a counterexample for the Pythagorean Theorem, thereby showing that the theorem is strictly Euclidean. Suppose that in that example it turned out to be true that $a^2 + b^2 = c^2$. Would that have provided us with sufficient evidence to conclude that the Pythagorean theorem is neutral? Explain.

3. Consider the following statement: A statement cannot be proved using examples but can be disproved using counterexamples. Is this statement true or false? How does it pertain to the main topic of this section?

4. Show that, in Figure 6.7.3, the coordinates of M_2 are $(0, 2 - \sqrt{3})$.

5. Show that, in Figure 6.7.3, the coordinates of the center of circle C' are $(-\frac{5}{3}, 2)$.

6. Show that the points $P'(0.463387, 0.886155)$ and $Q'(-0.95519, -0.29599)$ are the points of intersection of the circles $x^2 + y^2 = 1$ and $(x + \frac{5}{3})^2 + (y - 2)^2 = \frac{52}{9}$.

7. The following statement is neutral: The segment joining the midpoints of two sides of a triangle is parallel to the third side. Prove this statement, first, by showing that in Figure 6.7.3 that $\overline{M_1M_2}$ is contained in the base of a Saccheri quadrilateral having $\overline{AB}$ as its summit, and then by invoking Corollary 3.6.5.

For Problems 8 through 20, find a counterexample within the hyperbolic model that can be used to show that each of the following theorems is strictly Euclidean.

8. Theorem 4.2.1: The sum of the measures of the interior angles of a triangle is 180°.

9. Corollary 4.2.2: The measure of an exterior angle of a triangle is *equal to* the sum of the measures of the remote interior angles.

10. Theorem 4.2.3: The opposite sides of a parallelogram are congruent.

11. Theorem 4.2.4: If three parallel lines are crossed by a transversal in such a way as to make congruent segments between the parallels, then every transversal that crosses these parallel lines will contain, between the parallels, congruent segments.

12. Theorem 4.2.8: Two lines parallel to the same line are parallel to each other.

13. Theorem 4.2.9: If a line intersects one of two parallel lines, then it intersects the other.

14. Theorem 4.2.10: Each diagonal of a parallelogram partitions the parallelogram into a pair of congruent triangles.

15. Theorem 4.2.11: The diagonals of a parallelogram bisect each other.

16. Theorem 4.2.16: The median to the hypotenuse of a right triangle is one-half the length of the hypotenuse.

17. Theorem 4.2.17: If in a right triangle one of the angles measures 30°, then the side opposite this angle is one-half the length of the hypotenuse.

18. Theorem 4.2.18: If one leg of a right triangle is one-half the length of the hypotenuse, then the angle opposite that leg has a measure of 30°.

19. Theorem 4.2.19: The sum of the measures of the interior angles of a convex n-gon is $(n - 2)(180°)$.

20. Theorem 4.2.20: The sum of the exterior angles (one at each vertex) of a convex n-gon is 360°.

21. The proof in Chapter 4 of the median concurrence theorem (Theorem 4.2.5) makes extensive use of various consequences of the Euclidean parallel postulate. Keeping this in mind, consider $\triangle ABC$ in Figure 6.7.3. Are the three "hyperbolic medians" of this triangle concurrent? Does this show that the median concurrence theorem is strictly Euclidean, strictly hyperbolic, or neutral ? Explain.

22. Find a triangle, within the model of the hyperbolic plane, in which the perpendicular bisectors for the sides are not concurrent.

23. Do you think hyperbolic triangles have a nine-point circle? Explain why or why not.
24. Theorem 4.7.5 states that in Euclidean geometry the bisector of an interior angle of a triangle is concurrent with the bisectors of the remote exterior angles. Do you think this theorem is strictly Euclidean, strictly hyperbolic, or neutral? Explain.
25. Classify the theorem of Menalaus as strictly Euclidean, strictly hyperbolic, or neutral.
26. Classify Ceva's theorem as strictly Euclidean, strictly hyperbolic, or neutral.

6.8 ELLIPTIC GEOMETRY

A Geometry With No Parallels?

In Chapter 1 when we considered the question of the existence of parallel lines in incidence geometry, we determined that their existence must be a consequence of an additional axiom (see Section 1.4). At that time, we concluded that if we considered any line l and any point P, where P was not on line l, three possibilities exist for such a parallel axiom:

1. There exist no lines on P that are parallel to l,
2. There exists exactly one line on P that is parallel to l, or
3. There exists more than one line on P that is parallel to l.

Using axioms of neutral geometry as the basis of our incidence geometry, the additional assumption of a characteristic postulate equivalent to alternative 2 gave rise to what is commonly called Euclidean geometry. Some of the consequences of this axiom were discussed in Chapters 3 through 5. In the previous sections of this chapter we again used the axioms of neutral geometry as a basis, but this time replaced alternative 2 by a characteristic postulate equivalent to alternative 3, giving us the non-Euclidean geometry with multiple parallels referred to as hyperbolic geometry.

In this section we will attempt to briefly answer the remaining question, Can we, using axioms of neutral geometry as a basis together with the additional assumption of a characteristic postulate equivalent to alternative 1,[28] create a consistent but different non-Euclidean geometry? Unfortunately, the alternate interior angle theorem (Theorem 3.4.1) and its consequences, which were proven in Chapter 3, provide us with the existence of at least one line through P that is parallel to l. The existence of such a

---

[28] Note that this alternative is equivalent to the obtuse angle hypothesis made by Saccheri (see Section 3.5).

parallel line is a direct contradiction of alternative 1. Therefore the question must be answered with a resounding no. However, if we were willing to alter, slightly, the axioms of neutral geometry that give rise to the proof of the alternate interior angle theorem, perhaps we could still produce a consistent geometry that exhibits alternative 1 as its parallel axiom. As we shall see, this transition from neutral geometry to a geometry with no parallel lines is not as easily accomplished as was the transition to hyperbolic geometry. The assumption of the existence of no parallel lines will prove to be incompatible not only with the alternate interior angle theorem but, as we shall see, with several other properties as well. In addition, the modification of several preliminary results from neutral geometry would require us to critically evaluate all other results in order to ensure the consistency of this geometry. And, as discussed in earlier sections, determining which statements are theorems in neutral geometry is a difficult task in itself, and here the process is even more difficult. Such an extensive investigation is beyond the scope of this book and, for that reason, we shall present a more intuitive and significantly less rigorous discussion than we have in the previous developments of Euclidean and hyperbolic geometry.

Two Models

Since the alternate interior angle theorem provides us with an unwanted parallel line, we need to investigate its proof more thoroughly. It is evident that the proof of the alternate interior angle theorem relies heavily on the validity of the exterior angle theorem (Theorem 3.2.7). On closer inspection of its proof, we discover that the exterior angle theorem relies on three major components: triangle congruence, angle addition, and plane separation. Since triangle congruence and angle addition lie at the heart of geometry, alteration of these concepts seems inappropriate. Therefore plane separation[29] becomes the inevitable choice. It appears therefore that if we were to remove those axioms of neutral geometry that imply plane separation, we would be able to produce a consistent geometry that contains no parallel lines. To that end we must conceive of a geometry in which a line does not divide the plane into two distinct regions. In other words, we must conceive of a geometry in which we are able to pass from one side of a line to another without crossing that line. It is here that our intuition begins to fail us. We may find it difficult to comprehend such a line and such a plane. To assist us, the physical model of a Möbius band (Figure 6.8.1)[30] provides

[29] Inherent in this assumption is the fact that lines are infinite in extent. As we will see in a future example, we could accomplish the same result by replacing this assumption with that of the boundlessness of lines and allowing two distinct lines to intersect in more than one point.

[30] See H. E. Wolfe, *Introduction to Non-Euclidean Geometry* (New York: Holt, Rinehart and Winston, 1945), p. 179.

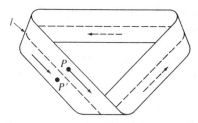

Figure 6.8.1

some insight into how it might be possible to pass from one side of a line to the other without crossing the line.

Before pursuing this problem further, it will prove to be advantageous for us to view it from a different perspective. This alternate perspective may indeed be the one that Bernhard Riemann (1826–1866) had in mind when, during a lecture to the faculty of the University of Göttingen in 1854, he formulated his theory of non-Euclidean geometry. Riemann posed the idea that space need not be infinite, as suggested by Euclid, but may simply be regarded as unbounded. It was, as you should recall, the assumption of the fact that straight lines could be produced infinitely that led us to rejection of the obtuse angle hypothesis (see Lemma 3.5.3). Riemann chose to replace Euclid's tacit assumption of the infinitude of a line by assuming that a line is boundless and not necessarily infinite. In this proposed geometry it will soon become evident that two other familiar geometric concepts must also be altered. As we will see in the following model, it becomes necessary to allow two lines to intersect in more than one point, and along with that, the concept of betweeness as it relates to points on a line must also be eliminated.

As a model of this geometry, we will choose as points, points on the surface of a sphere, and as lines, great circles[31] on the surface of that sphere (Figure 6.8.2a). In this model, to assist us in interpreting concepts like line segments, angles, and their measures, we will rely on the reader's familiarity with Euclidean concepts of major and minor arcs of a circle. With this understanding, line segment $\overline{AB}$ will be interpreted as either of the two arcs of the great circle containing points A and B. An angle shall be interpreted as the union of two distinct minor arcs that share a common endpoint (Figure 6.8.2b). In addition, the distance between two points can be interpreted as the length of the minor arc joining the two points, and the measure of an angle interpreted as the measure of the angle formed by the tangents to and in the planes of the minor arcs that form the angle.

It should be obvious that in this model all lines, that is, great circles, intersect in two distinct points and that no parallel lines exist. It should be equally as obvious that plane separation does exist, since a line separates the

---

[31] Great circles are circles on the surface of a sphere that have the same diameter as the sphere.

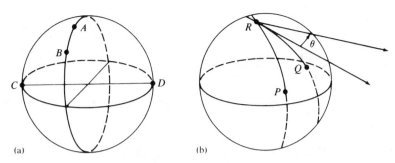

Figure 6.8.2   (a) Line $\overline{CD}$ and line segment $\overline{AB}$. (b) $\angle PRQ$, where $m\angle PRQ = \theta$.

surface of the sphere (i.e., plane) into two hemispheres (i.e., half-planes). Please note that since great circles clearly have finite length, it was indeed necessary, as suggested by Riemann, to replace the concept of the infinitude of a line by the fact that it is unbounded. This model is commonly referred to as *double elliptic geometry.*[32]

This spherical model of double elliptic geometry serves another very important role. Since each of its undefined terms has, as its interpretation, a concept from Euclidean spherical geometry,[33] we find that double elliptic geometry derives its consistency from the assumed consistency of Euclidean spherical geometry in the same way that the Poincaré model guarantees the consistency of hyperbolic geometry. Therefore, in the terminology of Chapter 1, double elliptic geometry and Euclidean spherical geometry are relatively consistent (see Section 1.2).

With the additional insight gained through our discussion of Riemann's double elliptic geometry, we can now return to our initial discussion of a non-Euclidean geometry in which plane separation fails to hold. Felix Klein (1849–1925), perhaps motivated either by the fact that two diametrically opposed points on the surface of a sphere do not determine a unique line (i.e., great circle) or by the fact that double elliptic geometry exhibits plane separation (the axiom whose denial we identified as a crucial issue in the development of this type of non-Euclidean geometry), derived what is commonly known as *single elliptic geometry.* Recognizing that since each point on the surface of a sphere determines a unique antipodal or diametrically

[32] The word "double" refers to the fact that two lines intersect in two distinct points. The name "elliptic" was coined by Felix Klein in reference to the fact that a straight line contains no infinitely distant points, under the obtuse angle hypothesis. (See Wolfe, *Non-Euclidean Geometry*, p. 63.)

[33] Riemann's spherical model of double elliptic geometry should not be confused with Euclidean spherical geometry. Double elliptic geometry is a type of non-Euclidean plane geometry which happens to have a model on the surface of a sphere. Euclidean spherical geometry, on the other hand, is the geometry on the surface of a three-dimensional sphere and is independent of any parallel axiom.

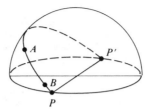

Figure 6.8.3 Line $\overleftrightarrow{AB}$ with antipodal points P and P'.

opposed point, Klein suggested that each pair of antipodal points could be abstractly identified as a "single point," hence the name "single elliptic geometry." With this interpretation, a line becomes a great circle with each pair of antipodal points identified or, if you will, a great semicircle with its endpoints identified (Figure 6.8.3). As a result, two distinct lines, which must intersect as a consequence of the elliptic characteristic postulate, intersect in only one point (i.e., point pair), and as we had hoped, a line no longer separates the plane. We should also note that in this model, as in double elliptic geometry, lines remain closed figures that are still unbounded and finite in length.

Some Results in Elliptic Geometry

Now that we can conceive of a consistent non-Euclidean geometry that has no parallel lines, we shall briefly look at some of its consequences. The intent here is not to provide a rigorous development of elliptic geometry, but to provide some results for the purpose of comparison with Euclidean and hyperbolic geometries. For ease in visualizing some of the following theorems, it might be helpful for the reader to consider the spherical model of double elliptic geometry and the hemispherical model of single elliptic geometry discussed previously.

THEOREM 6.8.1.   *The Polar Property Theorem.* If l is any line in elliptic geometry, then there exists at least one point P such that every line connecting P to a point on l is perpendicular to l, and P is equidistant from all points on l.

Proof. Let Q and R be any distinct points on l and then construct lines m and n which are perpendicular to l at Q and R, respectively (Figure 6.8.4). By the elliptic characteristic postulate, m and n must intersect at some point P. Therefore since P, Q, and R are noncollinear, figure PQR is a triangle with two congruent angles, and by the converse of the isosceles triangle theorem (Theorem 3.3.2), $\overline{PQ} \cong \overline{PR}$. Now suppose that S is the midpoint of $\overline{QR}$. If we draw $\overline{PS}$, we see that $\triangle PQS \cong \triangle PRS$, and consequently, $\overline{PS}$ is perpendicular to $\overline{QR}$ and $\overline{PS} \cong \overline{PQ} \cong \overline{PR}$. And since Q and R were chosen arbitrarily, the proof is complete.

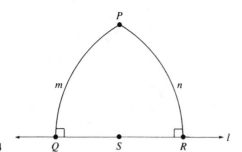

Figure 6.8.4

In the previous theorem point *P* is called the *pole* of line *l*, *l* is called the *polar* of point *P*, and the unique distance from *P* to *l* is called the *polar distance*.

Now suppose that in the previous proof we extend $\overline{PQ}$ to point *P′*, such that *QP′* is equal to the polar distance, and draw $\overline{P'R}$. It should be evident that $\triangle PQR \cong \triangle P'QR$, and therefore *P′* appears to be a "second" pole for line *l*. This would indeed be true if *l* separates the plane as it does in double elliptic geometry. But what happens in single elliptic geometry? As we should expect, in single elliptic geometry *P′* and *P* turn out to be a pair of antipodal points, thus they are not distinct, and lines *m* and *n* intersect in only one point.

THEOREM 6.8.2.   In any right triangle of elliptic geometry, each of the other two angles has a measure less than, equal to, or greater than a right angle dependent on whether the side opposite it has a measure that is less than, equal to, or greater than the polar distance.

Proof. Let $\triangle ABC$ have a right angle at *C*. On $\overline{CB}$, locate point *P* such that *CP* is equal to the polar distance (Figure 6.8.5). $\overline{AC}$ has point *P* as its pole, and therefore $\overline{AP}$ is perpendicular to $\overline{AC}$ and $\angle PAC$ is a right angle. Now if *P = B* (i.e., *CB* equals the polar distance), then $\angle CAB(P)$ is a right angle. Correspondingly, If *CB* is less than the polar distance, then $m\angle CAB < 90°$, and if *CB* is greater than the polar distance, then $m\angle CAB > 90°$.

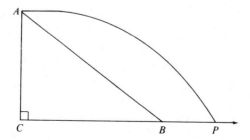

Figure 6.8.5   *C*

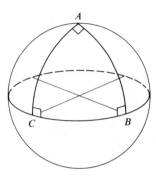

Figure 6.8.6

EXAMPLE 6.8.1.   Suppose that in $\triangle ABC$ where $\angle C$ is a right angle, we have $AB = BC = AC =$ the polar distance (Figure 6.8.6). Then, by the previous theorem, each of the angles is a right angle and the angle sum of $\triangle ABC$ is a somewhat surprising 270°.

Now if you recall our previous discussions of Saccheri and Lambert quadrilaterals and the triangle angle sum theorems in the sections on Euclidean and hyperbolic geometries, the following results should not be surprising.

THEOREM 6.8.3.   In elliptic geometry the summit angles of a Saccheri quadrilateral are congruent and obtuse.

Proof. Let $\square ABCD$ be a Saccheri quadrilateral where $\angle A$ and $\angle B$ are right angles and $\overline{AD} \cong \overline{BC}$. A theorem from neutral geometry (Theorem 3.6.2) tells us that the summit angles are congruent; therefore it suffices to show that $\angle BCD$ is obtuse. If we let E and F be the midpoints of $\overline{AB}$ and $\overline{CD}$, respectively, then by Theorem 3.6.4, $\overline{EF}$ is perpendicular to both $\overline{AB}$ and $\overline{CD}$ (Figure 6.8.7). Suppose that we now extend $\overrightarrow{EC}$ and $\overrightarrow{FB}$ until they meet at a point P. By definition, P is the pole of $\overleftrightarrow{EF}$, hence EP is the polar distance. Since $CP < EP$, the previous theorem tells us that $m\angle BCP < 90°$, and thus $m\angle BCD > 90°$ as required.

COROLLARY 6.8.4.   In elliptic geometry the fourth angle of a Lambert quadrilateral is obtuse. The proof is left as an exercise.

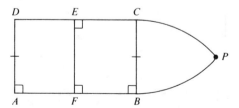

Figure 6.8.7

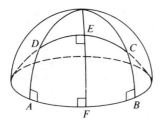

Figure 6.8.8

Figure 6.8.8 pictures Saccheri quadrilateral □*ABCD* and Lambert quadrilateral □*AFED* in the hemispherical model of single elliptic geometry.

THEOREM 6.8.5.   In elliptic geometry, the angle sum of any right triangle is greater than 180°.

Proof. Let △*ABC* have a right angle at *C*. We must show that S(△*ABC*) > 180°. As a result of Theorem 6.8.2, we need only consider the case where *AC* and *BC* are both less than the polar distance, that is, where ∠*A* and ∠*B* are both acute. We will begin our proof by constructing △*QAP* congruent to △*QBR* in the following manner. First, construct $\overrightarrow{AX}$ such that ∠*BAX* ≅ ∠*ABC*. Then, using *Q* as the midpoint of $\overline{AB}$, construct $\overrightarrow{QR}$ perpendicular to $\overline{BC}$ and $\overline{AP}$ ≅ $\overline{BR}$ (Figure 6.8.9). As a result of the congruence of △*QAP* and △*QBR*, ∠*APQ* is a right angle and *P*, *Q*, and *R* are collinear (see Exercise Set 6.8, Problem 6). Now by definition, □*ACRP* is a Lambert quadrilateral, hence ∠*CAR* is obtuse. From this observation, we can conclude that m∠*CAB* + m∠*ABC* > 90° and S(△*ABC*) > 180°.

THEOREM 6.8.6.   In elliptic geometry the angle sum of any triangle is greater than 180°. The proof is left as an exercise.

COROLLARY 6.8.7.   In elliptic geometry the sum of the measures of the interior angles of any convex quadrilateral is greater than 360°. The proof is left as an exercise.

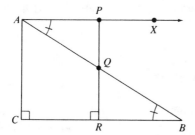

Figure 6.8.9

COROLLARY 6.8.8.   Rectangles do not exist in elliptic geometry. The proof is left as an exercise.

The final two results, whose proofs are beyond the scope of this discussion, are offered for the purpose of comparsion with analogous results in Euclidean and hyperbolic geometries.

THEOREM 6.8.9.   If three angles of one triangle are congruent, respectively, to the three angles of a second triangle, then the triangles are congruent.

THEOREM 6.8.10.   The area of a triangle is proportional to its excess; that is, area $\triangle ABC = k[S(\triangle ABC) - 180°]$, where k is a constant dependent on the unit of length choosen.

The following exercises are intended to provide additional insight into elliptic geometry and its relationship to Euclidean and hyperbolic geometry. The reader is encouraged to sketch diagrams in both double and single elliptic models and is cautioned to be aware of using results from Chapter 3 before checking their validity.

EXERCISE SET 6.8

1. Determine which of the theorems, proven in Chapter 3, are not true in elliptic geometry.
2. Prove Corollary 6.8.4.
3. Explain why, in the proof of Theorem 6.8.5, points P, Q, and R must be collinear.
4. Prove Theorem 6.8.6.
5. Prove Corollary 6.8.7.
6. Prove Corollary 6.8.8.
7. In elliptic geometry prove that the length of the summit of a Saccheri quadrilateral is less than the length of its base.
8. In elliptic geometry prove that in a Lambert quadrilateral the length of each side containing the obtuse angle is less than the length of the side opposite it.
9. In elliptic geometry prove that the length of the line joining the midpoints of the base and the summit of a Saccheri quadrilateral is greater than the length of its sides.

6.9 GEOMETRY IN THE REAL WORLD

In Chapter 4 we outlined Euclidean geometry, while in the previous sections of this chapter we investigated two distinct types of non-Euclidean geometry. To summarize our results and to assist us in comparing and contrasting these radically different axiomatic approaches to the geometry of the plane, we offer Table 6.9.1.

TABLE 6.9.1   A Comparison of Euclidean and Non-Euclidean Geometry

| | Type of Geometry | | | |
| | | | Elliptic | |
Property	Euclidean	Hyperbolic	Double	Single
Number of parallels to a line *l* through a point not on *l*	One	At least two	None	
$S(\triangle ABC)$	Equals 180°	Is less than 180°	Is greater than 180°	
The intersection of two distinct lines is	At most one point	At most one point	Two points	One point pair
Exhibits plane separation	Yes	Yes	Yes	No
Straight lines are	Unbounded and infinite in length	Infinite in length	Unbounded and finite in length	
Betweeness of points on a line	Yes	Yes	No	No
Two lines perpendicular to the same line	Are parallel	Are parallel	Intersect in two poles	Intersect in one pole
The summit angles of a Saccheri quadrilateral are	Right	Acute	Obtuse	
Rectangles	Exist	Do not exist	Do not exist	
Parallel lines	Are everywhere equidistant	Are not everywhere equidistant	Do not exist	
If a line intersects one of two parallel lines	It intersects the other	It may intersect the other	Parallels do not exist	

(Continued)

TABLE 6.9.1 *(Continued)*

| | Type of Geometry | | | |
| | | | Elliptic | |
Property	Euclidean	Hyperbolic	Double	Single
Two lines parallel to a third line are	Parallel	May be parallel	Parallels do not exist	
Two triangles with congruent corresponding angles are	Similar	Congruent	Congruent	
The area of a triangle	Is equal to $bh/2$	Is proportional to its defect	Is proportional to its excess	
A triangle of maximal area	Does not exist	Exists	Exists	

The previous comparison of the three distinctly different geometries leads us to the inevitable question, In spite of the fact that each of the geometries is a consistent deductive system, isn't it true that Euclidean geometry is an accurate description of physical space, while hyperbolic and elliptic geometries are merely logical exercises of value to mathematicians but with little applicability to the real world?

This controversy seems to center around our willingness to accept Euclidean geometry as the "absolute truth," much in the same way as Kant did when he pronounced that there could be no geometry other than Euclidean (see Section 6.6). But why is it that we seem so certain that Euclidean geometry is an accurate model of physical space? Is it possible that we could empirically determine its validity?

The answer to the first question is easy. We simply need to observe that Euclidean geometry is comfortable; it has been used for thousands of years; and for the relatively small physical space accessible to human beings during their day-to-day existence, it seems an accurate description.

The second question is much less easily answered. In 1827 C. F. Gauss suggested a criterion by which the "truth" of Euclidean geometry might be established. Recognizing that the angle sum of a triangle is one of the major distinguishing factors among the geometries, Gauss attempted to empirically validate the angle sum theorem. It had been established that in both non-Euclidean geometries, the larger the area of the triangle, the greater the difference between the angle sum and 180°. With this in mind, Gauss created a large triangle using distant mountain tops as its vertices.[34] After measuring the angles, he found that their sum differed from 180°, but not by a sufficient amount to eliminate error in measurement. And even if Gauss had had access to the most sophisticated measuring devices available today, and the sum proved to be 180°, the potential error in measurement, however small, would still have allowed for an actual sum of either greater than or less than 180°. It is also possible that any discrepancy in the angle sum could have been explained by the modification of some physical law, such as the law of optics; it is difficulties such as these that prompted Poincaré to declare the impropriety of asking which geometry is the "true one."[35] In spite of the fact that Gauss's experiment was inconclusive, it lends more credence to the claim that within a small frame of reference, each of the three geometries fit physical space equally well.[36]

From this perspective, the more important question is not which geometry is true, but in which geometry are certain real-world applications most appropriate? The answer is that for most applications to the physical world we choose the geometry that is most convenient. For example, most applications in the fields of surveying, navigation, and engineering deal with relatively small portions of the earth's surface, which when viewed as a plane allow for the simpler calculations of Euclidean geometry. Long-distance travel "on" the surface of the earth, as evidenced by the Arctic polar flight paths chosen by some commercial aircraft as the routes of least distance, seem best described by elliptic geometry. And, in addition, there appears to be scientific evidence that binocular visual space is perhaps best described by hyperbolic geometry. To further the case, Einstein, when developing his general theory of relativity, found that none of the geometries discussed in this chapter adequately described the orbits of the planets nor what appeared to happen to light rays as they traveled through space. As a consequence, he found that another non-Euclidean geometry, a complex form of Riemannian geometry, was more appropriate for his use.

It seems, therefore, perfect retribution that Kant's statement with regard to the incontrovertible truth of Euclidean geometry was made while "living on, if not in, a non-Euclidean world."[37]

---

[34] See R. J. Trudeau, *The Non-Euclidean Revolution.* (Boston: Birkhauser, 1987), pp. 147–148.

[35] See Eves, *History of Mathematics*, Problem Study 13.12, p. 396.

[36] For an interesting and more complete discussion of this problem, see Morris Kline, *Mathematics in Western Culture* (New York: Oxford University Press, 1964), pp. 417–427.

[37] Kline, *Mathematics*, p. 427.

CHAPTER 6 SUMMARY

6.2 A Return to Neutral Geometry

Definition. If line l and points P, Q, R, and S are as in Figure 6.2.1 and if $m\angle RPQ = d_0$ (where d_0 is the least upper bound of the set $D = \{d/d = m\angle \overrightarrow{RPQ}$ and $\overrightarrow{PQ} \cap \overleftrightarrow{RS} \neq \emptyset\}$), then $\angle RPQ$ is called the *angle of parallelism* for RS and P.

Theorem 6.2.1. If $\angle RPQ$ is the angle of parallelism for a line $\overleftrightarrow{RS}$ and a point P (see Figure 6.2.1), and if $m\angle RPQ = d_0$, we may conclude that (i) if $m\angle RPQ < d_0$, then $\overrightarrow{PQ} \cap \overleftrightarrow{RS} \neq \emptyset$, and (ii) if $m\angle RPQ < d_0$, then $\overrightarrow{PQ} \cap \overleftrightarrow{RS} \neq \emptyset$.

Theorem 6.2.2. If T-R-S and $\angle RPQ$ and $\angle RPQ'$ are the angles of parallelism for P and RS and for P and RT, respectively (see Figure 6.2.4), then $m\angle RPQ = m\angle RPQ'$.

Theorem 6.2.3. The angle of parallelism for a given line and point is less than or equal to 90°.

Corollary 6.2.4. If $\angle RPQ$ is the angle of parallelism for line l and point P (refer to Figure 6.2.1), then $\overleftrightarrow{PQ} \cap l = \emptyset$. (*Note:* This corollary involves *line* $\overleftrightarrow{PQ}$ rather than *ray* $\overrightarrow{PQ}$.)

Theorem 6.2.5. If the angle of parallelism for a point P and a line $\overleftrightarrow{RS}$ is less than 90°, then there exist at least two lines through P that are parallel to $\overleftrightarrow{RS}$.

Theorem 6.2.6. The Euclidean parallel postulate is equivalent to the following statement: The measure of the angle of parallelism is 90°.

6.3 The Hyperbolic Parallel Postulate

Definition. The distance between a line and a point not on the line is the length of the (unique) perpendicular from the point to the line. (The distance between a point and any line containing the point is, by definition, zero.)

Theorem 6.3.1. Given a line l and a point P not on l, there are at least two lines through P that are parallel to l.

Theorem 6.3.2. The hyperbolic parallel postulate is equivalent to the following statement. The summit angles of a Saccheri quadrilateral are acute.

Theorem 6.3.3. Rectangles do not exist in hyperbolic geometry.

Theorem 6.3.4. The fourth angle of a Lambert quadrilateral is acute.

Theorem 6.3.5.   In a Lambert quadrilateral, the sides contained by the sides of the acute angle are longer than the sides that they are opposite.

Theorem 6.3.6.   Parallel lines are *not* everywhere equidistant.

Theorem 6.3.7.   The summit of a Saccheri quadrilateral is longer than the base.

Theorem 6.3.8.   Two parallel lines that are crossed by a transversal have congruent alternate interior angles if and only if the transversal contains the midpoint of a segment perpendicular to both lines.

Theorem 6.3.9.   In hyperbolic geometry there exist triangles that cannot be circumscribed.

6.4 Some Hyperbolic Results Concerning Polygons

Definition.   *Equivalent Polygons:* Two polygons P_1 and P_2 are said to be equivalent if and only if each polygon can be partitioned into a finite set of triangles in such a way that (i) the two sets of triangles can be placed in a one-to-one correspondence, and (ii) the one-to-one correspondence associates triangles that are congruent.

Theorem 6.4.1.   If $\triangle ABC$ is a triangle, then its defect $d(\triangle ABC)$ is positive [i.e., $d(\triangle ABC) > 0$ or, equivalently, $S(\triangle ABC) < 180°$].

Theorem 6.4.2 (Corollary).   The sum of the measures of the interior angles of a convex quadrilateral is less than 360°.

Theorem 6.4.3.   If $\triangle ABC$ is partitioned into a pair of component triangles by a Cevian line, the defect of $\triangle ABC$ is equal to the sum of the defects of the component triangles.

Theorem 6.4.4.   If $\triangle ABC$ is partitioned into a triangle and a quadrilateral by a line, the defect of $\triangle ABC$ is equal to the sum of the defects of the component pieces (namely, the triangle and the quadrilateral).

Theorem 6.4.5 AAA Congruence Condition: Two triangles are congruent if their corresponding angles have the same measure.

Theorem 6.4.6.   If a convex polygon is partitioned into triangles in any manner, the defect of the polygon is equal to the sum of the defects of the component triangles.

Theorem 6.4.7.   Every triangle is equivalent to its associated Saccheri quadrilateral.

Theorem 6.4.8.   Any two congruent convex polygons are equivalent.

Theorem 6.4.9.   The equivalence relation as defined for polygons is reflexive, symmetric, and transitive.

Theorem 6.4.10 (Lemma).   Two Saccheri quadrilaterals are congruent if their summits and summit angles are congruent.

Theorem 6.4.11.   If two triangles have the same defect and a pair of congruent sides, then they are equivalent.

Theorem 6.4.12.   If two triangles have the same defect, then they are equivalent.

Theorem 6.4.13.   If two triangles are equivalent, then they have the same defect.

Theorem 6.4.14.   Two triangles have the same defect if and only if they are equivalent.

6.5 Area in Hyperbolic Geometry

Definition.   The *area* $A(P)$ for a simple polygon P in the hyperbolic plane is directly proportional to the polygon's defect, so that

$$A(P) = k \times d(P)$$

Theorem 6.5.1.   To every polygonal region there corresponds a unique positive area (SMSG Postulate 17).

Theorem 6.5.2.   If two polygons are congruent, then the polygonal regions have the same area (SMSG Postulate 18).

Theorem 6.5.3.   If a polygonal region P is partitioned into two subregions R_1 and R_2, then

$$A(P) = A(R_1) + A(R_2)$$

(SMSG Postulate 19).

Theorem 6.5.4.   If $\triangle ABC$ is a hyperbolic triangle, then

$$A(\triangle ABC) < k \times 180°$$

where degrees are the unit of angular measure.

6.7 Classifying Theorems

We have three means of classifying theorems as neutral, Euclidean, or hyperbolic.

1. A theorem can be classified as *neutral* if its proof makes no use of the Euclidean parallel postulate or a consequence of it. For example, the

ASA congruence condition was proved in Chapter 3 without any use of a parallel postulate.

2. A theorem can be classified as *strictly Euclidean* if a counterexample to it can be constructed within the model of the hyperbolic plane. For example, the Pythagorean theorem is strictly Euclidean since we were able to find, within the model of the hyperbolic plane, a triangle in which $a^2 + b^2 \neq c^2$.

3. A theorem can be classified as *strictly hyperbolic* if a counterexample to it can be found within the Euclidean plane. For example, Theorem 6.3.4, which states that parallel lines are *not* everywhere equidistant, is strictly hyperbolic since in the Euclidean plane we can prove (see Exercise Set 6.8, Problem 8) that all pairs of parallel lines are everywhere equidistant.

6.8 Elliptic Geometry

Theorem 6.8.1 *The Polar Property Theorem:*   If l is any line in elliptic geometry, then there exists at least one point P such that every line connecting P to a point on l is perpendicular to l and P is equidistant from all points on l.

In Theorem 6.8.1 the point P is called the *pole* of the line l, l is called the *polar* of point P, and the unique distance from P to l is called the *polar distance*.

Theorem 6.8.2.   In any right triangle of elliptic geometry (with a right angle at C), each of the angles other than $\angle C$ is less than, equal to, or greater than a right angle dependent on whether the side opposite it has a measure that is less than, equal to, or greater than the polar distance.

Theorem 6.8.3   In elliptic geometry the summit angles of a Saccheri quadrilateral are congruent and obtuse.

Corollary 6.8.4.   In elliptic geometry the fourth angle of a Lambert quadrilateral is obtuse.

Theorem 6.8.5.   In elliptic geometry the angle sum of any right triangle is greater than 180°.

Theorem 6.8.6.   In elliptic geometry the angle sum of any triangle is greater than 180°.

Corollary 6.8.7.   In elliptic geometry the sum of the measures of the interior angles of any convex quadrilateral is greater than 360°.

Corollary 6.8.8.   Rectangles do not exist in elliptic geometry.

Theorem 6.8.9.   If three angles of one triangle are congruent, respectively, to the three angles of a second triangle, then the triangles are congruent.

Theorem 6.8.10.   The area of a triangle is proportional to its excess; that is, $A(\triangle ABC) = k[S(\triangle ABC - 180°]$, where k is a constant dependent on the unit of length chosen.

ALL ROADS
LEAD TO . . .

Projective Geometry

7.1 INTRODUCTION

As the final leg of our journey down various "roads to geometry," we will investigate another distinct geometry—projective geometry—which began as an outgrowth of Euclidean geometry but has, during the last several centuries, taken on a spirit of its own. You may recall that in Section 1.4, where incidence geometries were discussed, two of the exercises posed involved the axioms of projective geometry. In this chapter we will explore some of the elementary results from the "projective plane." The object of this survey is only to whet the appetite of students who may choose to pursue this topic in further studies. A single chapter is not sufficient to do justice to the elegance of projective geometry, but perhaps this brief encounter will convey the flavor of the subject matter.

The roots of projective geometry can be traced back to the Renaissance period when artists, particularly painters, strove toward realism in their work. Prior to this time, works of art often portrayed scenes, such as landscapes, in ways that satisfied the brain but not the eye. For example, depth in scenes was often missing because objects that were parallel (e.g., the shoulders of roads) were portrayed on canvas as parallel lines. While this makes sense rationally, it is not the way in which these images meet and are seen by our eyes. The images processed by our brain are actually *projections* of the scene from the three-dimensional space in which they exist to a two-dimensional scene which our brain interprets. Figure 7.1.1 displays the contrast between a drawing that is "in perspective" and one that is not.

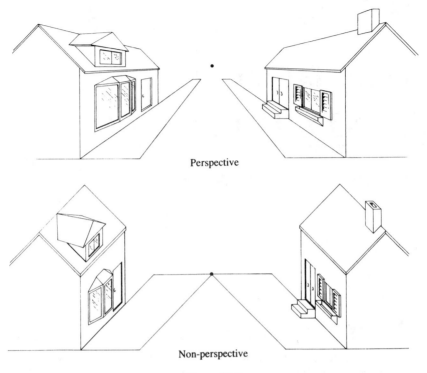

Perspective

Non-perspective

Figure 7.1.1

It should be clear by now that, at least in the Euclidean plane, parallel lines never intersect (in fact that is how we have defined parallel lines). Nevertheless, when we project a three-dimensional image onto a two-dimensional plane, some lines that are parallel appear to converge toward a common point. To visualize this, just imagine standing on a set of railroad tracks. The image that meets the eye is a pair of intersecting lines, even though our intellect tells us that these tracks neither intersect nor even get closer to one another in the distance.

Considerations such as these came to be studied by artists and, eventually, by mathematicians (often they were one and the same) during the Renaissance period. The outcomes of their studies led to the body of theorems we now call projective geometry.

As mentioned above, the origins of projective geometry are rooted in attempts to study, in a mathematically rigorous fashion, the way in which three-dimensional objects can be portrayed in a two-dimensional fashion. In order to visualize this process we need only look out of a window and think of the image we see as a "snapshot" appearing on the window. This snapshot can be thought of as a projection of the three-dimensional objects onto the window. In this projection, lines we know are parallel[1] will appear as intersecting lines in the snapshot. In fact, there will be no parallel lines in the projection. Nevertheless, projections preserve many Euclidean properties that are characteristic of the world our senses perceive (i.e., the world that has been projected onto the window). Renaissance artists and mathematicians faced the task of building an axiomatic system that preserved the essential Euclidean properties while avoiding the existence of parallel lines.

To accomplish this task we will return to our Euclidean axioms (i.e., the SMSG axioms) and modify them in such a way as to preserve the essential properties of incidence, while providing a framework within which parallel lines (of the Euclidean kind) can be allowed to have a point of intersection. In order to accomplish this objective, we will do what mathematicians do best—we will postulate the existence of additional points in the Euclidean plane, called ideal points, and logically deduce the consequences that result from expanding the Euclidean plane in this fashion. The resulting geometry, called *projective geometry,* is at once an example of mathematics applied to the solution of a real problem (namely, the desire of artists for realism) and mathematics as pure reason.

7.2 THE REAL PROJECTIVE PLANE

The introductory remarks in the preceding section have set the stage for an extension of Euclidean plane geometry, as discussed in Chapter 4, to a slightly larger plane, called the *real projective plane,* in which lines that are parallel in the Euclidean sense have a point of intersection. This extension brings with it, of course, a contradiction in terms, since the Euclidean notion of parallelism of lines implies that parallel lines have *no* points of intersection. In order to complete the projective plane, then, we must append to the Euclidean plane a set of points, known as *ideal points,* that will serve as the points of intersection of lines that are parallel in Euclidean geometry. The Euclidean points that constitute the remainder of the projective plane will be called *ordinary points.* With these new terms available, we can think of parallel lines as lines having no ordinary point of intersection.

In order to accomplish this extension successfully, we will need to address several issues that will ensure that the extended plane is, as much as possible, consistent with the axioms on which we built Euclidean geometry. In particular, we will want to preserve the incidence relations posited by SMSG Postulate 1 (or, more formally, by Hilbert's Axioms I-1 to I-4). Because of this, we will proceed cautiously as we extend the Euclidean plane to the projective plane.

---

[1] Since projective geometry evolved before the development of non-Euclidean geometries, we will assume that the three-dimensional world in which we live is Euclidean so that parallel lines exist.

Figure 7.2.1

One consideration worth noting concerns the number of ideal points that will need to be appended to the Euclidean plane. For instance, if line l is parallel to line m (Figure 7.2.1), then we will need to include in the projective plane a point I_l (the ideal point on line l) at which l and m intersect.

Since I_l is also on line m, it could as well be called I_m, so that $I_l = I_m$. Now consider any other line n that is parallel to l (and is also parallel to m since parallelism, within the Euclidean plane, is transitive). Will l and n share a point? If so, will it be I_l or a different ideal point? Since we wish all lines to intersect and we wish to append only one ideal point to each line, it follows that every line parallel to l should contain the same ideal point I_l. Consequently, it seems as if every collection of mutually parallel lines should intersect in a single ideal point, so that we will need one ideal point for each possible direction in the Euclidean plane. This provides us with an infinity of ideal points.

If we then choose any two of the ideal points, there should be a unique line containing these two points. (This is an incidence property that is characteristic of the Euclidean plane.) Can this line be an ordinary line? No, since each ordinary line is supposed to have just one ideal point. Consequently, it seems necessary to collect all of the ideal points into a single *ideal line*. With these considerations as the motivation, we will define the real projective plane in the following manner:

DEFINITION. *The Real Projective Plane.* Suppose that P is an ordinary point in the Euclidean plane. To each Euclidean line l that contains P, exactly one ideal point, denoted by I_l, will be appended. The set of all ideal points will be called the ideal line and will be denoted by L. The members of L have the properties that (i) if m is a line that does *not* contain P, and if $m \| l$, then $m \cap l = I_l$, and (ii) if m is a line not containing P, and if $m \cap l = I_l$, then $m \| l$ or $m = L$. The union $E \cup L$ consisting of all ordinary (i.e., Euclidean) points and all ideal points is called the real projective plane.

Note that in this definition the choice of P is arbitrary, since through every point in the Euclidean plane there is a line in every direction, so that every ideal point (and only those ideal points) appended using lines through

P would also be appended using lines through any other point P'. Verifying this aspect of the definition is included as an exercise.

With this definition in place, we begin the task of showing that the real projective plane, as defined above, does in fact exhibit the properties expected of it. We begin by examining some of the incidence relations mentioned above. The first result, which is helpful in proving the two theorems that follow it, establishes the fact that we have been successful in including just one ideal point per Euclidean line.

THEOREM 7.2.1.   In the real projective plane, every ordinary line x contains exactly one ideal point.

Proof. We must first show the existence of at least one ideal point on every line. First, if x is an ordinary line that contains point P, we have, by definition, appended an ideal point I_x to the line. If, on the other hand, x does *not* contain P, then the Euclidean parallel postulate (are we allowed to use it?) guarantees that there is a line l through P that is parallel to x. From this, it follows that $x \cap l = I_l$, so that the point I_l is on x. Together these arguments indicate that every ordinary line contains at least one ideal point.

Next we must show that no ordinary line contains more than one ideal point. If $P \in x$, we are assured by the definition that there is only one ideal point on x. So, assume that x is a line that does not contain P and that there are two ideal points I_l and I_m contained by x. From this, it follows that there are two lines through P, l, and m that are parallel to x, a conclusion that contradicts the Euclidean parallel postulate, so that there cannot be multiple ideal points on an ordinary line. From all of the above we conclude that each ordinary line contains *exactly* one ideal point.

Our next theorem guarantees that incidence in the real projective plane is consistent with SMSG Axiom 1.

THEOREM 7.2.2.   In the real projective plane, every two distinct points A and B determine a unique line.

Proof. We will examine the three possible cases that could occur.

1. A and B are both ordinary points: Under this hypothesis, the Euclidean axioms apply and a unique ordinary line contains the two points. The only other line that might contain A and B is the ideal line. However, the ideal line was constructed so that it contains no ordinary points, and so it cannot contain A and B. Therefore the line containing A and B is unique.

2. A and B are both ideal points: If A and B are both ideal points, then the ideal line L (and no other line) contains A and B, since the ideal line is unique and every ordinary line contains only one ideal point.

3. One point is ordinary and one is ideal: Without loss of generality, assume that $A = I_l$ and that B is ordinary. We know that A lies on just one line l that contains some arbitrary point P, and that there is a unique line m through B that is parallel to (or perhaps coincides with) l. Line m contains both A and B (why?) and is unique, since it is the only line through B that contains A (why?).

We see then that any choice of the two points A and B results in a unique line.

The next theorem results from the interchange of the words "point" and "line" in the previous one. As we shall see in the next section, the interchange of these terms, along with the relational terms involved in the statement, produces interesting (and valid) results in projective geometry.

THEOREM 7.2.3.   In the real projective plane every two distinct lines l and m determine a unique point.

The proof of this theorem can be approached in much the same way as the proof for Theorem 7.2.2. In fact, with an appropriate interchange of terms, the proofs follow an almost identical path. The details are left as an exercise.

Theorems 7.2.2 and 7.2.3 together show that there is no urgent need to distinguish between ordinary and ideal points in the real projective plane, since the incidence relations apply equally well to either type of point. Our axioms of incidence from Chapter 3 and 4 are valid, as theorems, for *all* points in the projective plane. It might seem therefore that we have succeeded in enlarging Euclidean geometry by including a new class of points (ideal points) that were previously missing. This idea, however, runs counter to the final axiom given by David Hilbert in *Grundlagen der Geometrie*:

AXIOM V-2.   *Axiom of Linear Completeness.* The system of points on a line with its order and congruence relations *cannot be extended* in such a way that the relations existing among its elements, as well as the basic properties of linear order and congruence resulting from Axioms I-III and V-1, remain valid.

An implication of this axiom is that by extending the Euclidean plane to the real projective plane we must, somewhere within the system, invalidate at least some of the Euclidean postulates stated by Hilbert. (These axioms were revised in the SMSG set, but this did not change their essential meaning.) This should make one wonder about where the predicted inconsistencies might appear. Which Euclidean properties will be preserved, and which will be invalidated?

Certainly, the Euclidean parallel postulate (Hilbert's Postulate IV-1) is no longer valid, since we have intentionally forced all pairs of lines to intersect. It is interesting to note that in Postulate V-2 Hilbert specifically excluded Postulate IV-1 from the list of postulates whose validity might be negated by an extension of the system.

This suggests that postulates other than the Euclidean parallel postulate are invalid in the real projective plane. Where will these problems surface? We begin with a look at betweenness.

Specifically, how do the ideal points, which are now full-fledged points of the plane, relate to ordinary points with respect to betweenness? Recall that the SMSG ruler postulate provided us with a mechanism that we could use to order points on a line. In particular, the ruler postulate guarantees a one-to-one correspondence between the points of ordinary lines and the real numbers. If this is so, what coordinate is given to point I_l on line l? It is tempting to think of the ideal point on l as the "point at infinity" so that its coordinate is ∞. But ∞ is not a number, and even if it were, there would still be a problem since, given a point A, there would be no point B such that A-∞-B, a violation of Hilbert's Axiom II-2.

From this it seems that ideal points cannot be endpoints of the line, so that perhaps we should place I_l between two ordinary points, say A and B. However, because of the one-to-one correspondence between the ordinary points of l and the real numbers, every point C between A and B has a unique coordinate. Placing I_l between A and B leaves some point C without a coordinate, and thus there is *no* one-to-one correspondence between the ordinary points of l and the real numbers. From this, we conclude that betweenness and all ideas derived from it (most notably congruence and measure[2]) have been sacrificed during the extension of the Euclidean plane to the real projective plane.

This eliminates, in the projective plane, a great many of the results developed in Chapter 4, since the axiom set for the real projective plane *does not include*

1. The ruler postulate
2. The ruler placement postulate
3. The protractor postulate
4. The SAS congruence postulate
5. The Euclidean parallel postulate, and
6. The SMSG postulates that concern area,

since all of these involve, either directly or indirectly, the notion of betweenness that has been lost in the real projective plane.

---

[2] If betweenness is lost, there is a major problem defining terms such as "line segment"—see Section 2.4—so that a unit of measure, that is, a line segment of unit length, cannot be established.

What then do we have to work with in this new geometry? We already
have three theorems (Theorems 7.2.1 through 7.2.3) involving incidence. In
addition, there are several other, rather weak, characteristics that have survived the "purge" discussed above. These will form the axioms for the real
projective plane.

Axioms for Real Projective Geometry

1. There exists at least one line.
2. Each line contains *at least* three points.[3]
3. Not all of the points are on the same line.
4. Two distinct points determine a unique line (Theorem 7.2.2).
5. Two distinct lines determine a unique point (Theorem 7.2.3).
6. There is a one-to-one correspondence between the real numbers and *all
   but one* of the points on each line (Clearly, this axiom overrides Axiom
   2. However, Axioms 1 through 5 define what is called a finite projective geometry; see Exercise Set 7.2, Problem 9 and 10.)

Very clearly, the axioms for projective geometry are heavily weighted
in favor of the incidence relations. Recall from Section 1.4 that projective
geometry is a special case of incidence geometry in which each line has at
least three points and no parallel lines exist. These properties are explicitly
included in the axioms for the real projective plane.

Some of the most elementary results from projective geometry, the
proofs for which have been included as exercises, are as follows:

THEOREM 7.2.4.   In the real projective plane there exist four distinct
points, no three of which are on the same line.

THEOREM 7.2.5.   In the real projective plane every point is contained
by at least three lines.

THEOREM 7.2.6.   In the real projective plane there exist four lines, no
three of which contain (or are concurrent with) the same point.

THEOREM 7.2.7.   If P is any point in the real projective plane, then
there exists a line that does not contain P.

---

[3] The motivation for including at least three points on each line is that in the Euclidean
plane each line has at least two points, to which a third point, the ideal point, was appended,
making three points on each projective line. There is, however, no longer a need to differentiate
between ordinary and ideal points since without the idea of betweenness there is no way to
distinguish the two.

THEOREM 7.2.8.   If P is any point in the real projective plane, and if l is any line in the real projective plane that does not contain P, then there is a one-to-one correspondence between the points on l and the lines that contain P.

THEOREM 7.2.9.   In the real projective plane the points of any line can be placed in one-to-one correspondence with the points of any other line.

EXERCISE SET 7.2

1. The definition of the real projective plane involves choosing an arbitrary point P in the Euclidean plane. One ideal point is appended to each line that contains P. The real projective plane consists of the ordinary Euclidean points along with all of the ideal points appended. Explain why any other choice of P, say P', would result in the same extension of the Euclidean plane, that is, explain why the choice of P is arbitrary.
2. Prove Theorem 7.2.3.
3. Prove Theorem 7.2.4.
4. Prove Theorem 7.2.5.
5. Prove Theorem 7.2.6.
6. Prove Theorem 7.2.7.
7. Prove Theorem 7.2.8.
8. Prove Theorem 7.2.9.

Problems 9 and 10 are the projective analogs of Problems 7 and 8 in Exercise Set 1.4.

9. A finite projective geometry is a geometry that uses only axioms 1 through 5 from the axiom set for real projective geometry. Prove that, in a finite projective geometry, if one line contains $n + 1$ points, then every line contains $(n + 1)$ points and every point is contained by $(n + 1)$ lines.
10. A finite projective plane is said to be of order n if one of its lines (and thus, using the result from Problem 9, all its lines) has $(n + 1)$ points. Prove that a finite projective plane of order n has, in all, $(n^2 + n + 1)$ points and the same number of lines.

In Problems 11 through 16 determine which of the geometries listed, each of which was discussed in Chapter 1, are finite projective geometries. Justify your choice for each.

11. The Fe-Fo geometry from Example 1.2.1.
12. The four-point geometry from Section 1.3.
13. The four-line geometry from Exercise Set 1.3, Problem 5.
14. Fano's geometry from Section 1.4.
15. Young's geometry from Section 1.4.
16. The incidence geometry derived from the incidence axioms in Section 1.4.

7.3 DUALITY

As mentioned in the previous section, Theorems 7.2.2 and 7.2.3 are closely related in that if, in the statement of each theorem, the primitive terms "point" and "line" are interchanged, the same two statements result, albeit in reversed order. For this reason, Theorems 7.2.2 and 7.2.3 are said to be *duals* of one another. Since both statements are theorems, we can say that Theorem 7.2.2 *and its dual* are both true in the projective plane.

Note, however, that in the Euclidean plane Theorem 7.2.3 is *not* a theorem. This is so because in Euclidean geometry parallel lines exist and certain choices of lines *l* and *m* would fail to determine a point. It is a curious consequence of the axioms for projective geometry that these dual statements are both valid. The main question to be investigated in this section concerns which other projective statements have duals that are valid. We begin by comparing each projective axiom to its dual.

Projective Axiom	Dual
1. There exists at least one *line*.	1'. There exists at least one *point*.
2. Each *line* contains at least three *points*.	2'. Each *point* is contained by at least three *lines*.
3. Not all of the *points* are on the same *line*.	3'. Not all of the *lines* contain the same *point*.
4. Two distinct *points* determine a unique *line*.	4'. Two distinct *lines* determine a unique *point*.
5. Two distinct *lines* determine a unique *point*.	5'. Two distinct *points* determine a unique *line*.
6. There is a one-to-one correspondence between the real numbers and all but one of the *points* on each *line*.	6'. There is a one-to-one correspondence between the real numbers and all but one of the *lines* containing each *point*.

In the dual of each axiom, the primitive terms "point" and "line" are exchanged *and* the incidence relations are restated accordingly. We know that duals 4' and 5' are true in the projective plane. We will now consider the other duals as logical consequences (i.e., theorems) of the axioms for the projective plane.

THEOREM 7.3.1.   *Dual of Projective Axiom 1.* There exists at least one point.

Proof. The validity of this statement follows immediately from projective Axioms 1 and 2.

THEOREM 7.3.2.   *Dual of Projective Axiom 2.* Each point is contained by at least three lines.

Proof. This statement can be proven by identifying three lines that must contain any point. The details are left as an exercise (see Exercise Set 7.3, Problem 1).

THEOREM 7.3.3.   *Dual of Projective Axiom 3.* Not all the lines contain the same point.

Proof. To prove this statement, we will choose an arbitrary point *P* and consider any two of the distinct lines that contain *P*. (We have at least three to choose from—why?). We will use *l* and *m* to denote these lines. We can find a point *Q* distinct from *P* on line *l*, and a point *R* distinct from *Q* and *P* on line *m*. (Why?) Consider the unique line *n* that contains *R* and *Q*. Line *n* cannot contain *P* since, otherwise, lines *l* and *n* would both contain *P* and *Q* so that lines *n* and *l* would be the same line. Similarly, if *P* is on *n*, *m* and *n* would also denote the same line. This would mean that *l* and *m* are not distinct lines, contrary to the hypothesis. We may conclude from this that *n* does not contain *P*. This means that no point is on all lines or, equivalently, not all the lines contain the same point.

THEOREM 7.3.4.   *Dual of Projective Axiom 4.* Two distinct lines determine a unique point.

Proof. This is Theorem 7.2.3.

THEOREM 7.3.5.   *Dual of Projective Axiom 5.* Two distinct points determine a unique line.

Proof. This is Theorem 7.2.2.

THEOREM 7.3.6.   *Dual of Projective Axiom 6.* There is a one-to-one correspondence between the real numbers and all but one of the lines containing each point.

Proof. Consider any point *P* in the real projective plane. There is at least one line *l* that does not contain *P*. (Why?) All but one of the points (we will call this "extra" point *Q*) of *l* can be placed in a one-to-one correspondence with the real numbers, and each of the points of this correspondence determines, with *P*, a unique line containing *P* that intersects *l*. From this, we may construct a one-to-one correspondence between the lines through *P*, except for line $\overrightarrow{PQ}$, and the real numbers. But line $\overrightarrow{PQ}$ has not been included in this pairing. Consequently, all but one of the lines through *P* are in a one-to-one correspondence with the real numbers, establishing the dual of projective Axiom 6.

The implications of establishing duals for the projective axioms are far-reaching. This means that each theorem that is subsequently proved has a dual statement that is also valid, and that the proof of the dual statement can be constructed by converting the terms in the proof of the theorem to their duals. In a sense then we have companion systems, each of which can be used to generate theorems that are true in the other. This can be advantageous for us, since now the proof of any statement from projective geometry can be approached in either of two ways (by proving the statement itself or by proving its dual). If one appears to be easier than the other, we can choose to proceed with the easier of the two.

Implicit in the notion of duality is the requirement that we translate all related terms used in the statement when phrasing the dual. Certainly, the exchange of the terms "point" and "line" does not present a problem, and even the translations concerning the incidence relations are generally easy to effect. However, there are other terms in projective geometry that need a measure of care, since they often have duals that are not familiar to us. To see this, consider a geometric figure we have discussed at length a number of times in previous chapters—the triangle.

To begin, we must note that our notion of the term "triangle" must be adjusted in projective geometry since triangles are generally thought of as a union of three line segments and, as mentioned earlier, the term "line segment" depends on the idea of betweenness, which has been lost in the projective plane. For this reason, in projective geometry a definition of the following sort is usually stated:

DEFINITION. A *triangle* is a figure consisting of three noncollinear points (called the vertices) and the three unique lines (called the sides) that the three points taken in pairs determine.

Note that the sides of a projective triangle are lines rather than line segments because of the considerations mentioned above. Note also that the triangle is defined, based on the points that constitute the vertices, rather than on the lines that comprise the sides. The dual of this definition, a figure called a trilateral, is obtained by interchanging the notions of point and line and by appropriately translating the incidence relations.

DEFINITION. A *trilateral* is a figure consisting of three nonconcurrent lines (called the sides) and the three distinct points (called the vertices) determined by the pairwise intersection of the sides.

Clearly, a triangle and a trilateral define the same figure, so that a triangle is a "self-dual" figure in projective geometry. This is not always the case, however, as we can see by advancing to projective figures with more than three points and/or lines.

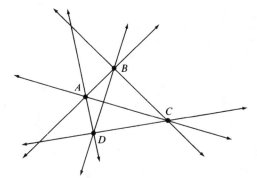

Figure 7.3.1

DEFINITION.   A *complete quadrangle*[4] is a figure consisting of four points (called the vertices) in the plane, no three being collinear, together with the six lines (called the sides) that result from joining the vertices pairwise (Figure 7.3.1).

The dual of this definition, stated as follows, results in a new projective figure that we shall call a complete quadrilateral.

DEFINITION.   A *complete quadrilateral*[5] is a figure consisting of four lines (called the sides), no three of which are concurrent, and the six points (called the vertices) that result from the pairwise intersection of the sides (Figure 7.3.2).

The quadrangle and the quadrilateral are, as duals, related figures, but they are *not* exactly the same. In particular, the complete quadrangle has four vertices, while the complete quadrilateral has six. There is a corre-

---

[4] Complete quadrangles contrast with *simple quadrangles,* which are figures consisting of four points (no three collinear) in a given order, such as A-B-C-D, and the four lines $\overline{AB}$, $\overline{BC}$, $\overline{CD}$, $\overline{DA}$.

[5] Simple quadrilaterals also exist and are defined so as to be duals of simple quadrangles. See Exercise Set 7.3, Problem 10.

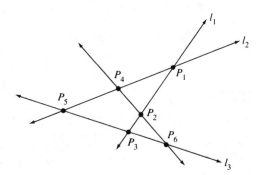

Figure 7.3.2

sponding contrast concerning the number of sides. The purpose of making this distinction is that now any theorem that we can prove concerning quadrangles will simultaneously result in a companion theorem concerning quadrilaterals, and conversely. This can be a useful tool to keep in mind as we proceed.

The comments concerning quadrangles and quadrilaterals can be generalized to sets of five (or more) points and/or lines, resulting in pentangles (or polyangles) and pentalaterals (or polylaterals).[6]

The distinctions between triangles and trilaterals, between quadrangles and quadrilaterals, and in general between polyangles and polylaterals should be kept clearly in mind as we proceed with our survey of the basic properties of the projective plane. These figures play important roles in the discussions of perspectivities and projectivities that are the main topics of the next section.

EXERCISE SET 7.3

1. Suppose that P is a point in a finite projective plane. In order to prove Theorem 7.3.2 we need to show that there are at least three lines that contain P. From Axiom 1 we know that there is at least one line containing P which we will call l_1. We also know that there is a point B that is not on l_1 (Axiom 3), so that there is a second line l_2 that contains P and B. Continue this process in order to prove Theorem 7.3.2.

2. State the dual of Theorem 7.2.4.   5. State the dual of Theorem 7.2.7.
3. State the dual of Theorem 7.2.5.   6. State the dual of Theorem 7.2.8.
4. State the dual of Theorem 7.2.6.   7. State the dual of Theorem 7.2.9.
8. Which of the dual statements given in Problems 2 through 7 are valid statements in projective geometry?
9. Which of the dual statements given in Problems 2 through 7 are valid statements in Euclidean geometry?
10. Propose a definition for the term "simple quadrilateral."
11. A hexangle is a polyangle with six vertices. How many sides does a hexangle have?
12. A hexalateral is a polylateral with six sides. How many vertices does a hexalateral have?
13. A polyangle with n vertices may be called an n-angle. How many sides does an n-angle have?
14. A polylateral with n sides may be called an n-lateral. How many vertices does an n-lateral have?

---

[6] Some authors prefer the term "complete n point in the projective plane" in place of "polyangle," and the term "complete n line in the projective plane" in place of "polylateral." This notation is particularly convenient when dealing in projective 3-space, but since we will generally remain in the projective plane the terms "polyangle" and "polylateral" will suffice.

7.4 PERSPECTIVITY

Our investigation of projective geometry began in Section 7.1 with a discussion of the origins of the subject, namely, artists' attempts to project three-dimensional scenes onto two-dimensional canvasses. It may have seemed in the last two sections that we have wandered far from that original direction. This is to some extent true. It is not unusual for mathematicians, while studying a topic, to drift away from the origins of the topic. It is exciting, however, when we find in seemingly unrelated places ideas and concepts that have a clear connection with the origins of the study. In this section we will examine such an idea, the notion of *perspectivity,* within the context of the axiomatic foundations of projective geometry that were posited in Sections 7.2 and 7.3.

So far in the discussion of projective geometry we have restricted ourselves to two-dimensional, or plane, projective space. This can naturally be generalized to three-dimensional projective spaces by appending to Euclidean 3-space an ideal plane comprised of ideal lines, one in the direction of each family of parallel Euclidean planes. We will forego the formalities of this extension[7] since the two-dimensional analog (namely, the real projective plane) is sufficiently similar to provide us with guidance through the relatively few three-dimensional cases we will encounter.

As a first example of a perspectivity, we will consider a rather simple projection—that of a triangle onto a "canvas" as shown in Figure 7.4.1.

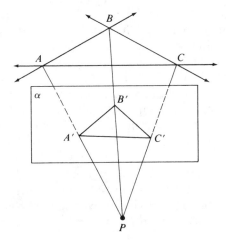

Figure 7.4.1

Plane α is our canvas, and $\triangle A'B'C'$ is the projection of $\triangle ABC$ onto α from the perspective of an eye positioned at point P. In a sense then,

[7] For a more formal discussion of projective space, see J. W. Young, *Projective Geometry* (Chicago: Open Court Publishing Co., 1930). pp. 75–78.

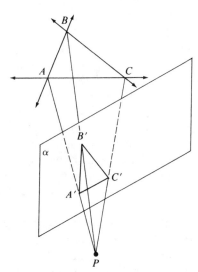

Figure 7.4.2

$\triangle A'B'C'$ is in perspective with $\triangle ABC$. In fact, in Figure 7.4.1 the position of plane α is such that it appears that the mapping relating $\triangle A'B'C'$ to $\triangle ABC$ is a homothety (or size transformation), so that the two triangles are similar in the sense of similarity defined in Chapter 4. Suppose, however, that plane α is positioned as shown in Figure 7.4.2.

In this figure there appears to be a distortion in the image of $\triangle ABC$. Of course, depending on the perspective of the artist painting this scene, Figure 7.4.2 might be a better portrayal of $\triangle ABC$ than Figure 7.4.1. In addition, the terms "similarity" and "position of plane α" both imply some type of measurement and therefore have no meaning in projective geometry, as discussed in previous sections. Thus a definition of what is meant by perspective in projective geometry must not involve measurements, but rather be dependent only on the incidence relations that comprise the axioms for the geometry. With these considerations in mind, we offer the following definition.

DEFINITION. *Perspectivity With Respect to a Point.* Two triangles ($\triangle ABC$ and $\triangle A'B'C'$) are said to be *perspective* with respect to a point P (called the point, or center, of perspectivity) providing the lines containing corresponding pairs of vertices ($\overrightarrow{AA'}$, $\overrightarrow{BB'}$, and $\overrightarrow{CC'}$) are concurrent at point P.

Very clearly, the pairs of triangles shown in Figures 7.4.1 and 7.4.2 are perspective with respect to point P in each figure. Other configurations of points and planes yield relationships that are not as obviously perspective. For example, in Figure 7.4.3a, $\triangle XYZ$ and $\triangle X'Y'Z'$ are perspective with

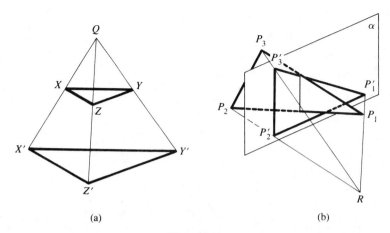

(a)   (b)

Figure 7.4.3

respect to point Q, while in Figure 7.4.3b, $\triangle P_1P_2P_3$ and $\triangle P_1'P_2'P_3'$ are perspective with respect to point R.

Another possible configuration for triangles that are perspective with respect to a point is shown in Figure 7.4.4. In this case $\triangle ABC$ and $\triangle A'B'C'$ are coplanar.

Triangles that are perspective with respect to a point are not as rare as one might expect. It is in fact quite easy to generate pairs of such triangles. One method of doing so proceeds in the following manner. To begin, choose a point P to serve as the center of perspectivity and draw three distinct lines (l_1, l_2, and l_3) that contain P (Figure 7.4.5a). Next arbitrarily place points A and A' on l_1, B and B' on l_2, and C and C' on line l_3 as shown in Figure 7.4.5b. Complete $\triangle ABC$ and $\triangle A'B'C'$ by sketching the sides (Figure 7.4.5c). It is easy to verify that these triangles are perspective with respect to point P. Sketch several more pairs of perspective triangles in this fashion using different configurations of points until you feel comfortable with the definition of perspectivity in this sense.

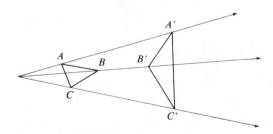

Figure 7.4.4

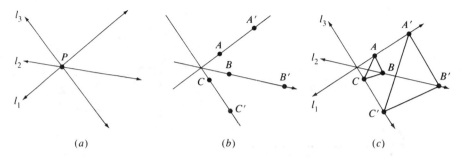

Figure 7.4.5

Next, since duality is a central theme of projective geometry, we will write the dual of the definition of perspectivity with respect to a point. (Before reading this definition you may wish to attempt to write one yourself.)

DEFINITION.   *Perspectivity With Respect to a Line.* Two trilaterals[8] ($\triangle ABC$ and $\triangle A'B'C'$) are said to be perspective with respect to a line l (called the line, or axis, of perspectivity) providing the points determined by the corresponding pairs of sides (i.e., points of intersection of the corresponding pairs of sides) are collinear on line l.

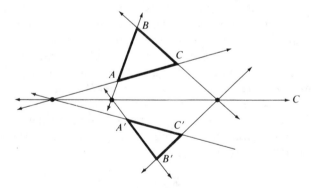

Figure 7.4.6

Figure 7.4.6 shows a pair of coplanar triangles $\triangle ABC$ and $\triangle A'B'C'$ that are perspective with respect to a line l.

Visualizing space triangles that are perspective with respect to a line is

---

[8] Technically we need to translate "triangle" to "trilateral" here because the terms are duals of one another. However, since the triangle and trilateral are self-dual, we can denote either term using the $\triangle$ notation.

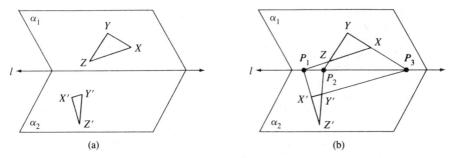

Figure 7.4.7

a bit more difficult than visualizing triangles that are perspective with respect to a point. However, doing so can provide an insight that is instructive and will be valuable as we proceed through this chapter.

To see this, consider $\triangle XYZ$ and $\triangle X'Y'Z'$ which are perspective with respect to line l in planes α_1 and α_2 as shown in Figure 7.4.7a.

Since l is the axis of perspectivity, the intersections of $\overline{XY}$, $\overline{XZ}$, and $\overline{YZ}$ with $\overline{X'Y'}$, $\overline{X'Z'}$, and $\overline{Y'Z'}$, respectively, must lie on line l. This means that the three points P_1, P_2, and P_3 are collinear on line l and contained by both planes α_1 and α_2. Therefore if planes α_1 and α_2 are distinct (which they clearly are here), then l is the (unique) line of intersection of the planes. This leads us to the following theorem.

THEOREM 7.4.1.   If $\triangle XYZ$ and $\triangle X'Y'Z'$ are noncoplanar projective triangles that are perspective with respect to a line l, then l is the intersection of the (distinct) planes that contain $\triangle XYZ$ and $\triangle X'Y'Z'$.

In the next section we will investigate the relationship between the two definitions of perspectivity. The following set of exercises may provide some hints concerning this relationship.

EXERCISE SET 7.4

1. Verify that $\triangle P_1P_2P_3$ and $\triangle P_1'P_2'P_3'$ are perspective with respect to point R in Figure 7.4.3b.
2. Verify that $\triangle ABC$ and $\triangle A'B'C'$ are perspective with respect to point P in Figure 7.4.5c.
3. In this section a method was described for drawing pairs of coplanar triangles that are perspective with respect to a point. Translate the statement of this method into its dual and use the resulting technique to draw a pair of coplanar triangles that are perspective with respect to a line.
4. Can two triangles in the Euclidean plane be perspective with respect to a point using the definition given in this section? Explain why or why not.

5. Can two triangles in the Euclidean plane be perspective with respect to a line using the definition given in this section? Explain why or why not.

6. Suppose that two Euclidean triangles are related by a translation. (See the definition of translation in Section 5.3.) Are these triangles perspective with respect to a Euclidean point? With respect to an ideal point? Explain.

7. Draw a pair of Euclidean triangles that are related by a homothety in which the constant of proportionality is -1. Are these triangles perspective with respect to a point? With respect to a line? Explain. Will your response change if the constant is -2? If it is $+2$? If it is an arbitrary constant k?

8. Consider $\triangle ABC$ and $\triangle A'B'C'$ shown in Figure 7.4.6. These triangles are by design perspective with respect to line l. Make a copy of this figure and determine if the triangles are perspective with respect to a point P.

9. Consider $\triangle ABC$ and $\triangle A'B'C'$ in Figure 7.4.4. These triangles are by design perspective with respect to point P. Make a copy of this figure and determine if they are perspective with respect to some line l.

10. Suppose that two triangles share a common side. Are these triangles perspective with respect to a line? If so, identify the axis of perspectivity. If not, explain why not.

11. Suppose that two triangles share a common side. Are the triangles perspective with respect to a point? If so, identify the center of the perspectivity. If not, explain why not.

12. Suppose that $\triangle ABC$ is a Euclidean triangle and that points P, Q, and R are the midpoints of the three sides (as shown in Figure 7.4.8). Are $\triangle ABC$ and $\triangle PQR$ perspective with respect to a line in the Euclidean plane? In the projective plane? Explain.

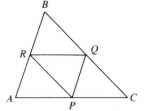

Figure 7.4.8

7.5 THE THEOREM OF DESARGUES

One of the earliest and most original contributors to the study of projective geometry was the French mathematician Gérard Desargues (1593–1660). In 1639 he published a paper entitled *Brouillon projet d'une atteinte aux événemens des rencontres d'un cone avec un plan*,[9] which was an early exposition on what are now called conic sections. Because other mathematicians, most notably René Descartes, dominated French mathematics during this period

---

[9] Roughly translated as *Draft of a Project Concerning the Incidence of a Cone With a Plane*.

and because Desargues' terminology was somewhat unorthodox, this treatise attracted little attention during Desargues' lifetime. In fact, it wasn't until 1845, when the French geometer Michel Chasles (1793–1880) happened across a copy of *Brouillon projet*, that this work took its rightful place in the history of projective geometry. Among Desargues' many contributions to the foundations of projective geometry is a theorem that today bears his name. The theorem relates the two aspects of projective perspectivity (namely, line perspectivity and point perspectivity) discussed in Section 7.4 and is the major topic of this section.

As suggested in Problems 8 and 9 in Exercise Set 7.4, triangles that are perspective with respect to a point (i.e., are point-perspective) may also be perspective with respect to a line (i.e., line-perspective). This is not surprising, given the discussion of duality that took place in Section 7.3. Desargues' theorem provides sure evidence that Problems 8 and 9 in Exercise Set 7.4 were not accidental and that in fact the two notions of perspectivity are equivalent in the sense that each implies the other.

The proof of Desargues' theorem, the formal statement of which will come later, requires that we first consider a pair of noncoplanar triangles that are point-perspective. It has been shown that the planar case cannot be proved otherwise. This means that we must once again deal with projective space (rather than in the projective plane), as we did in the proof of Theorem 7.4.1. Because of this it may be helpful, before proceeding further, to review SMSG Postulates 5 through 10, as they provide the structure for the incidence relationships in three dimensions.

THEOREM 7.5.1.   If two noncoplanar triangles are point-perspective, then they are also line-perspective.

Proof. Figure 7.5.1 shows two noncoplanar triangles $\triangle ABC$ and $\triangle A'B'C'$ that are perspective from point P. Points A, B, and C cannot be

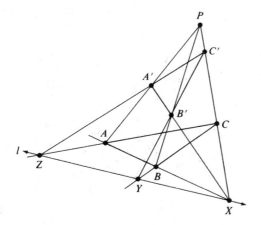

Figure 7.5.1

collinear (see the definition of a triangle in Section 7.3), and therefore these three points determine a unique plane that we will denote as α_1. Similarly, points A', B', and C' are contained in a unique plane α_2. Since all projective planes have a line of intersection, we can say that $\alpha_1 \cap \alpha_2 = l$ for some line l in the projective space. To prove the theorem we will demonstrate that $\triangle ABC$ and $\triangle A'B'C'$ are perspective with respect to line l. We will accomplish this by showing that the intersections $\overleftrightarrow{AB} \cap \overleftrightarrow{A'B'}$, $\overleftrightarrow{BC} \cap \overleftrightarrow{B'C'}$, and $\overleftrightarrow{AC} \cap \overleftrightarrow{A'C'}$ all occur on line l.

To see this, we first note that $\overleftrightarrow{AA'}$ and $\overleftrightarrow{BB'}$ are concurrent at point P. (Why?) Therefore the points A, A', B, B', and P are coplanar on a plane that we will call β_1 (a portion of β_1 can be seen as $\triangle APB$). Next note that $\overleftrightarrow{AB}$ and $\overleftrightarrow{A'B'}$ are coplanar (in β_1), so that there is a point X such that $\overleftrightarrow{AB} \cap \overleftrightarrow{A'B'} = X$.

Now consider the planes that contain X: (1) $X \in \beta_1$, by the preceding discussion; (2) $X \in \alpha_1$, since $X \in \overleftrightarrow{AB}$ and $\overleftrightarrow{AB}$ is contained in plane α_1; and (3) $X \in \alpha_2$, since $X \in \overleftrightarrow{A'B'}$ and $\overleftrightarrow{A'B'}$ is contained in plane α_2. Statements 2 and 3 indicate that $X \in \alpha_1 \cap \alpha_2$, and since $\alpha_1 \cap \alpha_2 = l$, we see that $X \in l$.

To complete the proof we need to show that the intersections $\overleftrightarrow{AC} \cap \overleftrightarrow{A'C'}$ (point Y in Figure 7.5.1) and $\overleftrightarrow{BC} \cap \overleftrightarrow{B'C'}$ (point Z) also lie on l. These arguments parallel the preceding one and have been left as exercises.

Next we consider the converse of Theorem 7.5.1.

THEOREM 7.5.2.   If two noncoplanar triangles are line-perspective, then they are also point-perspective.

Proof. Since this theorem is so closely related to the previous one, the proof will make use of several of the same lines and planes. This time, however, our hypotheses are that $\overleftrightarrow{AB} \cap \overleftrightarrow{A'B'} = X$, $\overleftrightarrow{BC} \cap \overleftrightarrow{B'C'} = Y$, and $\overleftrightarrow{AC} \cap \overleftrightarrow{A'C'} = Z$ (Figure 7.5.2). We will use these assumptions to show that $\overleftrightarrow{AA'}$, $\overleftrightarrow{BB'}$ and $\overleftrightarrow{CC'}$ share a common point P.

We begin by noting that the points A, B, A', B', and X are coplanar (why?) in a plane that we will call β_1. Therefore lines $\overleftrightarrow{AA'}$ and $\overleftrightarrow{BB'}$ are

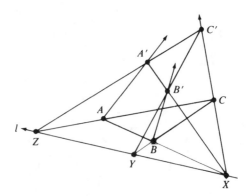

Figure 7.5.2

coplanar and, like all projective lines, are concurrent at a point we will call P_1.

In like fashion, we see that B, C, B', C', and Y are contained in a plane β_2, so that $\overleftrightarrow{BB'}$ and $\overleftrightarrow{CC'}$ share a common point P_2, and that A, C, A', C', and Y share a plane β_3, meaning that $\overleftrightarrow{AA'}$ and $\overleftrightarrow{CC'}$ intersect at some point P_3. We will show that P_1, P_2, and P_3 are really the same point.

First, it should be clear that β_1, β_2, and β_3 are distinct planes (see Exercise Set 7.5, Problem 3), so that there is a unique point P that is common to all three planes. This tells us that

$$\beta_1 \cap \beta_2 \cap \beta_3 = P \tag{1}$$

Next since $\overleftrightarrow{BB'}$ is on β_1 *and* β_2, we may say that

$$\beta_1 \cap \beta_2 = \overleftrightarrow{BB'} \tag{2}$$

Similarly, since $\overleftrightarrow{AA'}$ is on β_1 and β_3, we may say that

$$\beta_1 \cap \beta_3 = \overleftrightarrow{AA'} \tag{3}$$

Now, combining Equations (1) through (3) (and the fact that $\beta_1 \cap \beta_2 \cap \beta_3 = \beta_1 \cap \beta_2 \cap \beta_2 \cap \beta_3$), we may write

$$P = \beta_1 \cap \beta_2 \cap \beta_3 = \beta_1 \cap \beta_2 \cap \beta_2 \cap \beta_3$$
$$= (\beta_1 \cap \beta_2) \cap (\beta_2 \cap \beta_3) = BB' \cap AA' = P_1$$

By rearranging the terms of the intersection $\beta_1 \cap \beta_2 \cap \beta_3$ and making certain other minor adjustments, we may show also that $P = P_2$ and $P = P_3$. From this we can deduce that P is common to $\overleftrightarrow{AA'}$, $\overleftrightarrow{BB'}$, and $\overleftrightarrow{CC'}$ and conclude that $\triangle ABC$ and $\triangle A'B'C'$ are perspective from point P.

Theorems 7.5.1 and 7.5.2 establish Desargues' theorem for pairs of triangles that are *not* coplanar. In order to generalize this result to include pairs of coplanar triangles, we need several other preliminary results relating pairs of coplanar triangles to a third, noncoplanar, triangle. The first of these is given as the next theorem.

THEOREM 7.5.3.   If two coplanar triangles are point-perspective with the same noncoplanar triangle, then they are line-perspective with each other.

In Figure 7.5.3 $\triangle ABC$ and $\triangle A'B'C'$ are contained in plane α_1, while $\triangle QRS$ is contained in a different plane α_2. By hypothesis, $\triangle ABC$ and $\triangle QRS$ are point-perspective with respect to some point P, while $\triangle A'B'C'$ and $\triangle QRS$ are point-perspective with respect to some other point T.

With these hypotheses in mind, we may apply Theorem 7.5.1 and conclude that $\triangle ABC$ and $\triangle QRS$ are perspective with respect to some line l. Also, Theorem 7.4.1 indicates that $l = \alpha_1 \cap \alpha_2$. Using the same reasoning,

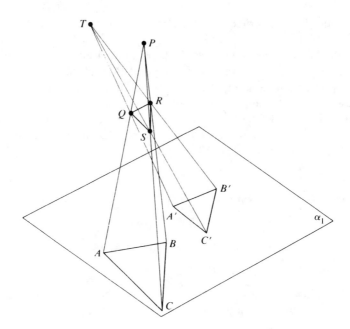

Figure 7.5.3

we may conclude that $\triangle A'B'C'$ and $\triangle QRS$ are perspective with respect to the same line l (see Exercise Set 7.5, Problem 6).

Consider the point X_1 on l where $\overrightarrow{AB}$ and $\overrightarrow{QR}$ meet, and the point X_2 on l where $\overrightarrow{A'B'}$ and $\overrightarrow{QR}$ meet. Since $\overrightarrow{QR}$ is not contained in α_1, we may conclude that $\overrightarrow{QR}$ intersects α_1 in a single point, so that $x_1 = x_2$. In like fashion, we may show that $\overrightarrow{AC}$, $\overrightarrow{A'C'}$, and $\overrightarrow{QS}$ are concurrent at some point Y, and that $\overrightarrow{BC}$, $\overrightarrow{B'C'}$, and $\overrightarrow{RS}$ are concurrent at some point Z such that Y and Z are both on l. Applying the definition of line perspectivity, we may say that $\triangle ABC$ and $\triangle A'B'C'$ are perspective with respect to line l.

The next theorem, which is closely related to the previous theorem, can be proved in a similar fashion. Its proof is included as an exercise.

THEOREM 7.5.4.   If two coplanar triangles are point-perspective with the same noncoplanar triangle, then they are point-perspective with each other.

One further result is needed in order to complete Desargues' theorem.

THEOREM 7.5.5.   If two coplanar triangles are point-perspective, then there is a noncoplanar triangle that is point-perspective with each.

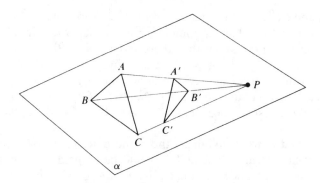

Figure 7.5.4

Proof. In Figure 7.5.4 we are given two triangles $\triangle ABC$ and $\triangle A'B'C'$ in plane α that are perspective with respect to point P. We wish to identify a noncoplanar triangle that is point-perspective with each.

To do this we will choose any line l that contains P but is not in plane α. We then choose any two points X and Y on l and draw $\overleftrightarrow{XC'}$ and $\overleftrightarrow{YC}$ as shown in Figure 7.5.5.

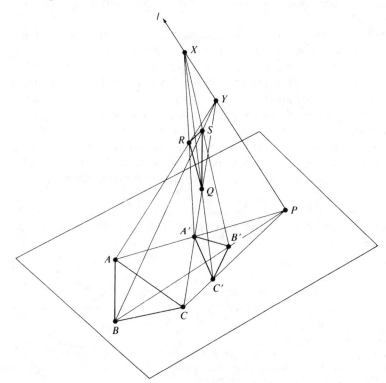

Figure 7.5.5

Since l and $\overrightarrow{CC'}$ are concurrent at point P, we know that $\overrightarrow{XC'}$ and $\overrightarrow{YC}$ are coplanar and therefore share a common point which we will call Q. In a similar manner, we may show that there exist points R and S such that $\overrightarrow{XA'} \cap \overrightarrow{YA} = R$ and $\overrightarrow{XB'} \cap \overrightarrow{YB} = S$. We claim three things: (1) that points Q, R, and S constitute a triangle (i.e., are noncollinear), (2) that $\triangle QRS$ is not contained in α, and (3) that each of $\triangle ABC$ and $\triangle A'B'C'$ is point-perspective with $\triangle QRS$. We consider these claims separately.

1. If Q, R, and S were collinear, the plane determined by the (unique, noncollinear) points Y, A, and B would be coincident with the plane determined by points Y, A, and C. Under this condition, $\triangle ABC$ would degenerate to AC, contrary to hypothesis. Therefore we may speak of $\triangle QRS$.
2. Consider point Q. If $Q \in \alpha$, then $\overrightarrow{YC}$ would be contained in α, since it would contain two points (Q and C) of α. This would imply that $Y \in \alpha$, contrary to the choice of Y. Therefore Q is not contained in α, from which it follows that $\triangle QRS$ isn't either.
3. Justifying that each of $\triangle ABC$ and $\triangle A'B'C'$ is point-perspective with $\triangle QRS$ is a relatively easy task and has been left as an exercise.

Summarizing the previous discussions, we see the close relationship between point perspectivity and line perspectivity. In particular, we have shown that noncoplanar triangles that are point-perspective are also line-perspective, and conversely (Theorems 7.5.1 and 7.5.2). Next Theorems 7.5.3 and 7.5.4 showed that coplanar triangles that are point-perspective with the same noncoplanar triangle are both point-perspective and line-perspective with each other. Finally, Theorem 7.5.5 established that *if* two coplanar triangles are point-perspective, then there exists a noncoplanar triangle with which they are both point-perspective. Combining these theorems, we arrive at the highly significant theorem of Desargues.

THEOREM 7.5.6.   *The Theorem of Desargues.* Two projective triangles are perspective with respect to a point if and only if they are perspective with respect to a line.

EXERCISE SET 7.5

1. Show that in Figure 7.5.1
   (a) Lines $\overrightarrow{AC}$ and $\overrightarrow{A'C'}$ must have a point of intersection
   (b) Point Z (which is the intersection of lines $\overrightarrow{AC}$ and $\overrightarrow{A'C'}$) is contained in line l (the line where α_1 and α_2 intersect).

2. Show that, in Figure 7.5.1,
   (a) Lines $\overleftrightarrow{BC}$ and $\overleftrightarrow{B'C'}$ have a point of intersection
   (b) Point Y (which is the intersection of lines $\overleftrightarrow{BC}$ and $\overleftrightarrow{B'C'}$) is contained in line l
   (the line where α_1 and α_2 intersect).
3. Explain why, in Figure 7.5.1, the plane that contains A, B, and P must be distinct
   from both the plane that contains B, C, and P *and* the plane that contains A, C,
   and P. (*Hint:* Assume otherwise and show that the triangles are then coplanar,
   contrary to hypothesis.)
4. Explain why, for any three planes β_1, β_2, and β_3, we may say

$$\beta_1 \cap \beta_2 \cap \beta_3 = \beta_1 \cap \beta_2 \cap \beta_2 \cap \beta_3$$

   (This equality was used in the proof of Theorem 7.5.2.)
5. In the proof of Theorem 7.5.2 we made the statement

$$\beta_1 \cap \beta_2 \cap \beta_2 \cap \beta_3 = (\beta_1 \cap \beta_2) \cap (\beta_2 \cap \beta_3)$$

   What property of set operations justifies this statement?
6. Suppose that $\triangle ABC$ and $\triangle A'B'C'$ are coplanar and that $\triangle QRS$ is contained in a
   different plane. Suppose also that $\triangle ABC$ and $\triangle A'B'C'$ are both line-perspective
   with respect to $\triangle QRS$. Explain why the axis must be the same for both perspectivities.
7. Prove Theorem 7.5.4.
8. Show that, in Figure 7.5.5, $\triangle ABC$ and $\triangle A'B'C'$ are both point-perspective with
   $\triangle QRS$. (*Hint:* Consider points X and Y candidates for the centers of the perspectivities.)

7.6 PROJECTIVE TRANSFORMATIONS

One of the major topics of Chapter 5 was transformational geometry. In that
chapter you studied ways in which the Euclidean plane could be mapped in a
one-to-one fashion onto itself. Two of the major classifications of transformations studied were isometries and homotheties (size transformations).
These transformations were particularly useful in Euclidean geometry because each preserves, among other things, betweenness.

Since the properties of betweenness were lost when the Euclidean
plane was extended to the real projective plane, the transformations discussed in Chapter 5 cannot be applied within projective geometry. Instead,
mappings within the projective plane will have to be defined in terms that are
meaningful within projective geometry. Since projective geometry is based
heavily on the notion of incidence of points and lines, it makes sense that
mappings relating points to other points should be based on incidence relations.

Suppose, for example, we wish to pair in a one-to-one manner the
points on line l with the points on line m in Figure 7.6.1.

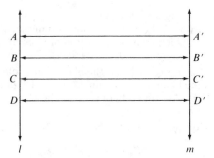

Figure 7.6.1

The mapping shown in the figure pairs the points of *l* with the points of *m* using the arrows (or vectors) connecting the corresponding pairs of points. Although this mapping does not explicitly mention betweenness, it does involve vectors, and vectors depend on direction and distance, two ideas that depend very much on betweenness.

An alternative method of pairing the points on *l* with the points on *m* is illustrated in Figure 7.6.2.

In this mapping points on *l* are paired with points on *m* by means of their common incidence with a line through *P*. Since every point on *l* shares a common line with *P* (why?), and since each of these lines must cross *m* at a unique point (why?), we see that each point on *l* is paired with exactly one point on *m*. Similarly, we see that each point on *m* is paired with exactly one point on *l*. Together these statements imply that the mapping shown is one-to-one. We formalize this with the following definition.

DEFINITION.   *Central Perspectivity.* Suppose that *l* and *m* are lines in the real projective plane and that *P* is a point that is not on either *l* or *m*. We define the mapping $T_p: l \rightarrow m$ in the following way: If $A \in l$ and $B \in m$, then $T_p(A) = B$ if and only if $B \in \overrightarrow{PA}$. The mapping T_p defined in this way is called a central perspectivity with center *P*.

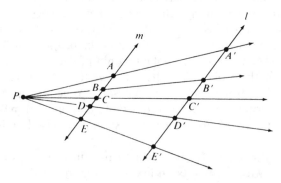

Figure 7.6.2

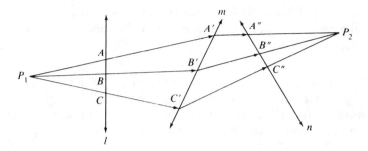

Figure 7.6.3

The definition of central perspectivity ensures that the mapping is one-to-one (see Exercise Set 7.6, Problems 1 and 2), so that T_p is a transformation relating the points of line l to the points of line m. In addition, since T_p is one-to-one, its inverse T_p^{-1} exists and maps the points of line m to the points of line l in such a way that $T_p^{-1}(T_p(A)) = A$.

In Chapter 5 we saw that transformations in the Euclidean plane could be composed to define new transformations. For example, a translation of the plane could be followed by a reflection of the plane to produce a new classification of transformations called glide reflections. Additionally, we saw that the composition of two reflections could be equivalent to either a translation or a rotation, depending on whether the lines of reflection were parallel or not.

Central perspectivities can be composed also. For example, Figure 7.6.3 shows the composition of $T_{p_1}: l \rightarrow m$ and $T_{p_2}: m \rightarrow n$.

Figure 7.6.4 shows the relationship among A, B, and C on line l and their "second images" A'', B'', and C'' on line n.

The natural question that now arises is, Is there a single central perspectivity that maps A, B, and C to A'', B'', and C'', respectively? The answer is no. As indicated by the dotted lines, there is a point X where $\overleftrightarrow{AA''}$ and $\overleftrightarrow{BB''}$ meet, so that $T_X(A) = A''$ and $T_X(B) = B''$. However, it is clear that $T_X(C) \neq T_X(C'')$, so that no single central perspectivity is equivalent to the composi-

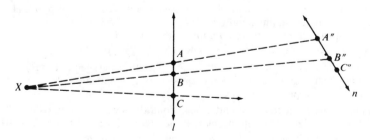

Figure 7.6.4

tion of T_{p_1} nd T_{p_2}. We should like, however, to speak of a single transformation, or mapping, that relates A, B, and C to A'', B'', and C''. This motivates the following definition.

DEFINITION.   *Projective Transformation.* A projective transformation (or *projectivity*) is a transformation of the points of one projective line to the points of another projective line, which may be expressed as the composition of a finite number of central perspectivities.

For example, in Figure 7.6.4, points A, B, and C are related to points A'', B'', and C'', respectively, by a projectivity composed of two central projections T_{p_1} and T_{p_2}. If we introduced a finite number of other central projectivities T_{p_3}, T_{p_4}, . . . , T_{p_n}, the composition would still be a projectivity.

A complete development of the important theorems relating to projective transformations is beyond the scope of this brief chapter. However, a few of the most important results appear below without proof. Listed in the bibliography are several excellent references that can provide the details for those interested in this subject.

THEOREM 7.6.1.   A projectivity relating the points of two distinct lines can be expressed as the product of two (or fewer) central projectivities.

THEOREM 7.6.2.   If A, B, and C are points on line l, and A', B', and C' are three points on a second line m, then there is exactly one distinct projectivity that maps A to A', B to B', and C to C'. (Theorem 7.6.2 is known as the fundamental theorem of projective geometry.)

THEOREM 7.6.3.   If P is a projectivity between distinct lines l and m, and if $P(A) = A$ for some point A, then P is a central perspectivity.

EXERCISE SET 7.6

1. Suppose that $T: l \to m$ is a central perspectivity centered at P. Suppose also that A is a point on line l. What projective postulate guarantees that $T(A)$ is unique?
2. Suppose that $T: l \to m$ is a central perspectivity centered at P. Suppose also that points A and B are on line l and that $T(A) = T(B) = C$, where C is a point on line m. What projective postulate guarantees that $A = B$?
3. Write a definition for $T_p^{-1}: B \to A$ such that $T_p^{-1}(T_p(A)) = A$.
4. Suppose that P is a projectivity composed of central perspectivities T_P, T_Q, and T_R. Does P have an inverse? If so, express it as a composition of the inverses of the three central perspectivities.
5. If P_1 and P_2 are projectivities, under what conditions will the composition $P_1 \circ P_2$ be defined? Under what conditions will the composition fail to be defined?
6. Reread the definition of a homothety transformation from Chapter 5. Explain the difference between a homothety and a central perspectivity.

CHAPTER 7 SUMMARY

7.2 The Real Projective Plane

The Real Projective Plane: Suppose that P is an ordinary point in the Euclidean plane. To each Euclidean line l that contains P exactly one ideal point, denoted by I_1, will be appended. The set of all ideal points will be called the ideal line and will be denoted by L. The members of L have the properties that (i) if m is a line that does *not* contain P, and if $m \parallel l$, then $m \cap l = I_1$, and (ii) if m is a line not containing P, and if $m \cap l = I_1$, then $m \parallel l$ or $m = L$. The union $E \cup L$ consisting of all ordinary (i.e., Euclidean) points and all ideal points is called the real projective plane.

Theorem 7.2.1.   In the real projective plane every ordinary line x contains exactly one ideal point.

Theorem 7.2.2.   In the real projective plane every two distinct points A and B determine a unique line.

Theorem 7.2.3.   In the real projective plane every two distinct lines l and m determine a unique point.

Theorem 7.2.4.   In the real projective plane there exist four distinct points, no three of which are on the same line.

Theorem 7.2.5.   In the real projective plane every point is contained by at least three lines.

Theorem 7.2.6.   In the real projective plane there exist four lines, no three of which contain (or are concurrent with) the same point.

Theorem 7.2.7.   If P is any point in the real projective plane, then there exists a line that does not contain P.

Theorem 7.2.8.   If P is any point in the real projective plane, and if l is any line in the real projective plane that does not contain P, then there is a one-to-one correspondence between the points on l and the lines that contain P.

Theorem 7.2.9.   In the real projective plane the points of any line can be placed in one-to-one correspondence with the points of any other line.

7.3 Duality

Triangle.   A figure consisting of three noncollinear points (called the vertices) and the three unique lines (called the sides) that the three points taken in pairs determine.

Trilateral.   A figure consisting of three nonconcurrent lines (called the sides) and the three distinct points (called the vertices) determined by the pairwise intersection of the sides.

Complete Quadrangle. A figure consisting of four points (called the vertices) in the plane, no three being collinear, together with the six lines (called the sides) that result from joining the vertices pairwise. (See Figure 7.3.1.)

Complete Quadrilateral. A figure consisting of four lines (called the sides), no three of which are concurrent, and the six points (called the vertices) that result from the pairwise intersection of the sides. (See Figure 7.3.2.)

Theorem 7.3.1.   *Dual of Projective Axiom 1:* There exists at least one point.

Theorem 7.3.2.   *Dual of Projective Axiom 2:* Each point is contained by at least three lines.

Theorem 7.3.3.   *Dual of Projective Axiom 3:* Not all of the lines contain the same point.

Theorem 7.3.4.   *Dual of Projective Axiom 4:* Two distinct lines determine a unique point.

Theorem 7.3.5.   *Dual of Projective Axiom 5:* Two distinct points determine a unique line.

Theorem 7.3.6.   *Dual of Projective Axiom 6:* There is a one-to-one correspondence between the real numbers and all but one of the lines containing each point.

7.4 Perspectivity

Perspectivity With Respect to a Point: Two triangles ($\triangle ABC$ and $\triangle A'B'C'$) are said to be perspective with respect to a point P (called the point, or center, of perspectivity) providing the lines containing corresponding pairs of vertices ($\overrightarrow{AA'}$, $\overrightarrow{BB'}$, and $\overrightarrow{CC'}$) are concurrent at point P.

Perspectivity With Respect to a Line: Two trilaterals ($\triangle ABC$ and $\triangle A'B'C'$) are said to be perspective with respect to a line l (called the line, or axis, of perspectivity) providing the points determined by the corresponding pairs of sides (i.e., points of intersection of the corresponding pairs of sides) are collinear on line l.

Theorem 7.4.1.   If $\triangle XYZ$ and $\triangle X'Y'Z'$ are noncoplanar projective triangles that are perspective with respect to a line l, then l is the intersection of the (distinct) planes that contain $\triangle XYZ$ and $\triangle X'Y'Z'$.

7.5 The Theorem of Desargues

Theorem 7.5.1.   If two noncoplanar triangles are point-perspective, then they are also line-perspective.

Theorem 7.5.2.   If two noncoplanar triangles are line-perspective, then they are also point-perspective.

Theorem 7.5.3.   If two coplanar triangles are point-perspective with the same noncoplanar triangle, then they are line-perspective with each other.

Theorem 7.5.4.   If two coplanar triangles are point-perspective with the same noncoplanar triangle, then they are point-perspective with each other.

Theorem 7.5.5.   If two coplanar triangles are point-perspective, then there is a noncoplanar triangle that is point-perspective with each.

Theorem 7.5.6.   *The Theorem of Desargues:* Two projective triangles are perspective with respect to a point if and only if they are perspective with respect to a line.

7.6 Projective Transformations

Central Perspectivity: Suppose that l and m are lines in the real projective plane and that P is a point that is not on either l or m. We define the mapping $T_p: l \rightarrow m$ in the following way: If $A \in l$ and $B \in m$, then $T_p(A) = B$ if and only if $B \in \overrightarrow{PA}$. The mapping T_p defined in this way is called a central perspectivity with center P.

Projective Transformation: (**or** *projectivity*)   A transformation of the points of one projective line to the points of another projective line which may be expressed as the composition of a finite number of central perspectivities.

Theorem 7.6.1.   A projectivity relating the points of two distinct lines can be expressed as the product of two (or fewer) central projectivities.

Theorem 7.6.2.   If A, B, and C are points of line l, and A', B', and C' are three points on a second line m, then there is exactly one distinct projectivity that maps A to A', B to B', and C to C'. (Theorem 7.6.2 is known as the fundamental theorem of projective geometry.)

Theorem 7.6.3.   If P is a projectivity between distinct lines l and m, and $P(A) = A$, then P is a central perspectivity.

APPENDIX A

Euclid's Definitions and Postulates Book I[1]

DEFINITIONS

1. A point is that which has no part.
2. A line is breadthless length.
3. The extremities of a line are points.
4. A straight line is a line which lies evenly with the points on itself.
5. A surface is that which has length and breadth only.
6. The extremities of a surface are lines.
7. A plane surface is a surface which lies evenly with the straight lines on itself.
8. A plane angle is the inclination to one another of two lines in a plane which meet one another and do not lie in a straight line.
9. And when the lines containing the angle are straight, the angle is called rectilineal.
10. When a straight line set up on a straight line makes the adjacent angles equal to one another, each of the angles is right, and the straight line standing on the other is called perpendicular to that on which it stands.
11. An obtuse angle is an angle greater than a right angle.
12. An acute angle is an angle less than a right angle.

[1] Reprinted with permission from Thomas L. Heath, *The Thirteen Books of Euclid's Elements* (New York: Dover Publications, Inc., 1956), pp. 153–315.

13. A boundary is that which is an extremity of anything.

14. A figure is that which is contained by any boundary or boundaries.

15. A circle is a plane figure contained by one line such that all the straight lines falling upon it from one point among those lying within the figure are equal to one another.

16. And the point is called the center of the figure.

17. A diameter of the circle is any straight line drawn through the center and terminated in both directions by the circumference of the circle, and such a straight line bisects the circle.

18. A semicircle is the figure contained by the diameter and the circumference cut off by it. And the center of the semicircle is the same as that of the circle.

19. Rectilineal figures are those that are contained by straight lines, trilateral figures being those contained by three, quadrilateral those contained by four, and multilateral those contained by more than four.

20. Of trilateral figures, an equilateral triangle is that which has three sides equal, an isosceles triangle that which has two of its sides alone equal, and a scalene triangle that which has its three sides unequal.

21. Further, of trilateral figures, a right-angled triangle is that which has a right angle, an obtuse triangle one which has an obtuse angle, and an acute-angled triangle that which has its three angles acute.

22. Of quadrilateral figures, a square is that which is both equilateral and right-angled; an oblong that which is right-angled but not equilateral; a rhombus that which is equilateral but not right-angled; and a rhomboid that which has its opposite sides and angles equal to one another but is neither equilateral nor right-angled. And let quadrilaterals other than these be called trapezia.

23. Parallel straight lines are straight lines which, being in the same plane and being produced indefinitely in both directions, do not meet one another in either direction.

THE POSTULATES

1. To draw a straight line from any point to any point.

2. To produce a finite straight line continuously in a straight line.

3. To describe a circle with any center and radius.

4. That all right angles are equal to one another.

5. That, if a straight line falling on two straight lines makes the interior angles on the same side less than two right angles, the straight lines, if produced indefinitely, meet on that side on which are the angles less than the two right angles.

THE COMMON NOTIONS

1. Things which are equal to the same things are equal to one another.
2. If equals be added to equals, the wholes are equal.
3. If equals be subtracted from equals, the remainders are equal.
4. Things which coincide with one another are equal to one another.
5. The whole is greater than the part.

THE FIRST TEN PROPOSITIONS OF BOOK I

1. On a given finite straight line, to construct an equilateral triangle.
2. To place at a given point (as an extremity) a straight line equal to a given straight line.
3. Given two unequal straight lines, to cut off from the greater a straight line equal to the lesser.
4. If two triangles have two sides equal to two sides respectively, and have the angles contained by the equal straight lines equal, they will also have the base equal to the base, the triangle will be equal to the triangle, and the remaining angles will be equal to the remaining angles respectively, namely, those which the equal sides subtend.
5. In isosceles triangles, the angles at the base are equal to one another and, if the equal straight lines be produced further, the angles under the base will be equal to one another.
6. If in a triangle two angles be equal to one another, the sides which subtend the equal angles will also be equal to one another.
7. Given two straight lines constructed on a straight line (from its extremities) and meeting at a point, there cannot be constructed on the same straight line (from its extremities) and on the same side of it, two other straight lines meeting in another point and equal to the former two respectively, namely, each to that which has the same extremity with it.
8. If two triangles have the two sides equal to two sides respectively, and have also the base equal to the base, they will also have the angles equal which are contained by the equal straight lines.
9. To bisect a rectilineal angle.
10. To bisect a given finite straight line.

APPENDIX B

David Hilbert's Axiom Set
For
Euclidean Plane Geometry[1]

UNDEFINED TERMS

1. Point
2. Line
3. Plane
4. Lie (incidence of point and line)
5. Between
6. Congruence

The precise and mathematically complete description of these relations follows from the *axioms of geometry*.

GROUP I: AXIOMS OF INCIDENCE (CONNECTION)

I-1. For every two points A, B, there exists a line that contains each of the points A, B.

I-2. For every two points A, B, there is no more than one line that contains each of the points A, B.

I-3. There exist at least two points on a line. There exist at least three points which do not lie on a line.

[1] Modified from D. Hilbert, *Foundations of Geometry*, Trans. L. Unger, 1971. (By permission of The Open Court Publishing Co, Chicago.)

I-4. For any three points *A*, *B*, *C* that do not lie on the same line, there exists a plane α that contains each of the points *A*, *B*, *C*. For every plane, there exists a point which it contains.

GROUP II: AXIOMS OF ORDER

II-1. If point *B* is between points *A* and *C*, then *A*, *B*, and *C* are distinct points on the same line and *B* is between *C* and *A*.

II-2. For any two distinct points *A* and *C* there is at least one point *B* on the line $\overleftrightarrow{AC}$ such that *C* is between *A* and *B*.

II-3. If *A*, *B*, and *C* are three points on the same line, then no more than one is between the other two.

II-4. Let *A*, *B*, and *C* be three points that are not on the same line and let *l* be a line in the plane containing *A*, *B*, and *C* that does not meet any of the points *A*, *B*, or *C*. Then, if *l* passes through a point of the segment $\overline{AB}$, it will also pass through a point of segment $\overline{AC}$ or a point of segment $\overline{BC}$.

GROUP III: AXIOMS OF CONGRUENCE

III-1. If *A* and *B* are two points on a line *a*, and if *A'* is a point on the same or on another line *a'*, then it is always possible to find a point *B'* on a given side of the line *a'* such that *AB* and *A'B'* are congruent.

III-2. If a segment $\overline{A'B'}$ and a segment $\overline{A''B''}$ are congruent to the same segment $\overline{AB}$, then segments $\overline{A'B'}$ and $\overline{A''B''}$ are congruent to each other.

III-3. On a line *a*, let $\overline{AB}$ and $\overline{BC}$ be two segments which, except for *B*, have no point in common. Furthermore, on the same or another line *a'*, let $\overline{A'B'}$ and $\overline{B'C'}$ be two segments which, except for B', have no points in common. In that case if $\overline{AB} \approx \overline{A'B'}$ and $\overline{BC} \approx \overline{B'C'}$, then $\overline{AC} \approx \overline{A'C'}$.

III-4. If $\angle ABC$ is an angle and if $\overrightarrow{B'C'}$ is a ray, then there is exactly one ray $B'A'$ on each "side" of $\overrightarrow{B'C'}$ such that $\angle A'B'C' \cong \angle ABC$. Furthermore, every angle is congruent to itself.

III-5. If for two triangles *ABC* and *A'B'C'* the congruences $\overline{AB} \approx \overline{A'B'}$, $\overline{AC} \approx \overline{A'C'}$, and $\angle BAC \approx \angle B'A'C'$ are valid, then the congruence $\angle ABC \approx \angle A'B'C'$ is also satisfied.

GROUP IV: AXIOM OF PARALLELS

IV-1. Let a be any line and A a point not on it. Then there is at most one line in the plane that contains a and A that passes through A and does not intersect a.

GROUP V: AXIOMS OF CONTINUITY

V-1. *Archimedes Axiom:* If $\overline{AB}$ and $\overline{CD}$ are any segments, then there exists a number n such that n copies of $\overline{CD}$ constructed contiguously from A along the ray $\overrightarrow{AB}$ will pass beyond the point B.

V-2. *Axiom of Line Completeness:* An extension of a set of points on a line with its order and congruence relations that would preserve the relations existing among the original elements as well as the fundamental properties of line order and congruence that follow from Axioms I-III and V-1 is impossible.

APPENDIX C

Birkhoff's Postulates for Euclidean Plane Geometry[1]

UNDEFINED TERMS AND RELATIONS

1. *Points*
2. Sets of points called *lines*
3. *Distance* between any two points A and B: a non-negative real number, $d(A,B)$, such that $d(A,B) = d(B,A)$
4. *Angle* formed by three ordered points A, O, B ($A \neq O$, $B \neq O$: $\angle AOB$ such that m$\angle AOB$ is real number (mod 2π)

DEFINITIONS

1. A point B is *between* A and C ($A \neq C$) if $d(A,B) + d(B,C) = d(A,C)$.
2. The points A and C together with all points B between A and C form *line segment AC*.
3. The *half-line m'* with *endpoint O* is defined by two points O, A in line $m(A \neq O)$ as the set of all points A' of m such that O is not between A and A'.
4. If two distinct lines have no points in common they are *parallel*. A line is always regarded as parallel to itself.

[1] Reprinted from G. D. Birkhoff, "A Set of Postulates for Plane Geometry (Based on Scale and Protractor)," *Annals of Mathematics*, 33, 1932, by permission of the *Annals of Mathematics*.

5. Two half-lines m, n through O are said to form a *straight angle* if $m\angle mOn = \pi$. Two half-lines m, n through O are said to form a *right angle* if $m\angle mOn = (+/-)\pi/2$, in which case we also say that m is *perpendicular* to n.

6. If A, B, C are three distinct points the three segments AB, BC, CA are said to form a *triangle ABC* with sides AB, BC, CA and *vertices* A, B, C. If A, B, and C are collinear then $\triangle ABC$ is said to be *degenerate*.

7. Any two geometric figures are *similar* if there exists a one-to-one correspondence between the points of the two figures such that all corresponding distances are in proportion and corresponding angles have equal measures (except, perhaps, for their sign). Any two geometric figures are *congruent* if they are similar with a constant of proportionality, $k = 1$.

POSTULATES

POSTULATE I: *Postulate of Line Measure.* The points $A, B, \ldots$, of any line m can be placed into a one-to-one correspondence with the real numbers r so that $|r_B - r_A| = d(AB)$ for all points A and B.

POSTULATE II: *Point-line Postulate.* One and only one line m contains two given points P and Q $(P \neq Q)$.

POSTULATE III: *Postulate of Angle Measure.* The halflines m, n, $\ldots$, through any point O can be placed into a one-to-one correspondence with real numbers $a \pmod{2\pi}$ so that if $A \neq O$ and $B \neq O$ are points of m and n respectively, the difference $(a_n - a_m) \pmod{2\pi}$ is $m\angle AOB$.

POSTULATE IV: *Postulate of Similarity.* If $\triangle ABC$ and $\triangle A'B'C'$, and for some positive constant, k, $d(A',B') = kd(A,B)$, $d(A',C') = kd(A,C)$, and also $m\angle BAC = (\pm)m\angle B'A'C'$, then also $d(B',C') = kd(B,C)$ and $m\angle C'B'A' = (\pm)m\angle CBA$ and $m\angle A'C'B' = (\pm)m\angle ABC$.

APPENDIX D

The SMSG Postulates for Euclidean Geometry[1]

UNDEFINED TERMS

1. *Point*
2. *Line*
3. *Plane*

POSTULATES

POSTULATE 1. Given any two distinct points there is exactly one line that contains them.

POSTULATE 2. *The Distance Postulate.* To every pair of distinct points there corresponds a unique positive number. This number is called the distance between the two points.

POSTULATE 3. *The Ruler Postulate.* The points of a line can be placed in a correspondence with the real numbers such that

(1) To every point of the line there corresponds exactly one real number,
(2) To every real number there corresponds exactly one point of the line, and

[1] Reprinted from School Mathematics Study Group, *Geometry*, by permission of Yale University Press, New Haven.

(3) The distance between two distinct points is the absolute value of the difference of the corresponding real numbers.

POSTULATE 4. *The Ruler Placement Postulate.* Given two points P and Q of a line, the coordinate system can be chosen in such a way that the coordinate of P is zero and the coordinate of Q is positive.

POSTULATE 5. (a) Every plane contains at least three non-collinear points. (b) Space contains at least four non-coplanar points.

POSTULATE 6. If two points lie in a plane, then the line containing these points lies in the same plane.

POSTULATE 7. Any three points lie in at least one plane, and any three non-collinear points lie in exactly one plane.

POSTULATE 8. If two planes intersect, then that intersection is a line.

POSTULATE 9. *The Plane Separation Postulate.* Given a line and a plane containing it, the points of the plane that do not lie on the line form two sets such that

(1) each of the sets is convex and
(2) if P is in one set and Q is in the other, then segment PQ intersects the line.

POSTULATE 10. *The Space Separation Postulate.* The points of space that do not lie in a given plane form two sets such that

(1) Each of the sets is convex, and
(2) If P is in one set and Q is in the other, then segment PQ intersects the plane.

POSTULATE 11. *The Angle Measurement Postulate.* To every angle there corresponds a real number between $0°$ and $180°$.

POSTULATE 12. *The Angle Construction Postulate.* Let AB be a ray on the edge of the half-plane H. For every r between 0 and 180 there is exactly one ray AP, with P in H such that $m\angle PAB = r$.

POSTULATE 13. *The Angle Addition Postulate.* If D is a point in the interior of $\angle BAC$, then $m\angle BAC = m\angle BAD + m\angle DAC$.

POSTULATE 14. *The Supplement Postulate.* If two angles form a linear pair, then they are supplementary.

POSTULATE 15. *The SAS Postulate.* Given a one-to-one correspondence between two triangles (or between a triangle and itself). If two sides and the included angle of the first triangle are congruent to the corresponding parts of the second triangle, then the correspondence is a congruence.

POSTULATE 16. *The Parallel Postulate.* Through a given external point there is at most one line parallel to a given line.

POSTULATE 17. To every polygonal region there corresponds a unique positive real number called its area.

POSTULATE 18. If two triangles are congruent, then the triangular regions have the same area.

POSTULATE 19. Suppose that the region R is the union of two regions R_1 and R_2. If R_1 and R_2 intersect at most in a finite number of segments and points, then the area of R is the sum of the areas of R_1 and R_2.

POSTULATE 20. The area of a rectangle is the product of the length of its base and the length of its altitude.

POSTULATE 21. The volume of a rectangular parallelpiped is equal to the product of the length of its altitude and the area of its base.

POSTULATE 22. *Cavalieri's Principle:* Given two solids and a plane. If for every plane that intersects the solids and is parallel to the given plane the two intersections determine regions that have the same area, then the two solids have the same volume.

BIBLIOGRAPHY

ADLER, CLAIRE F., *Modern Geometry*. New York: McGraw-Hill Book Co., 1967.

ALTSHILLER-COURT, NATHAN, *College Geometry*. New York: Barnes and Noble, 1952.

BALLARD, WILLIAM. *Geometry*. Philadelphia: W. B. Saunders Co., 1970.

BAUMGART, JOHN K. (ED.), *Historical Topics for the Mathematics Classroom*. Washington D.C.: National Council of Teachers of Mathematics, 1969.

BIRKHOFF, G. D. "A Set of Postulates for Plane Geometry (Based on scale and protractor)," *Annals of Mathematics*, Vol. 33, 1932.

BLUMENTHAL, LEONARD M., *A Modern View of Geometry*. San Francisco: W. H. Freeman & Co., 1961.

BONOLA, ROBERTO, *Geometry—A Critical and Historical Study of Its Developments*. New York: Dover Publications, Inc., 1955.

BORSUK, KAROL, AND W. SZMIELEW, *Foundations of Geometry*. Amsterdam: North-Holland Publishing Co., Inc., 1960.

COXETER, H. S. M., *Non-Euclidean Geometry*. Toronto: University of Toronto Press, 1961.

COXETER, H. S. M., *The Real Projective Plane*. Cambridge: Cambridge University Press, 1961.

DALTON, LEROY C., AND H. SNYDER (EDS.), *Topics for Mathematics Clubs*. Reston Va.: National Council of Teachers of Mathematics, 1983.

EVES, HOWARD, *Foundations and Fundamental Concepts of Mathematics*.Boston: PWS-Kent Publishing Co., Inc., 1990.

EVES, HOWARD, *A Survey of Geometry*. Boston: Allyn and Bacon, Inc., 1972.

EVES, HOWARD, *An Introduction to the History of Mathematics*, 3rd ed. New York: Holt, Rinehart and Winston, 1969.

FISHBACK, W. T., *Projective and Euclidean Geometry*. New York: John Wiley & Sons, Inc., 1962.

FORDER, HENRY GEORGE, *The Foundations of Euclidean Geometry*. New York: Dover Publications, Inc., 1958.

GANS, DAVID, *An Introduction to Non-Euclidean Geometry*. New York: Academic Press, 1973.

GANS, DAVID, *Transformations and Geometries*. New York: Appleton-Century-Crofts, 1969.

GARNER, LYNN E., *An Outline of Projective Geometry*. New York: Elsevier North-Holland, Inc., 1981.

GEMIGNANI, MICHAEL C., *Axiomatic Geometry*. Reading, Mass.: Addison-Wesley Publishing Co., Inc., 1971.

GREENBURG, MARVIN JAY, *Euclidean and Non-Euclidean Geometry*, 2nd ed. San Francisco: W. H. Freeman & Co., 1980.

GROZA, VIVIAN S., *A Survey of Mathematics*. New York: Holt, Rinehart and Winston, 1968.

HALSTEAD, GEORGE B. (TRANS.), *Girolamo Saccheri's Euclides Vindicatus*. New York: Chelsea Publishing Co., Inc., 1986.

HEATH, THOMAS L., *The Thirteen Books of Euclid's Elements*. New York: Dover Publications, Inc., 1956.

HILBERT, DAVID, AND S. COHN-VOSSEN, *Geometry and the Imagination*. New York: Chelsea Publishing Co., Inc., 1952.

KLINE, MORRIS, *Mathematics in Western Culture*. New York: Oxford University Press, 1964.

KULCZYCKI, STEFAN, *Non-Euclidean Geometry*, Trans. S. Knaowski. New York: Pergamon Press, 1961.

LOOMIS, ELISHA SCOTT, *The Pythagorean Proposition*. Washington D.C.: National Council of Teachers of Mathematics, 1968.

MANNING, HENRY PARKER, *Non-Euclidean Geometry*. New York: Dover Publications, Inc., 1963.

MARTIN, GEORGE E., *The Foundations of Geometry and the Non-Euclidean Plane*. New York: Springer-Verlag, 1975.

MOISE, EDWIN E., *Elementary Geometry From an Advanced Standpoint* 2nd ed. Reading, Mass.: Addison-Wesley Publishing Co., Inc., 1974.

MOISE, EDWIN E., AND FLOYD L. DOWNS, *Geometry*. Menlo Park, Ca.: Addison-Wesley Publishing Co., Inc., 1982.

MORROW, GLENN R. (TRANS.), *Proclus' A Commentary on the First Book of Euclid's Elements*. Princeton: Princeton University Press, 1970.

PRENOWITZ, WALTER, AND M. JORDAN, *Basic Concepts of Geometry*. New York: Ardsley House Publishing Co., Inc., 1989.

ROBINSON, GILBERT DE B., *The Foundations of Geometry*. Toronto: University of Toronto Press, 1940.

RUNION, GARTH E., AND J. R. LOCKWOOD, *Deductive Systems—Finite and Non-Euclidean Geometries*. Reston, Va.: National Council of Teachers of Mathematics, 1978.

School Mathematics Study Group. *Geometry*. New Haven: Yale University Press, 1961.

SMART, JAMES R. *Modern Geometries*, 3rd ed. Pacific Grove, Ca.: Brooks/Cole, 1988.

SMITH, DAVID EUGENE, *History of Mathematics*, Vols. I and II. New York: Dover Publications, Inc., 1951.

TAYLOR, E. H., AND BARTOO, G. C., *An Introduction to College Geometry*. New York: Macmillan Co., 1949.

TORRETTI, ROBERTO, *Philosophy of Geometry from Riemann to Poincaré*. Boston: D. Reidel Publishing Co., Inc., 1978.

TRUDEAU, RICHARD, *The Non-Euclidean Revolution*. Boston: Birkhauser, 1986.

TULLER, ANNITA, *A Modern Introduction to Geometries*. Princeton: D. Van Nostrand Co., Inc., 1967.

WOLFE, HAROLD E., *Introduction to Non-Euclidean Geometry*. New York: Holt, Rinehart and Winston, 1945.

WYLIE, C. R., JR., *Foundations of Geometry*. New York: McGraw Hill Book Company, 1964.

WYLIE, C. R., JR., *Introduction to Projective Geometry*. New York: McGraw Hill Book Company, 1970.

YOUNG, JOHN WESLEY, *Projective Geometry*. Chicago: Open Court Publishing Co., 1930.

INDEX

The SMSG Postulates for Euclidean Geometry[1]

UNDEFINED TERMS

1. *Point*
2. *Line*
3. *Plane*

POSTULATES

POSTULATE 1.   Given any two distinct points there is exactly one line that contains them.

POSTULATE 2.   *The Distance Postulate*. To every pair of distinct points there corresponds a unique positive number. This number is called the distance between the two points.

POSTULATE 3.   *The Ruler Postulate*. The points of a line can be placed in a correspondence with the real numbers such that

(1) To every point of the line there corresponds exactly one real number,
(2) To every real number there corresponds exactly one point of the line, and
(3) The distance between two distinct points is the absolute value of the difference of the corresponding real numbers.

POSTULATE 4.   *The Ruler Placement Postulate*. Given two points P and Q of a line, the coordinate system can be chosen in such a way that the coordinate of P is zero and the coordinate of Q is positive.

POSTULATE 5.   (a) Every plane contains at least three non-collinear points. (b) Space contains at least four non-coplanar points.

POSTULATE 6.   If two points lie in a plane, then the line containing these points lies in the same plane.

POSTULATE 7.   Any three points lie in at least one plane, and any three non-collinear points lie in exactly one plane.

POSTULATE 8.   If two planes intersect, then that intersection is a plane.

POSTULATE 9.   *The Plane Separation Postulate*. Given a line and a plane containing it, the points of the plane that do not lie on the line form two sets such that

[1] Reprinted from School Mathematics Study Group, *Geometry*, by permission of Yale University Press, New Haven.

(1) each of the sets is convex and

(2) if P is in one set and Q is in the other, then segment PQ intersects the line.

POSTULATE 10. *The Space Separation Postulate.* The points of space that do not lie in a given plane form two sets such that

(1) Each of the sets is convex, and

(2) If P is in one set and Q is in the other, then segment PQ intersects the plane.

POSTULATE 11. *The Angle Measurement Postulate.* To every angle there corresponds a real number between 0° and 180°.

POSTULATE 12. *The Angle Construction Postulate.* Let AB be a ray on the edge of the half-plane H. For every r between 0 and 180 there is exactly one ray AP, with P in H such that $m\angle PAB = r$.

POSTULATE 13. *The Angle Addition Postulate.* If D is a point in the interior of $\angle BAC$, then $m\angle BAC = m\angle BAD + m\angle DAC$.

POSTULATE 14. *The Supplement Postulate.* If two angles form a linear pair, then they are supplementary.

POSTULATE 15. *The SAS Postulate.* Given a one-to-one correspondence between two triangles (or between a triangle and itself). If two sides and the included angle of the first triangle are congruent to the corresponding parts of the second triangle, then the correspondence is a congruence.

POSTULATE 16. *The Parallel Postulate.* Through a given external point there is at most one line parallel to a given line.

POSTULATE 17. To every polygonal region there corresponds a unique positive real number called its area.

POSTULATE 18. If two triangles are congruent, then the triangular regions have the same area.

POSTULATE 19. Suppose that the region R is the union of two regions R_1 and R_2. If R_1 and R_2 intersect at most in a finite number of segments and points, then the area of R is the sum of the areas of R_1 and R_2.

POSTULATE 20. The area of a rectangle is the product of the length of its base and the length of its altitude.

POSTULATE 21. The volume of a rectangular parallelpiped is equal to the product of the length of its altitude and the area of its base.

POSTULATE 22. *Cavalieri's Principle:* Given two solids and a plane. If for every plane that intersects the solids and is parallel to the given plane the two intersections determine regions that have the same area, then the two solids have the same volume.